ASSESSMENT OF FUEL ECONOMY TECHNOLOGIES FOR LIGHT-DUTY VEHICLES

Committee on the Assessment of Technologies for Improving
Light-Duty Vehicle Fuel Economy

Board on Energy and Environmental Systems

Division on Engineering and Physical Sciences

NATIONAL RESEARCH COUNCIL
OF THE NATIONAL ACADEMIES

THE NATIONAL ACADEMIES PRESS
Washington, D.C.
www.nap.edu

THE NATIONAL ACADEMIES PRESS **500 Fifth Street, N.W.** **Washington, DC 20001**

NOTICE: The project that is the subject of this report was approved by the Governing Board of the National Research Council, whose members are drawn from the councils of the National Academy of Sciences, the National Academy of Engineering, and the Institute of Medicine. The members of the committee responsible for the report were chosen for their special competences and with regard for appropriate balance.

This study was supported by Contract No. DTNH22-07-H-00155 between the National Academy of Sciences and the Department of Transportation. Any opinions, findings, conclusions, or recommendations expressed in this publication are those of the author(s) and do not necessarily reflect the views of the organizations or agency that provided support for the project.

International Standard Book Number-13: 978-0-309-15607-3
International Standard Book Number-10: 0-309-15607-6
Library of Congress Control Number: 2011927639

Copies of this report are available from the National Academies Press, 500 Fifth Street, N.W., Lockbox 285, Washington, DC 20055; (800) 624-6242 or (202) 334-3313 (in the Washington metropolitan area); Internet, http://www.nap.edu.

Printed in the United States of America

THE NATIONAL ACADEMIES

Advisers to the Nation on Science, Engineering, and Medicine

The **National Academy of Sciences** is a private, nonprofit, self-perpetuating society of distinguished scholars engaged in scientific and engineering research, dedicated to the furtherance of science and technology and to their use for the general welfare. Upon the authority of the charter granted to it by the Congress in 1863, the Academy has a mandate that requires it to advise the federal government on scientific and technical matters. Dr. Ralph J. Cicerone is president of the National Academy of Sciences.

The **National Academy of Engineering** was established in 1964, under the charter of the National Academy of Sciences, as a parallel organization of outstanding engineers. It is autonomous in its administration and in the selection of its members, sharing with the National Academy of Sciences the responsibility for advising the federal government. The National Academy of Engineering also sponsors engineering programs aimed at meeting national needs, encourages education and research, and recognizes the superior achievements of engineers. Dr. Charles M. Vest is president of the National Academy of Engineering.

The **Institute of Medicine** was established in 1970 by the National Academy of Sciences to secure the services of eminent members of appropriate professions in the examination of policy matters pertaining to the health of the public. The Institute acts under the responsibility given to the National Academy of Sciences by its congressional charter to be an adviser to the federal government and, upon its own initiative, to identify issues of medical care, research, and education. Dr. Harvey V. Fineberg is president of the Institute of Medicine.

The **National Research Council** was organized by the National Academy of Sciences in 1916 to associate the broad community of science and technology with the Academy's purposes of furthering knowledge and advising the federal government. Functioning in accordance with general policies determined by the Academy, the Council has become the principal operating agency of both the National Academy of Sciences and the National Academy of Engineering in providing services to the government, the public, and the scientific and engineering communities. The Council is administered jointly by both Academies and the Institute of Medicine. Dr. Ralph J. Cicerone and Dr. Charles M. Vest are chair and vice chair, respectively, of the National Research Council.

www.national-academies.org

COMMITTEE ON THE ASSESSMENT OF TECHNOLOGIES FOR IMPROVING LIGHT-DUTY VEHICLE FUEL ECONOMY

TREVOR O. JONES, NAE,[1] ElectroSonics Medical, Cleveland, Ohio, *Chair*
THOMAS W. ASMUS, NAE, DaimlerChrysler Corporation (retired), Oakland, Michigan
RODICA BARANESCU, NAE, NAVISTAR, Warrenville, Illinois
JAY BARON, Center for Automotive Research, Ann Arbor, Michigan
DAVID FRIEDMAN, Union of Concerned Scientists, Washington, D.C.
DAVID GREENE, Oak Ridge National Laboratory, Oak Ridge, Tennessee
LINOS JACOVIDES, NAE, Delphi Research Laboratory (retired), Grosse Pointe Farms, Michigan
JOHN H. JOHNSON, Michigan Technological University, Houghton
JOHN G. KASSAKIAN, NAE, Massachusetts Institute of Technology, Cambridge
ROGER B. KRIEGER, University of Wisconsin-Madison
GARY W. ROGERS, FEV, Inc., Auburn Hills, Michigan
ROBERT F. SAWYER, NAE, University of California, Berkeley

Staff

K. JOHN HOLMES, Study Director
ALAN CRANE, Senior Program Officer
LaNITA JONES, Administrative Coordinator
MADELINE WOODRUFF, Senior Program Officer
E. JONATHAN YANGER, Senior Project Assistant
JAMES J. ZUCCHETTO, Director, Board on Energy and Environmental Systems

[1] NAE, National Academy of Engineering.

BOARD ON ENERGY AND ENVIRONMENTAL SYSTEMS

ANDREW BROWN, JR., *Chair,* NAE,[1] Delphi Corporation, Troy, Michigan
RAKESH AGRAWAL, NAE, Purdue University, West Lafayette, Indiana
WILLIAM BANHOLZER, NAE, The Dow Chemical Company, Midland, Michigan
MARILYN BROWN, Georgia Institute of Technology, Atlanta
MICHAEL CORRADINI, NAE, University of Wisconsin-Madison
PAUL DeCOTIS, Long Island Power Authority, Albany, New York
CHRISTINE EHLIG-ECONOMIDES, NAE, Texas A&M University, College Station
WILLIAM FRIEND, NAE, Bechtel Group, Inc., McLean, Virginia
SHERRI GOODMAN, CNA, Alexandria, Virginia
NARAIN HINGORANI, NAE, Independent Consultant, Los Altos Hills, California
ROBERT HUGGETT, Independent Consultant, Seaford, Virginia
DEBBIE NIEMEIER, University of California, Davis
DANIEL NOCERA, NAS,[2] Massachusetts Institute of Technology, Cambridge
MICHAEL OPPENHEIMER, Princeton University, Princeton, New Jersey
DAN REICHER, Stanford University, Stanford, California
BERNARD ROBERTSON, NAE, DaimlerChrysler (retired), Bloomfield Hills, Michigan
ALISON SILVERSTEIN, Consultant, Pflugerville, Texas
MARK THIEMENS, NAS, University of California, San Diego
RICHARD WHITE, Oppenheimer & Company, New York City

Staff

JAMES ZUCCHETTO, Director
DANA CAINES, Financial Associate
ALAN CRANE, Senior Program Officer
JONNA HAMILTON, Program Officer
K. JOHN HOLMES, Senior Program Officer and Associate Board Director
LaNITA JONES, Administrative Coordinator
ALICE WILLIAMS, Senior Program Assistant
MADELINE WOODRUFF, Senior Program Officer
JONATHAN YANGER, Senior Program Assistant

---

[1] National Academy of Engineering.
[2] National Academy of Sciences.

DEDICATION

This report is dedicated to Dr. Patrick Flynn, a very active and contributing committee member and a member of the National Academy of Engineering, who passed away on August 21, 2008, while this report was being prepared.

Acknowledgments

As a result of the considerable time and effort contributed by the members of the Committee on the Assessment of Technologies for Improving Light-Duty Vehicle Fuel Economy, whose biographies are presented in Appendix A, this report identifies and estimates the effectiveness of technologies for improving fuel economy in light-duty vehicles, and the related costs. The committee's statement of task (Appendix B) clearly presented substantial challenges, which the committee confronted with fair and honest discussion supported with data from the National Highway Traffic Safety Administration (NHTSA), the Environmental Protection Agency (EPA), and the DOT-Volpe Research Laboratory. I appreciate the members' efforts, especially those who chaired the subgroups and led the compilation of the various chapters.

The data and conclusions presented in the report have benefited from a substantial amount of information provided by global automobile manufacturers, suppliers, and others in the regulatory communities and in non-governmental organizations. Appendix C lists the presentations provided to the committee. Members of the committee also visited industry organizations in North America, Europe, and Japan. In addition, the National Research Council contracted with outside organizations to develop and evaluate a number of technological opportunities.

The committee greatly appreciates and thanks the dedicated and committed staff of the National Research Council (NRC), and specifically the Board on Energy and Environmental Systems (BEES) under the direction of James Zucchetto (director of BEES). The committee particularly wishes to recognize the outstanding leadership of K. John Holmes, study director, and his staff. Thanks and recognition are due to the following BEES staff: Alan Crane, senior program officer; Madeline Woodruff, senior program officer; LaNita Jones, administrative coordinator; Jonathan Yanger, senior program assistant; and Aaron Greco, Mirzayan Policy Fellow, as well as consultants K.G. Duleep of Energy and Environmental Analysis, Inc.; Ricardo, Inc.; and IBIS, Inc. The committee also thanks Christopher Baillie, FEV, Inc., an unpaid consultant to the committee, for his many efforts, dedication, and hard work.

This report has been reviewed in draft form by individuals chosen for their diverse perspectives and technical expertise, in accordance with procedures approved by the Report Review Committee of the NRC. The purpose of this independent review is to provide candid and critical comments that will assist the institution in making its published report as sound as possible and to ensure that the report meets institutional standards for objectivity, evidence, and responsiveness to the study charge. The review comments and draft manuscript remain confidential to protect the integrity of the deliberative process.

We wish to thank the following individuals for their review of this report:

Tom Austin, Sierra Research Corporation,
Paul Blumberg, Consultant,
Andrew Brown, Delphi Corporation,
Wynn Bussmann, DaimlerChrysler Corporation (retired),
Laurence Caretto, California State University,
Coralie Cooper, NESCAUM,
James Fay, Massachusetts Institute of Technology,
Larry Howell, Consultant,
David Japikse, Concepts NREC,
Orron Kee, National Highway Traffic Safety Administration (retired),
Steven Plotkin, Argonne National Laboratory,
Priyaranjan Prasad, Prasad Consulting, and
Lee Schipper, Berkeley Transportation Center.

Although the reviewers listed above have provided many constructive comments and suggestions, they were not asked to endorse the conclusions or recommendations, nor did they see the final draft of the report before its release. The review of this report was overseen by Elisabeth M. Drake, Massachusetts Institute of Technology (retired), and Dale Stein, Michigan Technological University (retired). Appointed by the NRC, they were responsible for making certain that an independent examination of this report was carried out in accordance with institutional procedures and that all review comments were carefully considered. Responsibility for the final content of this report rests entirely with the authoring committee and the institution.

Trevor O. Jones, *Chair*
Committee on the Assessment of Technologies
for Improving Light-Duty Vehicle Fuel Economy

Contents

APPENDIXES

Summary

In 2007 the National Highway Traffic Safety Administration (NHTSA) requested that the National Academies provide an objective and independent update of the technology assessments for fuel economy improvements and incremental costs contained in the 2002 National Research Council (NRC) report *Effectiveness and Impact of Corporate Average Fuel Economy (CAFE) Standards*. The NHTSA also asked that the NRC add to its assessment technologies that have emerged since that report was prepared. To address this request, the NRC formed the Committee on the Assessment of Technologies for Improving Light-Duty Vehicle Fuel Economy. The statement of task, shown in Appendix B, directed the committee to estimate the efficacy, cost, and applicability of technologies that might be used over the next 15 years.

FINDINGS AND RECOMMENDATIONS

Overarching Finding

A significant number of technologies exist that can reduce the fuel consumption of light-duty vehicles while maintaining similar performance, safety, and utility. Each technology has its own characteristic fuel consumption benefit and estimated cost. Although these technologies are often considered independently, there can be positive and negative interactions among individual technologies, and so the technologies must be integrated effectively into the full vehicle system. Integration requires that other components of the vehicle be added or modified to produce a competitive vehicle that can be marketed successfully. Thus, although the fuel consumption benefits and costs discussed here are compared against those of representative base vehicles, the actual costs and benefits will vary by specific model. Further, the benefits of some technologies are not completely represented in the tests used to estimate corporate average fuel economy (CAFE). The estimate of such benefits will be more realistic using the new five-cycle tests that display fuel economy data on new vehicles' labels, but improvements to test procedures and additional analysis are warranted. Given that the ultimate energy savings are directly related to the amount of fuel consumed, as opposed to the distance that a vehicle travels on a gallon of fuel, consumers also will be helped by addition to the label of explicit information that specifies the number of gallons typically used by the vehicle to travel 100 miles.

Technologies for Reducing Fuel Consumption

Tables S.1 and S.2 show the committee's estimates of fuel consumption benefits and costs for technologies that are commercially available and can be implemented within 5 years. The cost estimates represent estimates for the current (2009/2010) time period to about 5 years in the future. The committee based these estimates on a variety of sources, including recent reports from regulatory agencies and other sources on the costs and benefits of technologies; estimates obtained from suppliers on the costs of components; discussions with experts at automobile manufacturers and suppliers; detailed teardown studies of piece costs for individual technologies; and comparisons of the prices for and amount of fuel consumed by similar vehicles with and without a particular technology.

Some longer-term technologies have also demonstrated the potential to reduce fuel consumption, although further development is required to determine the degree of improvement, cost-effectiveness, and expected durability. These technologies include camless valve trains, homogeneous-charge compression ignition, advanced diesel, plug-in hybrids, diesel hybrids, electric vehicles, fuel cell vehicles, and advanced materials and body designs. Although some of these technologies will see at least limited commercial introduction over the next several years, it is only in the 5- to 15-year time frame and beyond that they are expected to find widespread commercial application. Further, it will not be possible for some of these technologies to become solutions for significant technical and economic challenges, and thus some of these technologies will remain perennially 10 to 15 years out beyond a moving reference. Among its provisions,

TABLE S.1 Committee's Estimates of Effectiveness (shown as a percentage) of Near-Term Technologies in Reducing Vehicle Fuel Consumption

Technologies		Incremental values - A preceding technology must be included								
		I4			V6			V8		
Spark Ignition Techs	Abbreviation	Low	High	AVG	Low	High	AVG	Low	High	AVG
Low Friction Lubricants	LUB	0.5	0.5	**0.5**	0.5	0.5	**0.5**	0.5	0.5	**0.5**
Engine Friction Reduction	EFR	0.5	2.0	**1.3**	0.5	2.0	**1.3**	1.0	2.0	**1.5**
VVT- Coupled Cam Phasing (CCP), SOHC	CCP	1.5	3.0	**2.3**	1.5	3.5	**2.5**	2.0	4.0	**3.0**
Discrete Variable Valve Lift (DVVL), SOHC	DVVL	1.5	3.0	**2.3**	1.5	3.0	**2.3**	2.0	3.0	**2.5**
Cylinder Deactivation, SOHC	DEAC	NA	NA	NA	4.0	6.0	**5.0**	5.0	10.0	**7.5**
VVT - In take Cam Phasing (ICP)	ICP	1.0	2.0	**1.5**	1.0	2.0	**1.5**	1.5	2.0	**1.8**
VVT - Dual Cam Phasing (DCP)	DCP	1.5	2.5	**2.0**	1.5	3.0	**2.3**	1.5	3.0	**2.3**
Discrete Variable Valve Lift (DVVL), DOHC	DVVL	1.5	3.0	**2.3**	1.5	3.5	**2.5**	2.0	4.0	**3.0**
Continuously Variable Valve Lift (CVVL)	CVVL	3.5	6.0	**4.8**	3.5	6.5	**5.0**	4.0	6.5	**5.3**
Cylinder Deactivation, OHV	DEAC	NA	NA	NA	4.0	6.0	**5.0**	5.0	10.0	**7.5**
VVT - Coupled Cam Phasing (CCP), OHV	CCP	1.5	3.0	**2.3**	1.5	3.5	**2.5**	2.0	4.0	**3.0**
Discrete Variable Valve Lift (DVVL), OHV	DVVL	1.5	2.5	**2.0**	1.5	3.0	**2.3**	2.0	3.0	**2.5**
Stoichiometric Gasoline Direct Injection (GDI)	SGDI	1.5	3.0	**2.3**	1.5	3.0	**2.3**	1.5	3.0	**2.3**
Turbocharging and Downsizing	TRBDS	2.0	5.0	**3.5**	4.0	6.0	**5.0**	4.0	6.0	**5.0**
Diesel Techs										
Conversion to Diesel	DSL	15.0	35.0	**25.0**	15.0	35.0	**25.0**	NA	NA	NA
Conversion to Advanced Diesel	ADSL	7.0	13.0	**10.0**	7.0	13.0	**10.0**	22.0	38.0	**30.0**
Electrification/Accessory Techs										
Electric Power Steering (EPS)	EPS	1.0	3.0	**2.0**	1.0	3.0	**2.0**	1.0	3.0	**2.0**
Improved Accessories	IACC	0.5	1.5	**1.0**	0.5	1.5	**1.0**	0.5	1.5	**1.0**
Higher Voltage/Improved Alternato	HVIA	0.0	0.5	**0.3**	0.0	0.5	**0.3**	0.0	0.5	**0.3**
Transmission Techs										
Continuously Variable Transmission (CVT)	CVT	1.0	7.0	**4.0**	1.0	7.0	**4.0**	1.0	7.0	**4.0**
5-spd Auto. Trans. w/ Improved Internals		2.0	3.0	**2.5**	2.0	3.0	**2.5**	2.0	3.0	**2.5**
6-spd Auto. Trans. w/ Improved Internals		1.0	2.0	**1.5**	1.0	2.0	**1.5**	1.0	2.0	**1.5**
7-spd Auto. Trans. w/ Improved Internals		2.0		**2.0**	2.0		**2.0**	2.0		**2.0**
8-spd Auto. Trans. w/ Improved Internals		1.0		**1.0**	1.0		**1.0**	1.0		**1.0**
6/7/8-spd Auto. Trans. w/ Improved Internals	NAUTO	3.0	8.0	**5.5**	3.0	8.0	**5.5**	3.0	8.0	**5.5**
6/7-spd DCT from 4-spd AT	DCT	6.0	9.0	**7.5**	6.0	9.0	**7.5**	6.0	9.0	**7.5**
6/7-spd DCT from 6-spd AT	DCT	3.0	4.0	**3.5**	3.0	4.0	**3.5**	3.0	4.0	**3.5**
Hybrid Techs										
12V BAS Micro-Hybrid	MHEV	2.0	4.0	**3.0**	2.0	4.0	**3.0**	2.0	4.0	**3.0**
Integrated Starter Generator	ISG	29.0	39.0	**34.0**	29.0	39.0	**34.0**	29.0	39.0	**34.0**
Power Split Hybrid	PSHEV	24.0	50.0	**37.0**	24.0	50.0	**37.0**	24.0	50.0	**37.0**
2-Mode Hybrid	2MHEV	25.0	45.0	**35.0**	25.0	45.0	**35.0**	25.0	45.0	**35.0**
Plug-in hybrid	PHEV	NA	NA	NA	NA	NA	NA	NA	NA	NA
Vehicle Techs										
Mass Reduction - 1%	MR1	0.3		**0.3**	0.3		**0.3**	0.3		**0.3**
Mass Reduction - 2%	MR2	1.4		**1.4**	1.4		**1.4**	1.4		**1.4**
Mass Reduction - 5%	MR5	3.0	3.5	**3.3**	3.0	3.5	**3.3**	3.0	3.5	**3.3**
Mass Reduction - 10%	MR10	6.0	7.0	**6.5**	6.0	7.0	**6.5**	6.0	7.0	**6.5**
Mass Reduction - 20%	MR20	11.0	13.0	**12.0**	11.0	13.0	**12.0**	11.0	13.0	**12.0**
Low Rolling Resistance Tires	ROLL	1.0	3.0	**2.0**	1.0	3.0	**2.0**	1.0	3.0	**2.0**
Low Drag Brakes	LDB	1.0		**1.0**	1.0		**1.0**	1.0		**1.0**
Aero Drag Reduction 10%	AERO	1.0	2.0	**1.5**	1.0	2.0	**1.5**	1.0	2.0	**1.5**

NOTE: Some of the benefits (highlighted in green) are incremental to those obtained with preceding technologies shown in the technology pathways described in Chapter 9.

the Energy Independence and Security Act (EISA) of 2007 requires periodic assessments by the NRC of automobile vehicle fuel economy technologies, including how such technologies might be used to meet new fuel economy standards. Follow-on NRC committees will be responsible for responding to the EISA mandates, including the periodic evaluation of emerging technologies.

Testing and Reporting of Vehicle Fuel Use

Fuel economy is a measure of how far a vehicle will travel with a gallon of fuel, whereas fuel consumption is the amount of fuel consumed in driving a given distance. Although each is simply the inverse of the other, fuel consumption is the fundamental metric by which to judge absolute improvements in fuel efficiency, because what is important is gallons of fuel saved in the vehicle fleet. The amount of fuel saved directly relates not only to dollars saved on fuel purchases but also to quantities of carbon dioxide emissions avoided. Fuel economy data cause consumers to undervalue small increases (1-4 mpg) in fuel economy for vehicles in the 15-30 mpg range, where large decreases in fuel consumption can be realized with small increases in fuel economy. The percentage decrease in fuel consumption is approximately

TABLE S.2 Committee's Estimates of Technology Costs in U.S. Dollars (2008)

NRC 2009 Costs													
Technologies		Incremental Values - A preceding technology must be included											
		I4				V6				V8			
Spark Ignition Techs	Abbreviation	Low	High	AVG	AVG w/1.5 RPE	Low	High	AVG	AVG w/1.5 RPE	Low	High	AVG	AVG w/1.5 RPE
Low Friction Lubricants	LUB	3	5	**4**	**6**	3	5	**4**	**6**	3	5	**4**	**6**
Engine Friction Reduction	EFR	32.0	52.0	**42**	**63**	48	78	**63**	**94.5**	64	104	**84**	**126**
VVT- Coupled Cam Phasing (CCP), SOHC	CCP	35		**35**	**52.5**	70		**70**	**105**	70		**70**	**105**
Discrete Variable Valve Lift (DVVL), SOHC	DVVL	130	160	**145**	**217.5**	180	210	**195**	**292.5**	280	320	**300**	**450**
Cylinder Deactivation, SOHC	DEAC	NA	NA	**NA**	**NA**	340	400	**370**	**555**	357	420	**388.5**	**582.75**
VVT - In take Cam Phasing (ICP)	ICP	35		**35**	**52.5**	70		**70**	**105**	70		**70**	**105**
VVT - Dual Cam Phasing (DCP)	DCP	35		**35**	**52.5**	70		**70**	**105**	70		**70**	**105**
Discrete Variable Valve Lift (DVVL), DOHC	DVVL	130	160	**145**	**217.5**	180	220	**200**	**300**	260	300	**280**	**420**
Continuously Variable Valve Lift (CVVL)	CVVL	159	205	**182**	**273**	290	310	**300**	**450**	350	390	**370**	**555**
Cylinder Deactivation, OHV	DEAC	NA	NA	**NA**	**NA**	220	250	**235**	**352.5**	255		**255**	**382.5**
VVT - Coupled Cam Phasing (CCP), OHV	CCP	35		**35**	**52.5**	35		**35**	**52.5**	35		**35**	**52.5**
Discrete Variable Valve Lift (DVVL), OHV	DVVL	130	160	**145**	**218**	210	240	**225**	**338**	280	320	**300**	**450**
Stoichiometric Gasoline Direct Injection (GDI)	SGDI	117	195	**156**	**234**	169	256	**213**	**319**	295	351	**323**	**485**
Turbocharging and Downsizing	TRBDS	370	490	**430**	**645**	-144	205	**31**	**46**	525	790	**658**	**986**
Diesel Techs													
Conversion to Diesel	DSL	2154	2632	**2393**	**3590**	2857	3491	**3174**	**4761**	NA	NA	**NA**	**NA**
Conversion to Advanced Diese	ADSL	520	520	**520**	**780**	683	683	**683**	**1025**	3513	4293	**3903**	**5855**
Electrification/Accessory Techs													
Electric Power Steering (EPS)	EPS	70	120	**95**	**143**	70	120	**95**	**143**	70	120	**95**	**143**
Improved Accessories	IACC	70	90	**80**	**120**	70	90	**80**	**120**	70	90	**80**	**120**
Higher Voltage/Improved Alternator	HVIA	15	55	**35**	**53**	15	55	**35**	**53**	15	55	**35**	**53**
Transmission Techs													
Continuously Variable Transmission (CVT)	CVT	150	170	**160**	**240**	243	263	**253**	**380**	243	263	**253**	**380**
5-spd Auto. Trans. w/ Improved Internals		133		**133**	**200**	133		**133**	**200**	133		**133**	**200**
6-spd Auto. Trans. w/ Improved Internals		133	215	**174**	**262**	133	215	**174**	**262**	133	215	**174**	**262**
7-spd Auto. Trans. w/ Improved Internals		170	300	**235**	**353**	170	300	**235**	**353**	170	300	**235**	**353**
8-spd Auto. Trans. w/ Improved Internals		425		**425**	**638**	425		**425**	**638**	425		**425**	**638**
6/7/8-Speed Auto. Trans. with Improved Internals	NAUTO	137	425	**281**	**422**	137	425	**281**	**422**	137	425	**281**	**422**
6/7-spd DCT from 6-spd AT	DCT	-147	185	**19**	**29**	-147	185	**19**	**29**	-147	185	**19**	**29**
6/7-spd DCT from 4-spd AT	DCT	-14	400	**193**	**290**	-14	400	**193**	**290**	-14	400	**193**	**290**
Hybrid Techs													
12V BAS Micro-Hybrid	MHEV	450	550	**500**	**665**	585	715	**650**	**865**	720	880	**800**	**1064**
Integrated Starter Generator	ISG	1760	2640	**2200**	**2926**	2000	3000	**2500**	**3325**	3200	4800	**4000**	**5320**
Power Split Hybrid	PSHEV	2708	4062	**3385**	**4502**	3120	4680	**3900**	**5187**	4000	6000	**5000**	**6650**
2-Mode Hybrid	2MHEV	5200	7800	**6500**	**8645**	5200	7800	**6500**	**8645**	5200	7800	**6500**	**8645**
Series PHEV 40	PHEV	8000	12000	**10000**	**13300**	9600	14400	**12000**	**15960**	13600	20400	**17000**	**22610**
Vehicle Techs													
Mass Reduction - 1%	MR1	37	45	**41**	**61**	48	58	**53**	**80**	68	82	**75**	**113**
Mass Reduction - 2%	MR2	77	93	**85**	**127**	100	121	**111**	**166**	142	170	**156**	**234**
Mass Reduction - 5%	MR5	217	260	**239**	**358**	283	339	**311**	**467**	399	479	**439**	**659**
Mass Reduction - 10%	MR10	520	624	**572**	**859**	679	815	**747**	**1120**	958	1150	**1054**	**1581**
Mass Reduction - 20%	MR20	1600	1700	**1650**	**2475**	1600	1800	**1700**	**2550**	1600	1900	**1750**	**2625**
Low Rolling Resistance Tires	ROLL	30	40	**35**	**53**	30	40	**35**	**53**	30	40	**35**	**53**
Aero Drag Reduction 10%	AERO	40	50	**45**	**68**	40	50	**45**	**68**	40	50	**45**	**68**

equal to the percentage increase in fuel economy for values less than 10 percent (for example, a 9.1 percentage decrease in fuel consumption equals a 10 percent increase in fuel economy), but the differences increase progressively: for example, a 33.3 percent decrease in fuel consumption equals a 50 percent increase in fuel economy.

Recommendation: Because differences in the fuel consumption of vehicles relate directly to fuel savings, the labeling on new cars and light-duty trucks should include information on the gallons of fuel consumed per 100 miles traveled in addition to the already-supplied data on fuel economy so that consumers can become familiar with fuel consumption as a fundamental metric for calculating fuel savings.

Fuel consumption and fuel economy are evaluated by the U.S. Environmental Protection Agency (EPA) for the two driving cycles: the urban dynamometer driving schedule (city cycle) and the highway dynamometer driving schedule (highway cycle). In the opinion of the committee, the schedules used to compute CAFE should be modified so that vehicle test data better reflect actual fuel consumption. Excluding some driving conditions and accessory loads in determining CAFE discourages the introduction of certain technologies into the vehicle fleet. The three additional schedules recently adopted by the EPA for vehicle labeling purposes—ones that capture the effects of higher speed and acceleration, air conditioner use, and cold weather—represent a positive step forward, but further study is needed to assess to what degree the new test procedures can fully characterize changes in in-use vehicle fuel consumption.

Recommendation: The NHTSA and the EPA should review and revise fuel economy test procedures so that they better reflect in-use vehicle operating conditions and also provide the proper incentives to manufacturers to produce vehicles that reduce fuel consumption.

Cost Estimation

Large differences in technology cost estimates can result from differing assumptions. These assumptions include whether costs are long- or short-term costs; whether learning by doing is included in the cost estimate; whether the cost estimate represents direct in-house manufacturing costs or the cost of purchasing a component from a supplier; and which of the other changes in vehicle design that are required to maintain vehicle quality have been included in the cost estimate. Cost estimates also depend greatly on assumed production volumes.

In the committee's judgment, the concept of incremental retail price equivalent (RPE) is the most appropriate indicator of cost for the NHTSA's purposes because it best represents the full, long-run economic costs of decreasing fuel consumption. The RPE represents the average additional price consumers would pay for a fuel economy technology. It is intended to reflect long-run, substantially learned, industry-average production costs that incorporate rates of profit and overhead expenses. A critical issue is choice of the RPE markup factor, which represents the ratio of total cost of a component, taking into account the full range of costs of doing business, to only the direct cost of the fully manufactured component. For fully manufactured components purchased from a Tier 1 supplier,[1] a reasonable average RPE markup factor is 1.5. For in-house manufactured components, a reasonable average RPE markup factor over variable manufacturing costs is 2.0. In addition to the costs of materials and labor and the fixed costs of manufacturing, the RPE factor for components from Tier 1 suppliers includes profit, warranty, corporate overhead, and amortization of certain fixed costs, such as research and development. The RPE factor for in-house manufactured components from automobile manufacturers includes the analogous components of the Tier 1 markup for the manufacturing operations, plus additional fixed costs for vehicle integration design and vehicle installation, corporate overhead for assembly operations, additional product warranty costs, transportation, marketing, dealer costs, and profits. RPE markup factors clearly vary depending on the complexity of the task of integrating a component into a vehicle system, the extent of the changes required to other components, the novelty of the technology, and other factors. However, until empirical data derived via rigorous estimation methods are available, the committee prefers the use of average markup factors.

Available cost estimates are based on a variety of sources: component cost estimates obtained from suppliers, discussions with experts at automobile manufacturers and suppliers, publicly available transaction prices, and comparisons of the prices of similar vehicles with and without a particular technology. However, there is a need for cost estimates based on a teardown of all the elements of a technology and a detailed accounting of materials and capital costs and labor time for all fabrication and assembly processes. Such teardown studies are costly and are not feasible for advanced technologies whose designs are not yet finalized and/or whose system integration impacts are not yet fully understood. Estimates based on the more rigorous method of teardown analysis would increase confidence in the accuracy of the costs of reducing fuel consumption.

Technology cost estimates are provided by the committee for each fuel economy technology discussed in this report. Except as indicated, the cost estimates represent the price an automobile manufacturer would pay a supplier for a finished component. Thus, on average, the RPE multiplier of 1.5 would apply to the direct, fully manufactured cost to obtain the average additional price consumers would pay for a technology. Again, except where indicated otherwise, the

[1] A Tier 1 supplier is one that contracts directly with automobile manufacturers to supply technologies.

cost estimates provided are based on current conditions and do not attempt to estimate economic conditions and hence predict prices 5, 10, or 15 years into the future.

Spark-Ignition Gasoline Engine Technologies

Spark-ignition (SI) engines are expected to continue to be the primary source of propulsion for light-duty vehicles in the United States over the time frame of this report. There have been and continue to be significant improvements in reducing the fuel consumption of SI engines in the areas of friction reduction, reduced pumping losses through advanced valve-event modulation, thermal efficiency improvements, cooled exhaust gas recirculation, and improved overall engine architecture, including downsizing. An important attribute of improvements in SI engine technologies is that they offer a means of reducing fuel consumption in relatively small, incremental steps. This approach allows automobile manufacturers to create packages of technologies that can be tailored to meet specific cost and effectiveness targets, as opposed to developing diesel or full hybrid alternatives that offer a single large benefit, but at a significant cost increase. Because of the flexibility offered by this approach, and given the size of the SI engine-powered fleet, the implementation of SI engine technologies will continue to play a large role in reducing fuel consumption.

Of the technologies currently available, cylinder deactivation is one of the more effective in reducing fuel consumption. This feature is most cost-effective when applied to six-cylinder (V6) and eight-cylinder (V8) overhead valve engines, and typically reduces fuel consumption by 4 to 10 percent at an incremental RPE increase of about \$550. Stoichiometric direct injection typically affords a 1.5 to 3 percent reduction in fuel consumption at an incremental RPE increase of \$230 to \$480, depending on cylinder count and noise abatement requirements. Turbocharging and downsizing can also yield fuel consumption reductions. Downsizing—reducing engine displacement while maintaining vehicle performance—is an important strategy applicable in combination with technologies that increase engine torque, such as turbocharging or supercharging. Downsizing simultaneously reduces throttling and friction losses because downsized engines generally have smaller bearings and either fewer cylinders or smaller cylinder bore friction surfaces. Reductions in fuel consumption can range from 2 to 6 percent with turbocharging and downsizing, depending on many details of implementation. This technology combination is assumed to be added after direct injection, and its fuel consumption benefits are incremental to those from direct injection. Based primarily on an EPA teardown study, the committee's estimates of the costs for turbocharging and downsizing range from close to zero additional cost, when converting from a V6 to a four-cylinder (I4) engine, to almost \$1,000, when converting from a V8 to a V6 engine. Valve-event modulation (VEM) can further reduce fuel consumption and can also cause a slight increase in engine performance, which offers a potential opportunity for engine downsizing. There are many different implementations of VEM, and the costs and benefits depend on the specific engine architecture. Fuel consumption reduction can range from 1 percent with only intake cam phasing, to about 7 percent with a continuously variable valve lift and timing setup. The incremental RPE increase for valve-event modulation ranges from about \$50 to \$550, with the amount depending on the implementation technique and the engine architecture.

Variable compression ratio, camless valve trains, and homogeneous-charge compression ignition were all given careful consideration during the course of this study. Because of questionable benefits, major implementation issues, or uncertain costs, it is uncertain whether any of these technologies will have any significant market penetration in the next 10 to 15 years.

Compression-Ignition Diesel Engine Technologies

Light-duty compression-ignition (CI) engines operating on diesel fuels have efficiency advantages over the more common SI gasoline engines. Although light-duty diesel vehicles are common in Europe, concerns over the ability of such engines to meet emission standards for nitrogen oxides and particulates have slowed their introduction in the United States. However, a joint effort between automobile manufacturers and suppliers has resulted in new emissions control technologies that enable a wide range of light-duty CI engine vehicles to meet federal and California emissions standards. The committee found that replacing a 2007 model year SI gasoline power train with a base-level CI diesel engine with an advanced 6-speed dual-clutch automated manual transmission (DCT) and more efficient accessories packages can reduce fuel consumption by about 33 percent on an equivalent vehicle performance basis. The estimated incremental RPE cost of conversion to the CI engine is about \$3,600 for a four-cylinder engine and \$4,800 for a six-cylinder engine. Advanced-level CI diesel engines, which are expected to reach market in the 2011-2014 time frame, with DCT (7/8 speed) could reduce fuel consumption by about an additional 13 percent for larger vehicles and by about 7 percent for small vehicles. Part of the gain from advanced-level CI diesel engines comes from downsizing. The estimated incremental RPE cost of the conversion to the package of advanced diesel technologies is about \$4,600 for small passenger cars and \$5,900 for intermediate and large passenger cars.

An important characteristic of CI diesel engines is that they provide reductions in fuel consumption over the entire vehicle operating range, including city driving, highway driving, hill climbing, and towing. This attribute of CI diesel engines is an advantage when compared with other technology options that in most cases provide fuel consumption benefits for only part of the vehicle operating range.

The market penetration of CI diesel engines will be strongly influenced by both the incremental cost of CI diesel power trains above the cost of SI gasoline power trains and by diesel and gasoline fuel prices. Further, while technology improvements to CI diesel engines are expected to reach market in the 2011-2014 time frame, technology improvements to SI gasoline and hybrid engines will also enter the market. Thus, competition between these power train systems will continue with respect to reductions in fuel consumption and to cost. For the period 2014-2020, further potential reductions in fuel consumption by CI diesel engines may be offset by increases in fuel consumption as a result of changes in engines and emissions systems required to meet potentially stricter emissions standards.

Hybrid Vehicle Technologies

Because of their potential to eliminate energy consumption when the vehicle is stopped, permit braking energy to be recovered, and allow more efficient use of the internal combustion engine, hybrid technologies are one of the most active areas of research and deployment. The degree of hybridization can vary from minor stop-start systems with low incremental costs and modest reductions in fuel consumption to complete vehicle redesign and downsizing of the SI gasoline engine at a high incremental cost but with significant reductions in fuel consumption. For the most basic systems that reduce fuel consumption by turning off the engine while the vehicle is at idle, the fuel consumption benefit may be up to about 4 percent at an estimated incremental RPE increase of $670 to $1,100. The fuel consumption benefit of a full hybrid may be up to about 50 percent at an estimated incremental RPE cost of $3,000 to $9,000 depending on vehicle size and specific hybrid technology. A significant part of the improved fuel consumption of full hybrid vehicles comes from the complete vehicle redesign that can incorporate modifications such as low-rolling-resistance tires, improved aerodynamics, and the use of smaller, more efficient SI engines.

In the next 10 to 15 years, improvements in hybrid vehicles will occur primarily as a result of reduced costs for hybrid power train components and improvements in battery performance such as higher power per mass and volume, increased number of lifetime charges, and wider allowable state-of-charge ranges. During the past decade, significant advances have been made in lithium-ion battery technology. When the cost and safety issues associated with them are resolved, lithium-ion batteries will replace nickel-metal-hydride batteries in hybrid electric vehicles and plug-in hybrid electric vehicles. A number of different lithium-ion chemistries are being studied, and it is not yet clear which ones will prove most beneficial. Given the high level of activity in lithium-ion battery development, plug-in hybrid electric vehicles will be commercially viable and will soon enter at least limited production. The practicality of full-performance battery electric vehicles (i.e., with driving range, trunk space, volume, and acceleration comparable to those of vehicles powered with internal-combustion engines) depends on a battery cost breakthrough that the committee does not anticipate within the time horizon considered in this study. However, it is clear that small, limited-range, but otherwise full-performance battery electric vehicles will be marketed within that time frame. Although there has been significant progress in fuel cell technology, it is the committee's opinion that fuel cell vehicles will not represent a significant fraction of on-road light-duty vehicles within the next 15 years.

Non-engine Technologies for Reducing Vehicle Fuel Consumption

There is a range of non-engine technologies with varying costs and impacts. Many of these technologies are continually being introduced to new vehicle models based on the timing of the product development process. Coordinating the introduction of many technologies with the product development process is critical to maximizing impact and minimizing cost. Relatively minor changes that do not involve reengineering the vehicle or that require recertification for fuel economy, emissions, and/or safety can be implemented within a 2- to 4-year time frame. These changes could include minor reductions in mass (achieved by substitution of materials), improving aerodynamics, or switching to low-rolling-resistance tires. More substantive changes, which require longer-term coordination with the product development process because of the need for reengineering and integration with other subsystems, could include resizing the engine and transmission or aggressively reducing vehicle mass, such as by changing the body structure. The time frame for substantive changes for a single model is approximately 4 to 8 years.

Two important technologies impacting fuel consumption are those for light-weighting and for improving transmissions. Light-weighting has significant potential because vehicles can be made very light with exotic materials, albeit at potentially high cost. The incremental cost to reduce a pound of mass from the vehicle tends to increase as the total amount of reduced mass increases, leading to diminishing returns. About 10 percent of vehicle mass can be eliminated at a cost of roughly $800 to $1,600 and can provide a fuel consumption benefit of about 6 to 7 percent. Reducing mass much beyond 10 percent requires attention to body structure design, such as considering an aluminum-intensive car, which increases the cost per pound. A 10 percent reduction in mass over the next 5 to 10 years appears to be within reach for the typical automobile.

Transmission technologies have improved significantly and, like other vehicle technologies, show a similar trend of diminishing returns. Planetary-based automatic transmissions can have 5, 6, 7, and 8 speeds, but with incremental costs increasing faster than reductions in fuel consumption. DCTs are in production by some automobile manufacturers, and new production capacity for this transmission type has been announced. It is expected that the predominant trend in transmission design is conversion to 6- to 8-speed planetary-based automatics and to DCTs, with continuously variable transmissions remaining a niche application. Given the close linkage between the effects of fuel-consumption-reducing engine technologies and transmission technologies, the present study has for the most part considered the combined effects of engines and transmission combinations rather than potential separate effects.

Accessories are also being introduced to new vehicles to reduce the power load on the engine. Higher-efficiency air conditioning systems are available that more optimally match cooling with occupant comfort. Electric and electric/hydraulic power steering also reduces the load on an engine by demanding power only when the operator turns the wheel. An important motivating factor affecting the introduction of these accessories is whether or not their impact is measured during the EPA driving cycles used to estimate fuel consumption.

Modeling Reductions in Fuel Consumption Obtained from Vehicle Technologies

The two primary methods for modeling technologies' reduction of vehicle fuel consumption are full system simulation (FSS) and partial discrete approximation (PDA). FSS is the state-of-the-art method because it is based on integration of the equations of motion for the vehicle carried out over the speed-time representation of the appropriate driving or test cycle. Done well, FSS can provide an accurate assessment (within +/–5 percent or less) of the impacts on fuel consumption of implementing one or more technologies. The validity of FSS modeling depends on the accuracy of representations of system components. Expert judgment is also required at many points and is critical to obtaining accurate results. Another modeling approach, the PDA method, relies on other sources of data for estimates of the impacts of fuel economy technologies and relies on mathematical summation or multiplication methods to aggregate the effects of multiple technologies. Synergies among technologies can be represented using engineering judgment and lumped parameter models[2] or can be synthesized from FSS results. Unlike FSS, the PDA method cannot be used to generate estimates of the impacts of individual technologies on fuel consumption. Thus, the PDA method by itself, unlike FSS, is not suitable for estimating the fuel consumption impacts of technologies that have not already been tested in actual vehicles or whose fuel consumption benefits have not been estimated by means of FSS.

Comparisons of FSS modeling and PDA estimation supported by lumped parameter modeling have shown that the two methods produce similar results when similar assumptions are used. In some instances, comparing the estimates made by the two methods has enhanced the overall validity of estimated fuel consumption impacts by uncovering inadvertent errors in one or the other method. In the committee's judgment both methods are valuable, especially when used together, with one providing a check on the other. However, more work needs to be done to establish the accuracy of both methods relative to actual motor vehicles.

The Department of Transportation's Volpe National Transportation Systems Center has developed a model for the NHTSA to estimate how manufacturers can comply with fuel economy regulations by applying additional fuel savings technologies to the vehicles they plan to produce. The model employs a PDA algorithm that includes estimates of the effects of interactions among technologies applied. The validity of the Volpe model could be improved by taking into account main and interaction effects produced by the FSS methodology described in Chapter 8 of this report. In particular, modeling work done for the committee by an outside consulting firm has demonstrated a practical method for using data generated by FSS models to accurately assess the fuel consumption potentials of combinations of dozens of technologies on thousands of vehicle configurations. A design-of-experiments statistical analysis of FSS model runs demonstrated that main effects and first-order interaction effects alone could predict FSS model outputs with an R^2 of 0.99. Using such an approach could appropriately combine the strengths of both the FSS and the PDA modeling methods. However, in the following section, the committee recommends an alternate approach that uses FSS to better assess the contributory effects of the technologies applied in the reduction of energy losses and to better couple the modeling of fuel economy technologies to the testing of such technologies on production vehicles.

Application of Multiple Vehicle Technologies to Vehicle Classes

Figures 9.1 to 9.5 in Chapter 9 of this report display the technology pathways developed by the committee for eight classes of vehicles and the aggregated fuel consumption benefits and costs for the SI engine, CI engine, and hybrid power train pathways. The results of the committee's analysis are that, for the intermediate car, large car, and unibody standard truck classes, the average reduction in fuel consumption for the SI engine path is about 29 percent at a cost of approximately $2,200; the average reduction for the CI engine path is about 37 percent at a cost of approximately $5,900; and the average reduction for the hybrid power train path is about 44 percent at a cost of $6,000. These values are approximate and are provided here as rough estimates that can be used for qualitative comparison of SI engine-related technologies and

[2]Lumped parameter models are simplified analytical tools for estimating vehicle energy use based on a small set of energy balance equations and empirical relationships. With a few key vehicle parameters, these methods can explicitly account for the sources of energy loss and the tractive force required to move the vehicle.

other candidates for the reduction of vehicle fuel consumption, such as light-duty diesel or hybrid vehicles.

Improvements to Modeling of Multiple Fuel Economy Technologies

Many vehicle and power train technologies that improve fuel consumption are currently in or entering production or are in advanced stages of development in European or Asian markets where high consumer fuel prices have made commercialization of the technologies cost-effective. Depending on the intended vehicle use or current state of energy-loss reduction, the application of incremental technologies will produce varying levels of improvement in fuel consumption. Data made available to the committee from automobile manufacturers, Tier 1 suppliers, and other published studies also suggest a very wide range in estimated incremental cost. As noted above in this Summary, estimates based on teardown cost analysis, currently being utilized by the EPA in its analysis of standards for regulating light-duty-vehicle greenhouse gas emissions, should be expanded for developing cost impact analyses. The committee notes, however, that cost estimates are always more uncertain than estimates of fuel consumption.

FSS modeling that is based on empirically derived power train and vehicle performance and on fuel consumption data maps offers what the committee believes is the best available method to fully account for system energy losses and to analyze potential improvements in fuel consumption achievable by technologies as they are introduced into the market. Analyses conducted for the committee show that the effects of interactions between differing types of technologies for reducing energy loss can and often do vary greatly from vehicle to vehicle.

Recommendation: The committee proposes a method whereby FSS analyses are used on class-characterizing vehicles, so that synergies and effectiveness in implementing multiple fuel economy technologies can be evaluated with what should be greater accuracy. This proposed method would determine a characteristic vehicle that would be defined as a reasonable average representative of a class of vehicles. This representative vehicle, whether real or theoretical, would undergo sufficient FSS, combined with experimentally determined and vehicle-class-specific system mapping, to allow a reasonable understanding of the contributory effects of the technologies applied to reduce vehicle energy losses. Data developed under the United States Council for Automotive Research (USCAR) Benchmarking Consortium should be considered as a source for such analysis and potentially expanded. Under the USCAR program, actual production vehicles are subjected to a battery of vehicle, engine, and transmission tests in sufficient detail to understand how each candidate technology is applied and how they contribute to the overall performance and fuel consumption of light-duty vehicles. Combining the results of such testing with FSS modeling, and thereby making all simulation variables and subsystem maps transparent to all interested parties, would allow the best opportunity to define a technical baseline against which potential improvements could be analyzed more accurately and openly than is the case with the current methods employed.

The steps in the recommended process would be as follows:

1. Develop a set of baseline vehicle classes from which a characteristic vehicle can be chosen to represent each class. The vehicle may be either real or theoretical and will possess the average attributes of that class as determined by sales-weighted averages.
2. Identify technologies with a potential to reduce fuel consumption.
3. Determine the applicability of each technology to the various vehicle classes.
4. Estimate each technology's preliminary impact on fuel consumption and cost.
5. Determine the optimum implementation sequence (technology pathway) based on cost-effectiveness and engineering considerations.
6. Document the cost-effectiveness and engineering judgment assumptions used in step 5 and make this information part of a widely accessible database.
7. Utilize modeling software (FSS) to progress through each technology pathway for each vehicle class to obtain the final incremental effects of adding each technology.

If such a process were adopted as part of a regulatory rule-making procedure, it could be completed on 3-year cycles to allow regulatory agencies sufficient lead time to integrate the results into future proposed and enacted rules.

CONCLUDING COMMENTS

A significant number of approaches are currently available to reduce the fuel consumption of light-duty vehicles, ranging from relatively minor changes to lubricants and tires to large changes in propulsion systems and vehicle platforms. Technologies such as all-electric propulsion systems have also demonstrated the potential to reduce fuel consumption, although further development is required to determine the degree of improvement, cost-effectiveness, and durability. The development and deployment of vehicles that consume less fuel will be influenced not only by technological factors but also by economic and policy factors whose examination is beyond the scope of this study. Future NRC committees will be responsible for periodic assessments of the cost and benefits of technologies that reduce vehicle fuel consumption, including how such technologies might be used to meet new fuel economy standards.

1

Introduction

The impacts of fuel consumption by light-duty vehicles are profound, influencing economic prosperity, national security, and Earth's environment. Increasing energy efficiency has been a continuing and central objective for automobile manufacturers and regulators pursuing objectives that range from reducing vehicle operating costs and improving performance to reducing dependence on petroleum and limiting greenhouse gas emissions. Given heightened concerns about the dangers of global climate change, the needs for energy security, and the volatility of world oil prices, attention has again been focused on reducing the fuel consumption of light-duty vehicles. A wide array of technologies and approaches exist for reducing fuel consumption. These improvements range from relatively minor changes with low costs and small fuel consumption benefits—such as use of new lubricants and tires—to large changes in propulsion systems and vehicle platforms that have high costs and large fuel consumption benefits.

CURRENT POLICY CONTEXT AND MOTIVATION

The rapid rise in gasoline and diesel fuel prices experienced during 2006-2008 and growing recognition of climate-change issues have helped make vehicle fuel economy an important policy issue once again. These conditions have motivated several recent legislative and regulatory initiatives. The first major initiative was the mandate for increased CAFE standards under the Energy Independence and Security Act of 2007. This legislation requires the National Highway Traffic Safety Administration (NHTSA) to raise vehicle fuel economy standards, starting with model year 2011, until they achieve a combined average fuel economy of at least 35 miles per gallon (mpg) for model year 2020. The policy landscape has also been significantly altered by separate Supreme Court decisions related to the regulation of carbon dioxide as an air pollutant and the California greenhouse gas vehicle standards. These decisions helped spur the Obama administration to direct the U.S. Environmental Protection Agency (EPA) and the NHTSA to develop a joint fuel economy/greenhouse gas emission standard for light-duty vehicles that mirrors the stringency of the California emissions standard. Finalized on April 1, 2010, the rule requires that fleet-averaged fuel economy reach an equivalent of 35.4 mpg by model year 2016.

The significant downturn in the United States and world economies that occurred during the course of this study has had substantial negative impacts on the global automobile industry. Most manufacturers have experienced reduced sales and suffered losses. The automobile industry is capital intensive and has a very steep curve on profits around the break-even point: a small increase in sales beyond the break-even point can results in large profits, while a small decrease can result in large losses. Consumer spending decreased markedly due to lack of confidence in the economy as well as difficulties in the credit markets that typically finance a large portion of vehicle purchases. The U.S. market for light-duty vehicles decreased from about 16 million vehicles annually for the last few years to about 10 million in 2009. The overall economic conditions resulted in Chrysler and GM deciding to file for Chapter 19 bankruptcy and in Ford excessively leveraging its assets. GM and Chrysler have recently exited bankruptcy, and the U.S. government is now the major shareholder of GM. Fiat Automobiles has become a 20 percent shareholder in Chrysler, with the potential to expand its ownership to 35 percent, and the newly formed Voluntary Employee Beneficiary Association has a 55 percent stake.

These economic conditions will impact automotive companies' and suppliers' ability to fund in a timely manner the R&D necessary for fuel economy improvements and the capital expenditures required. Although addressing the impact of such conditions on the adoption of vehicle fuel economy technologies is not within the purview of this committee, these conditions do provide an important context for this study. Manufacturers will choose fuel economy technologies based on what they think will be most effective and best received by consumers. Customers also will have a central role in what technologies are actually chosen and will make those choices based partly on initial and operating costs.

Subsidies and other incentives also can significantly impact the market acceptance rate of technologies that reduce fuel consumption. Finally, adoption of these technologies must play out in a sometimes unpredictable marketplace and policy setting, with changing standards for emissions and fuel economy, government incentives, consumer preferences, and other events impacting their adoption. Thus, the committee acknowledges that technologies downplayed here may play a bigger role than anticipated, or that technologies covered in this report may never emerge in the marketplace.

The timing for introducing new fuel consumption technologies may have a large influence on cost and risk. The individual vehicle models produced by automobile manufacturers pass through a product cycle that includes introduction, minor refreshments of design and features, and then full changes in body designs and power trains. To reduce costs and quality concerns, changes to reduce fuel consumption normally are timed for implementation in accordance with this process. Further, new technologies are often applied first in lower-volume, higher-end vehicles because such vehicles are better able to absorb the higher costs, and their lower volumes reduce exposure to risk. In general, 2 to 3 years is considered the quickest time frame for bringing a new vehicle model to market or for modifying an existing model. Significant carryover technology and engineering from other models or previous vehicle models are usually required to launch a new model this quickly, and the ability to significantly influence fuel consumption is thus smaller. More substantial changes to a model occur over longer periods of time. Newly styled, engineered, and redesigned vehicles can take from 4 to 8 years to produce, each with an increasing amount of new content. Further, the engine development process often follows a path separate from that for other parts of a vehicle. Engines have longer product lives, require greater capital investment, and are not as critical to the consumer in differentiating one vehicle from another as are other aspects of a car. The normal power train development process evolves over closer to a 15-year cycle, although refinements and new technologies will be implemented throughout this period. It should be noted that there are significant differences among manufacturers in their approaches to introducing new models and, due to regulatory and market pressures, product cycles have tended to become shorter over time.

Although it is not a focus of this study, the global setting for the adoption of these fuel economy technologies is critical. The two main types of internal combustion engines, gasoline spark-ignition (SI) and diesel compression-ignition (CI), are not necessarily fully interchangeable. Crude oil (which varies in composition) contains heavier fractions that go into diesel production and lighter fractions that go into gasoline. A large consumer of diesel, Europe diverts the remaining gasoline fraction to the United States or elsewhere. China is now using mostly gasoline, and so there is more diesel available globally. And automobile manufacturers and suppliers worldwide are improving their capabilities in hybrid-electric technologies. Further, policy incentives may help favor one technology over another in individual countries.

STATEMENT OF TASK

The NHTSA has a mandate to keep up-to-date on the potential for technological improvements as it moves into planned vehicular regulatory activities. It was as part of its technology assessment that the NHTSA asked the National Academies to update the 2002 National Research Council report *Effectiveness and Impact of Corporate Average Fuel Economy (CAFE) Standards* (NRC, 2002) and add to its assessment other technologies that have emerged since that report was prepared. The statement of task (see Appendix B) directed the Committee on the Assessment of Technologies for Improving Light-Duty Vehicle Fuel Economy to estimate the efficacy, timing, cost, and applicability of technologies that might be used over the next 15 years. The list of technologies includes diesel and hybrid electric power trains, which were not considered in the 2002 NRC report. Weight and power reductions also were to be included, but not size or power-to-weight ratio reductions. Updating the fuel economy-cost relationships for various technologies and different vehicle size classes as represented in Chapter 3 of the 2002 report was central to the study request.

The current study focuses on technology and does not consider CAFE issues related to safety, economic effects on industry, or the structure of fuel economy standards; those issues were addressed in the 2002 report. The new study looks at lowering fuel consumption by reducing power requirements through such measures as reduced vehicle weight, lower tire rolling resistance, or improved vehicle aerodynamics and accessories; by reducing the amount of fuel needed to produce the required power through improved engine and transmission technologies; by recovering some of the exhaust thermal energy with turbochargers and other technologies; and by improving engine performance and recovering energy through regenerative braking in hybrid vehicles. Additionally, the committee was charged with assessing how ongoing changes to manufacturers' refresh and redesign cycles for vehicle models affect the incorporation of new fuel economy technologies. The current study builds on information presented in the committee's previously released interim report (NRC, 2008).

CONTENTS OF THIS REPORT

The committee organized its final report according to broad topics related to the categories of technologies important for reducing fuel consumption, the costs and issues associated with estimating the costs and price impacts of these technologies, and approaches to estimating the fuel consumption benefits possible with combinations of these technologies. Chapter 2 describes fundamentals of determining vehicle fuel consumption, tests for regulating fuel economy, and basic energy balance concepts, and it discusses why this report presents primarily fuel consumption data. Chapter 3 describes cost estimation for vehicle technologies, including methods for estimating the costs of a new technology and issues related to translating those costs into impacts on the retail price of a vehicle. Chapters 4 through 7 describe technologies for improving fuel consumption in spark-ignition gasoline engines (Chapter 4), compression-ignition diesel engines (Chapter 5), and hybrid-electric vehicles (Chapter 6). Chapter 7 covers non-engine technologies for reducing light-duty vehicle fuel consumption. Chapter 8 provides a basic overview of and discusses the attributes of two different approaches for estimating fuel consumption benefits—the discrete approximation and the full-system simulation modeling approaches. Chapter 9 provides an estimate of the costs and the fuel consumption benefits of multiple technologies for an array of vehicle classes. The appendixes provide information related to conducting the study (Appendixes A through C), a list of the acronyms used in the report (Appendix D), and additional information supplementing the individual chapters (Appendixes E through K).

REFERENCES

NRC (National Research Council). 2002. Effectiveness and Impact of Corporate Average Fuel Economy (CAFE) Standards. Washington, D.C.: National Academy Press.

NRC. 2008. Interim Report of the Committee on the Assessment of Technologies for Improving Light-Duty Vehicle Fuel Economy. Washington, D.C.: The National Academies Press.

2

Fundamentals of Fuel Consumption

INTRODUCTION

This chapter provides an overview of the various elements that determine fuel consumption in a light-duty vehicle (LDV). The primary concern here is with power trains that convert hydrocarbon fuel into mechanical energy using an internal combustion engine and which propel a vehicle though a drive train that may be a combination of a mechanical transmission and electrical machines (hybrid propulsion). A brief overview is given here of spark-ignition (SI) and compression-ignition (CI) engines as well as hybrids that combine electric drive with an internal combustion engine; these topics are discussed in detail in Chapters 4 through 6. The amount of fuel consumed depends on the engine, the type of fuel used, and the efficiency with which the output of the engine is transmitted to the wheels. This fuel energy is used to overcome (1) rolling resistance primarily due to flexing of the tires, (2) aerodynamic drag as the vehicle motion is resisted by air, and (3) inertia and hill-climbing forces that resist vehicle acceleration, as well as engine and drive line losses. Although modeling is discussed in detail in later chapters (Chapters 8 and 9), a simple model to describe tractive energy requirements and vehicle energy losses is given here as well to understand fuel consumption fundamentals. Also included is a brief discussion of customer expectations, since performance, utility, and comfort as well as fuel consumption are primary objectives in designing a vehicle.

Fuel efficiency is a historical goal of automotive engineering. As early as 1918, General Motors Company automotive pioneer Charles Kettering was predicting the demise of the internal combustion engine within 5 years because of its wasteful use of fuel energy: "[T]he good Lord has tolerated this foolishness of throwing away 90 percent of the energy in the fuel long enough" (Kettering, 1918). And indeed, in the 1920s through the 1950s peak efficiencies went from 10 percent to as much as 40 percent, with improvements in fuels, combustion system design, friction reduction, and more precise manufacturing processes. Engines became more powerful, and vehicles became heavier, bigger, and faster. However, by the late 1950s, fuel economy had become important, leading to the first large wave of foreign imports. In the wake of the 1973 oil crisis, the issue of energy security arose, and Congress passed the Energy Policy and Conservation Act of 1975 as a means of reducing the country's dependence on imported oil. The act established the Corporate Average Fuel Economy (CAFE) program, which required automobile manufacturers to increase the average fuel economy of passenger cars sold in the United States in 1990 to a standard of 27.5 miles per gallon (mpg) and allowed the U.S. Department of Transportation (DOT) to set appropriate standards for light trucks. The standards are administered in DOT by the National Highway Traffic Safety Administration (NHTSA) on the basis of U.S. Environmental Protection Agency (EPA) city-highway dynamometer test procedures.

FUEL CONSUMPTION AND FUEL ECONOMY

Before proceeding, it is necessary to define the terms *fuel economy* and *fuel consumption*; these two terms are widely used, but very often interchangeably and incorrectly, which can generate confusion and incorrect interpretations:

- *Fuel economy* is a measure of how far a vehicle will travel with a gallon of fuel; it is expressed in miles per gallon. This is a popular measure used for a long time by consumers in the United States; it is used also by vehicle manufacturers and regulators, mostly to communicate with the public. As a metric, fuel economy actually measures distance traveled per unit of fuel.
- *Fuel consumption* is the inverse of fuel economy. It is the amount of fuel consumed in driving a given distance. It is measured in the United States in gallons per 100 miles, and in liters per 100 kilometers in Europe and elsewhere throughout the world. Fuel consumption is a fundamental engineering measure that is directly related to fuel consumed per 100 miles and is useful because it can be employed as a direct measure of volumetric fuel savings. It is actually fuel consumption

that is used in the CAFE standard to calculate the fleet average fuel economy (the sales weighted average) for the city and highway cycles. The details of this calculation are shown in Appendix E. Fuel consumption is also the appropriate metric for determining the yearly fuel savings if one goes from a vehicle with a given fuel consumption to one with a lower fuel consumption.

Because fuel economy and fuel consumption are reciprocal, each of the two metrics can be computed in a straightforward manner if the other is known. In mathematical terms, if fuel economy is X and fuel consumption is Y, their relationship is expressed by $XY = 1$. This relationship is not linear, as illustrated by Figure 2.1, in which fuel consumption is shown in units of gallons per 100 miles, and fuel economy is shown in units of miles per gallon. Also shown in the figure is the decreasing influence on fuel savings that accompanies increasing the fuel economy of high-mpg vehicles. Each bar represents an increase of fuel economy by 100 percent or the corresponding decrease in fuel consumption by 50 percent. The data on the graph show the resulting decrease in fuel consumption per 100 miles and the total fuel saved in driving 10,000 miles. The dramatic decrease in the impact of increasing miles per gallon by 100 percent for a high-mpg vehicle is most visible in the case of increasing the miles per gallon rating from 40 mpg to 80 mpg, where the total fuel saved in driving 10,000 miles is only 125 gallons, compared to 500 gallons for a change from 10 mpg to 20 mpg. Likewise, it is instructive to compare the same absolute value of fuel economy changes—for example, 10-20 mpg and 40-50 mpg. The 40-50 mpg fuel saved in driving 10,000 miles would be 50 gallons, as compared to the 500 gallons in going from 10-20 mpg. Appendix E discusses further implications of the relationship between fuel consumption and fuel economy for various fuel economy values, and particularly for those greater than 40 mpg.

Figure 2.2 illustrates the relationship between the percentage of fuel consumption decrease and that of fuel economy increase. Figures 2.1 and 2.2 illustrate that the amount of fuel saved by converting to a more economical vehicle depends on where one is on the curve.

Because of the nonlinear relationship in Figure 2.1, consumers can have difficulty using fuel economy as a measure of fuel efficiency in judging the benefits of replacing the most inefficient vehicles (Larrick and Soll, 2008). Larrick and Soll further conducted three experiments to test whether people reason in a linear but incorrect manner about fuel economy. These experimental studies demonstrated a systemic misunderstanding of fuel economy as a measure of fuel efficiency. Using linear reasoning about fuel economy leads people to undervalue small improvements (1-4 mpg) in lower-fuel-economy (15-30 mpg range) vehicles where there are large decreases in fuel consumption (Larrick and Soll, 2008) in this range, as shown in Figure 2.1. Fischer (2009) further discusses the potential benefits of utilizing a metric based on fuel consumption as a means to aid consumers in calculating fuel and cost savings resulting from improved vehicle fuel efficiency.

Throughout this report, fuel consumption is used as the metric owing to its fundamental characteristic and its suitability for judging fuel savings by consumers. In cases where the committee has used fuel economy data from the

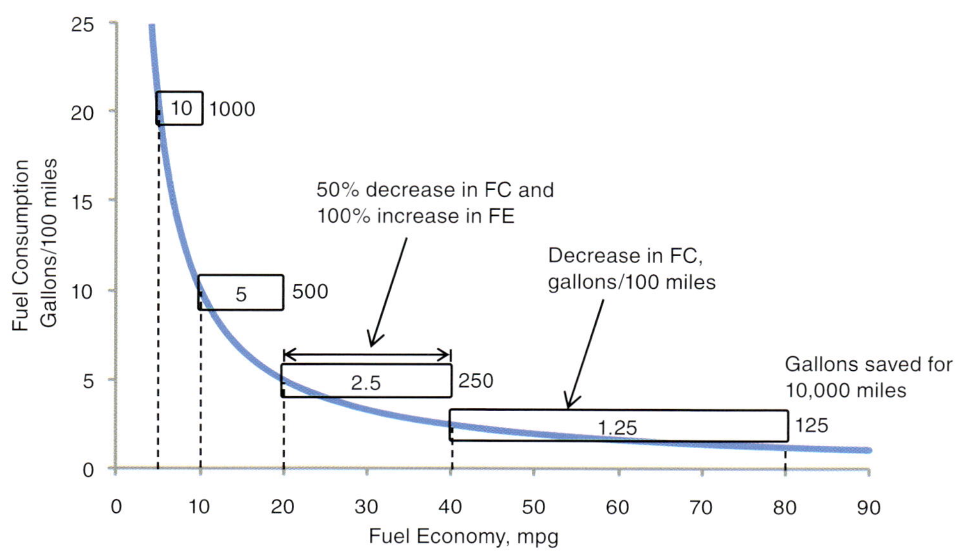

FIGURE 2.1 Relationship between fuel consumption (FC) and fuel economy (FE) illustrating the decreasing reward of improving fuel economy (miles per gallon [mpg]) for high-mile-per-gallon vehicles. The width of each rectangle represents a 50 percent decrease in FC or a 100 percent increase in FE. The number within the rectangle is the decrease in FC per 100 miles, and the number to the right of the rectangle is the total fuel saved over 10,000 miles by the corresponding 50 percent decrease in FC.

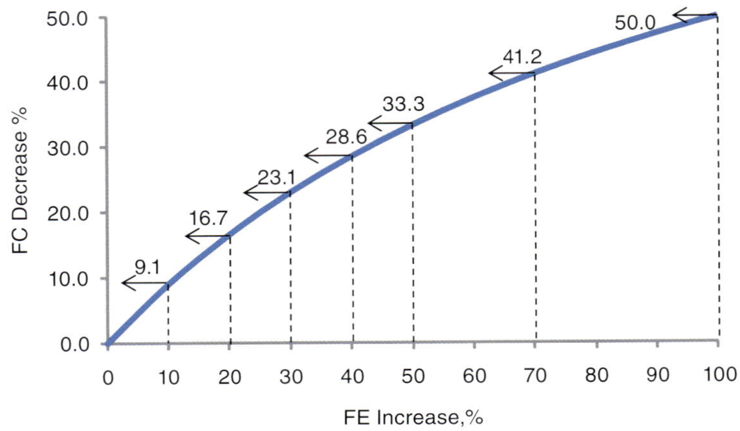

FIGURE 2.2 Percent decrease in fuel consumption (FC) as a function of percent increase in fuel economy (FE), illustrating the decreasing benefit of improving the fuel economy of vehicles with an already high fuel economy.

literature, the data were converted to fuel consumption, using the curve of either Figure 2.1 or 2.2 for changes in fuel economy. Because of this, the committee recommends that the fuel economy information sticker on new cars and trucks should include fuel consumption data in addition to the fuel economy data so that consumers can be familiar with this fundamental metric since fuel consumption difference between two vehicles relates directly to fuel savings. The fuel consumption metric is also more directly related to overall emissions of carbon dioxide than is the fuel economy metric.

ENGINES

Motor vehicles have been powered by gasoline, diesel, steam, gas turbine, and Stirling engines as well as by electric and hydraulic motors. This discussion of engines is limited to power plants involving the combustion of a fuel inside a chamber that results in the expansion of the air/fuel mixture to produce mechanical work. These internal combustion engines are of two types: gasoline spark-ignition and diesel compression-ignition. The discussion also addresses alternative power trains, including hybrid electrics.

Basic Engine Types

Gasoline engines, which operate on a relatively volatile fuel, also go by the name Otto cycle engines (after the person who is credited with building the first working four-stroke internal combustion engine). In these engines, a spark plug is used to ignite the air/fuel mixture. Over the years, variations of the conventional operating cycle of gasoline engines have been proposed. A recently popular variation is the Atkinson cycle, which relies on changes in valve timing to improve efficiency at the expense of lower peak power capability. Since in all cases the air/fuel mixture is ignited by a spark, this report refers to gasoline engines as spark-ignition engines.

Diesel engines—which operate on "diesel" fuels, named after inventor Rudolf Diesel—rely on compression heating of the air/fuel mixture to achieve ignition. This report uses the generic term compression-ignition engines to refer to diesel engines.

The distinction between these two types of engines is changing with the development of engines having some of the characteristics of both the Otto and the diesel cycles. Although technologies to implement homogeneous charge compression ignition (HCCI) will most likely not be available until beyond the time horizon of this report, the use of a homogeneous mixture in a diesel cycle confers the characteristic of the Otto cycle. Likewise the present widespread use of direct injection in gasoline engines confers some of the characteristics of the diesel cycle. Both types of engines are moving in a direction to utilize the best features of both cycles' high efficiency and low particulate emissions.

In a conventional vehicle propelled by an internal combustion engine, either SI or CI, most of the energy in the fuel goes to the exhaust and to the coolant (radiator), with about a quarter of the energy doing mechanical work to propel the vehicle. This is partially due to the fact that both engine types have thermodynamic limitations, but it is also because in a given drive schedule the engine has to provide power over a range of speeds and loads; it rarely operates at its most efficient point.

This is illustrated by Figure 2.3, which shows what is known as an engine efficiency map for an SI engine. It plots the engine efficiency as functions of torque and speed. The plot in Figure 2.3 represents the engine efficiency contours in units of brake-specific fuel consumption (grams per kilowatt-hour) and relates torque in units of brake mean effective pressure (kilopascals). For best efficiency, the engine should operate over the narrow range indicated by the roughly round contour in the middle; this is also referred to later in the chapter as the maximum engine brake thermal efficiency ($\eta_{b,max}$). In conventional vehicles, however, the engine needs to cover

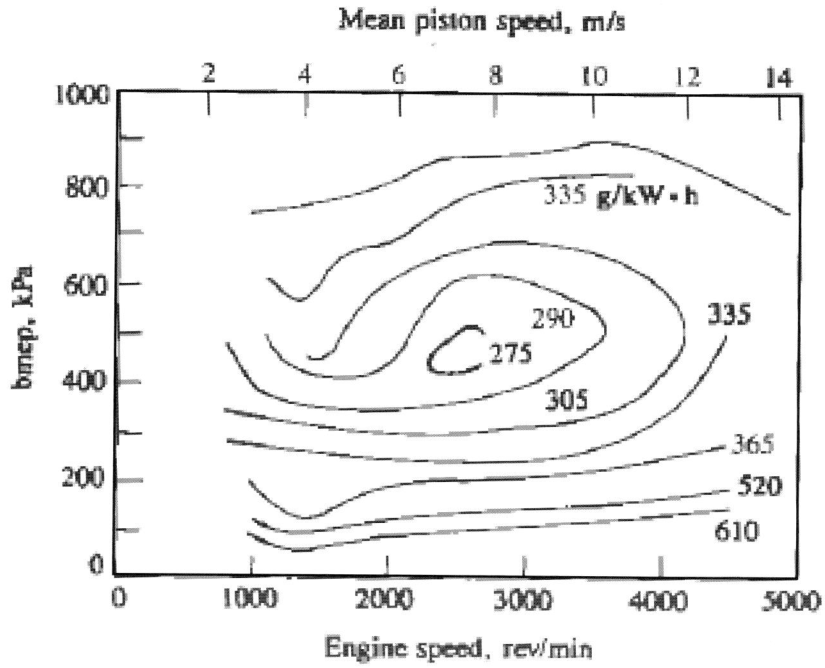

FIGURE 2.3 An example of an engine efficiency map for a spark-ignition engine. SOURCE: Reprinted with permission from Heywood (1988). Copyright 1988 by the McGraw-Hill Companies, Inc.

the entire range of torque and speeds, and so, on average, the efficiency is lower. One way to improve efficiency is to use a smaller engine and to use a turbocharger to increase its power output back to its original level. This reduces friction in both SI and CI engines as well as pumping losses.[1] Increasing the number of gear ratios in the transmission also enables the engine to operate closer to the maximum engine brake thermal efficiency. Other methods to expand the high-efficiency operating region of the engine, particularly in the lower torque region, are discussed in Chapters 4 and 5. As discussed in Chapter 6, part of the reason that hybrid electric vehicles show lower fuel consumption is that they permit the internal combustion engine to operate at more efficient speed-load points.

Computer control, first introduced to meet the air/fuel mixture ratio requirements for reduced emissions in both CI and SI engines, now allows the dynamic optimization of engine operations, including precise air/fuel mixture control, spark timing, fuel injection, and valve timing. The monitoring of engine and emission control parameters by the onboard diagnostic system identifies emission control system malfunctions.

A more recent development in propulsion systems is to add one or two electrical machines and a battery to create a hybrid vehicle. Such vehicles can permit the internal combustion engine to shut down when the vehicle is stopped and allow brake energy to be recovered and stored for later use. Hybrid systems also enable the engine to be downsized and to operate at more efficient operating points. Although there were hybrid vehicles in production in the 1920s, they could not compete with conventional internal combustion engines. What has changed is the greater need to reduce fuel consumption and the developments in controls, batteries, and electric drives. Hybrids are discussed in Chapter 6, but it is safe to say that the long-term future of motor vehicle propulsion may likely include advanced combustion engines, combustion engine-electric hybrids, electric plug-in hybrids, hydrogen fuel cell electric hybrids, battery electrics, and more. The challenge of the next generation of propulsion systems depends not only on the development of the propulsion technology but also on the associated fuel or energy infrastructure. The large capital investment in manufacturing capacity, the motor vehicle fleet, and the associated fuel infrastructure all constrain the rate of transition to new technologies.

Combustion-Related Traits of SI Versus CI Engines

The combustion process within internal combustion engines is critical for understanding the performance of SI versus CI engines. SI-engine combustion occurs mainly by turbulent flame propagation, and as turbulence intensity

[1] "Pumping loss" refers to the energy dissipated through fluid friction and pressure gradients developed from the air flow through the engine. A more detailed explanation is provided in Chapter 4 of this report.

tends to scale with engine speed, the combustion interval in the crank-angle domain remains relatively constant throughout the speed range (at constant intake-manifold pressure and engines having a conventional throttle). Thus, combustion characteristics have little effect on the ability of this type of engine to operate successfully at high speeds. Therefore, this type of engine tends to have high power density (e.g., horsepower per cubic inch or kilowatts per liter) compared to its CI counterpart. CI engine combustion is governed largely by means of the processes of spray atomization, vaporization, turbulent diffusion, and molecular diffusion. Therefore, CI combustion, in comparison with SI combustion, is less impacted by engine speed. As engine speed increases, the combustion interval in the crank-angle domain also increases and thus delays the end of combustion. This late end of combustion delays burnout of the particulates that are the last to form, subjecting these particulates to thermal quenching. The consequence of this quenching process is that particulate emissions become problematic at engine speeds well below those associated with peak power in SI engines. This ultimately limits the power density (i.e., power per unit of displacement) of CI diesel engines.

While power density gets much attention, torque density in many ways is more relevant. Thermal auto ignition in SI engines is the process that limits torque density and fuel efficiency potential. Typically at low to moderate engine speeds and high loads, this process yields combustion of any fuel/air mixture not yet consumed by the desired flame-propagation process. This type of combustion is typically referred to as engine knock, or simply knock. If this process occurs prior to spark ignition, it is referred to as pre-ignition. (This is typically observed at high power settings.) Knock and pre-ignition are to be avoided, as they both lead to very high rates of combustion pressure and ultimately to component failure. While approaches such as turbocharging and direct injection of SI engines alter this picture somewhat, the fundamentals remain. CI diesel engines, however, are not knock limited and have excellent torque characteristics at low engine speed. In the European market, the popularity of turbocharged CI diesel engines in light-duty vehicle segments is not only driven by the economics of fuel economy but also by the "fun-to-drive" element. That is, at equal engine displacement, the turbocharged diesel tends to deliver superior vehicle launch performance as compared with that of its naturally aspirated SI engine counterpart.

FUELS

The fuels and the SI and CI engines that use them have co-evolved in the past 100 years in response to improved technology and customer demands. Engine efficiencies have improved due to better fuels, and refineries are able to provide the fuels demanded by modern engines at a lower cost. Thus, the potential for fuel economy improvement may depend on fuel attributes as well as on engine technology. Implementing certain engine technologies may require changes in fuel properties, and vice versa. Although the committee charge is not to assess alternative liquid fuels (such as ethanol or coal-derived liquids) that might replace gasoline or diesel fuels, it is within the committee charge to consider fuels and the properties of fuels as they pertain to implementing the fuel economy technologies discussed within this report.

Early engines burned coal and vegetable oils, but their use was very limited until the discovery and exploitation of inexpensive petroleum. The lighter, more volatile fraction of petroleum, called gasoline, was relatively easy to burn and met the early needs of the SI engine. A heavier, less volatile fraction, called distillate, which was slower to burn, met the early needs of the CI engine. The power and efficiency of early SI engines were limited by the low compression ratios required for resistance to pre-ignition or knocking. This limitation had been addressed by adding a lead additive commonly known as tetraethyl lead. With the need to remove lead because of its detrimental effect on catalytic aftertreatment (and the negative environmental and human impacts of lead), knock resistance was provided by further changing the organic composition of the fuel and initially by reducing the compression ratio and hence the octane requirement of the engine. Subsequently, a better understanding of engine combustion and better engine design and control allowed increasing the compression ratios back to and eventually higher than the pre-lead-removal levels. The recent reduction of fuel sulfur levels to less than 15 parts per million (ppm) levels enabled more effective and durable exhaust aftertreatment devices on both SI and CI engines.

The main properties that affect fuel consumption in engines are shown in Table 2.1. The table shows that, on a volume basis, diesel has a higher energy content, called heat of combustion, and higher carbon content than gasoline; thus, on a per gallon basis diesel produces almost 15 percent more CO_2. However, on a weight basis the heat of combustion of diesel and gasoline is about the same, and so is the carbon content. One needs to keep in mind that this difference in energy content is one of the reasons why CI engines have lower fuel consumption when measured in terms of gallons rather than in terms of weight. Processing crude oil into fuels for vehicles is a complex process that uses hydrogen to break

TABLE 2.1 Properties of Fuels

	Lower Heat of Combustion (Btu/gal)	Lower Heat of Combustion (Btu/lb)	Density (lb/gal)	Carbon Content (g/gal)	Carbon Content (g/lb)
Gasoline	116,100	18,690	6.21	2,421	392
Diesel	128,500	18,400	6.98	2,778	392
Ethanol (E85)	76,300	11,580	6.59	1,560	237

SOURCE: After GREET Program, Argonne National Laboratory, http://www.transportation.anl.gov/modeling_simulation/GREET/.

down heavy hydrocarbons into lighter fractions. This is commonly called cracking. Diesel fuel requires less "molecular manipulation" for the conversion of crude oil into useful fuel. So if one wants to minimize the barrels of crude oil used per 100 miles, diesel would be a better choice than gasoline.

Ethanol as a fuel for SI engines is receiving much attention as a means of reducing dependence on imported petroleum and also of producing less greenhouse gas (GHG). Today ethanol is blended with gasoline at about 10 percent. Proponents of ethanol would like to see the greater availability of a fuel called E85, which is a blend of 85 percent ethanol and 15 percent gasoline. The use of 100 percent ethanol is widespread in Brazil, but it is unlikely to be used in the United States because engines have difficulty starting in cold weather with this fuel.

The effectiveness of ethanol in reducing GHG is a controversial subject that is not addressed here, since it generally does not affect the technologies discussed in this report. It is interesting to note that in a very early period of gasoline shortage, it was touted as a fuel of the future (Foljambe, 1916).

Ethanol has about 65 percent of the heat of combustion of gasoline, so the fuel consumption is roughly 50 percent higher as measured in gallons per 100 miles. Ethanol has a higher octane rating than that of gasoline, and this is often cited as an advantage. Normally high octane enables increases in the compression ratio and hence efficiency. To take advantage of this form of efficiency increase, the engine would need to be redesigned to accommodate an increased combustion ratio. For technical reasons the improvement with ethanol is very small. Also, during any transition period, vehicles that run on 85 to 100 percent ethanol must also run on gasoline, and since the compression ratio cannot be changed after the engine is built, the higher octane rating of ethanol fuel has not led to gains in efficiency. A way to enable this efficiency increase is to modify the SI engine so that selective ethanol injection is allowed. This technology is being developed and is further discussed in Chapter 4 of this report.

FUEL ECONOMY TESTING AND REGULATIONS

The regulation of vehicle fuel economy requires a reproducible test standard. The test currently uses a driving cycle or test schedule originally developed for emissions regulation, which simulated urban-commute driving in Los Angeles in the late 1960s and the early 1970s. This cycle is variously referred to as the LA-4, the urban dynamometer driving schedule (UDDS), and the city cycle. The U.S. Environmental Protection Agency (EPA) later added a second cycle to better capture somewhat higher-speed driving: this cycle is known as the highway fuel economy test (HWFET) driving schedule, or the highway cycle. The combination of these two test cycles (weighted using a 55 percent city cycle and 45 percent highway cycle split) is known as the Federal Test Procedure (FTP). This report focuses on fuel consumption data that reflect legal compliance with the CAFE requirements and thus do not include EPA's adjustments for its labeling program, as described below. Also discussed below are some technologies—such as those that reduce air-conditioning power demands or requirements—that improve on-road fuel economy but are not directly captured in the FTP.

Compliance with the NHTSA's CAFE regulation depends on the city and highway vehicle dynamometer tests developed and conducted by the EPA for its exhaust emission regulatory program. The results of the two tests are combined (harmonic mean) with a weighting of 55 percent city and 45 percent highway driving. Manufacturers self-certify their vehicles using preproduction prototypes representative of classes of vehicles and engines. The EPA then conducts tests in its laboratories of 10 to 15 percent of the vehicles to verify what the manufacturers report. For its labeling program, the EPA adjusts the compliance values of fuel economy in an attempt to better reflect what vehicle owners actually experience. The certification tests yield fuel consumption (gallons per 100 miles) that is about 25 percent better (less than) EPA-estimated real-world fuel economy. Analysis of the 2009 EPA fuel economy data set for more than 1,000 vehicle models yields a model-averaged difference of about 30 percent.

The certification test fails to capture the full array of driving conditions encountered during vehicle operations. Box 2.1 provides some of the reasons why the certification test does not reflect actual driving. Beginning with model year 2008, the EPA began collecting data on three additional test cycles to capture the effect of higher speed and acceleration, air-conditioner use, and cold weather. These data are part of air pollution emission compliance testing but not fuel economy or proposed greenhouse gas compliance. However, the results from these three test cycles will be used with the two FTP cycles to report the fuel economy on the vehicle label. Table 2.2 summarizes the characteristics of the five test schedules. This additional information guides the selection of a correction factor, but an understanding of fuel consumption based on actual in-use measurement is lacking.

The unfortunate consequence of the disparity between the official CAFE (and proposed greenhouse gas regulation) certification tests and how vehicles are driven in use is that manufacturers have a diminished incentive to design vehicles to deliver real-world improvements in fuel economy if such improvements are not captured by the official test. Some examples of vehicle design improvements that are not completely represented in the official CAFE test are more efficient air conditioning; cabin heat load reduction through heat-resistant glazing and heat-reflective paints; more efficient power steering; efficient engine and drive train operation at all speeds, accelerations, and road grades; and reduced drag to include the effect of wind. The certification tests give no incentive to provide information to the driver that would improve operational efficiency or to reward control strategies that compensate for driver characteristics that increase fuel consumption.

BOX 2.1
Shortcomings of Fuel Economy Certification Test

- *Dynamometer test schedules.* The UDDS and HWFET test schedule (driving cycles) were adopted in 1975 to match driving conditions and dynamometer limitations of that period. Maximum speed (56.7 mph) and acceleration (3.3 mph/sec, or 0-60 mph in 18.2 sec) are well below typical driving. The 55 percent city and 45 percent highway split may not match actual driving. Recent estimates indicate that a weighting of 57 percent highway and 43 percent city is a better reflection of current driving patterns in a number of geographic areas.
- *Test vehicles.* The preproduction prototypes do not match the full range of vehicles actually sold.
- *Driver behavior.* The unsteady driving characteristic of many drivers increases fuel consumption.
- *Fuel.* The test fuel does not match current pump fuel.
- *Air conditioning.* Air conditioning is turned off during the certification test. In addition to overestimating mileage, there is no regulatory incentive for manufacturers to increase air-conditioning efficiency. However, there is substantial market incentive for original equipment manufacturers both to increase air-conditioning efficiency and to reduce the sunlight-driven heating load for customer comfort benefits.
- *Hills.* There are no hills in the EPA certification testing.
- *Vehicle maintenance.* Failure to maintain vehicles degrades fuel economy.
- *Tires and tire pressure.* Test tires and pressures do not generally match in-use vehicle operation.
- *Wind.* There is no wind in the EPA certification testing.
- *Cold start.* There is no cold start in the EPA CAFE certification testing.
- *Turns.* There is no turning in the EPA certification testing.

The measurement of the fuel economy of hybrid, plug-in hybrid, and battery electric vehicles presents additional difficulties in that their performance on the city versus highway driving cycles differs from that of conventional vehicles. Regenerative braking provides a greater gain in city driving than in highway driving. Plug-in hybrids present an additional complexity in measuring fuel economy since this requires accounting of the energy derived from the grid. The Society of Automotive Engineers (SAE) is currently developing recommendations for measuring the emissions and fuel economy of hybrid-electric vehicles, including plug-in and battery electric vehicles. General Motors Company recently claimed that its Chevrolet Volt extended-range electric vehicle achieved city fuel economy of at least 230 miles per gallon, based on development testing using a draft EPA federal fuel economy methodology for the labeling of plug-in electric vehicles (General Motors Company press release, August 11, 2009).

CUSTOMER EXPECTATIONS

The objective of this study is to evaluate technologies that reduce fuel consumption without significantly reducing customer satisfaction. Although each vehicle manufacturer has a proprietary way of defining very precisely how its vehicle must perform, it is assumed here that the following parameters will remain essentially constant as the technologies that reduce fuel consumption are considered:

- Interior passenger volume;
- Trunk space, except for hybrids, where trunk space may be compromised;
- Acceleration, which is measured in a variety of tests, such as time to accelerate from 0 to 60 mph, 0 to 30, 55 to 65 (passing), 30 to 45, entrance ramp to highway, etc.;

TABLE 2.2 Test Schedules Used in the United States for Mileage Certification

	Test Schedule				
Driving Schedule Attributes	Urban (UDDS)	Highway (HWFET)	High Speed (US06)	Air Conditioning (SC03)	Cold Temperature UDDS
Trip type	Low speeds in stop-and-go urban traffic	Free-flow traffic at highway speeds	Higher speeds; harder acceleration and braking	Air conditioning use under hot ambient conditions	City test with colder outside temperature
Top speed	56.7 mph	59.9 mph	80.3 mph	54.8 mph	56.7 mph
Average speed	19.6 mph	48.2 mph	48 mph	21.4 mph	19.6 mph
Maximum acceleration	3.3 mph/sec	3.2 mph/sec	8.40 mph/sec	5.1 mph/sec	3.3 mph/sec
Simulated distance	7.45 mi.	10.3 mi.	8 mi.	3.58	7.45 mi.
Time	22.8 min	12.75 min	10 min	10 min	22.8 min
Stops	17	None	5	5	17
Idling time	18% of time	None	7% of time	19% of time	18% of time
Lab temperature	68-86°F			95°F	20°F
Vehicle air conditioning	Off	Off	Off	On	Off

SOURCE: After http://www.fueleconomy.gov/feg/fe_test_schedules.shtml.

TABLE 2.3 Average Characteristics of Light-Duty Vehicles for Four Model Years

	1975	1987	1998	2008
Adjusted fuel economy (mpg)	13.1	22	20.1	20.8
Weight	4,060	3,220	3,744	4,117
Horsepower	137	118	171	222
0 to 60 acceleration time (sec)	14.1	13.1	10.9	9.6
Power/weight (hp/ton)	67.5	73.3	91.3	107.9

SOURCE: EPA (2008).

- Safety and crashworthiness; and
- Noise and vibration.

These assumptions are very important. It is obvious that reducing vehicle size will reduce fuel consumption. Also, the reduction of vehicle acceleration capability allows the use of a smaller, lower-power engine that operates closer to its best efficiency. These are not options that will be considered.

As shown in Table 2.3, in the past 20 or so years, the net result of improvements in engines and fuels has been increased vehicle mass and greater acceleration capability while fuel economy has remained constant (EPA, 2008). Presumably this tradeoff between mass, acceleration, and fuel consumption was driven by customer demand. Mass increases are directly related to increased size, the shift from passenger cars to trucks, the addition of safety equipment such as airbags, and the increased accessory content. Note that although the CAFE standards for light-duty passenger cars have been for 27.5 mpg since 1990, the fleet average remains much lower through 2008 due to lower CAFE standards for light-duty pickup trucks, sport utility vehicles (SUVs), and passenger vans.

TRACTIVE FORCE AND TRACTIVE ENERGY

The mechanical work produced by the power plant is used to propel the vehicle and to power the accessories. As discussed by Sovran and Blaser (2006), the concepts of tractive force and tractive energy are useful for understanding the role of vehicle mass, rolling resistance, and aerodynamic drag. These concepts also help evaluate the effectiveness of regenerative braking in reducing the power plant energy that is required. The analysis focuses on test schedules and neglects the effects of wind and hill climbing. The instantaneous tractive force (F_{TR}) required to propel a vehicle is

$$F_{TR} = R + D + \left[M + 4\left(\frac{I_W}{r_W^2}\right)\right]\frac{dV}{dt} = r_0 Mg + C_D A \frac{V^2}{2}\rho + \left[M + 4\left(\frac{I_W}{r_W^2}\right)\right]\frac{dV}{dt} \tag{2.1}$$

where R is the rolling resistance, D is the aerodynamic drag with C_D representing the aerodynamic drag coefficient, M is the vehicle mass, V is the velocity, dV/dt is the rate of change of velocity (i.e., acceleration or deceleration), A is the frontal area, r_o is the tire rolling resistance coefficient, g is the gravitational constant, I_w is the polar moment of inertia of the four tire/wheel/axle rotating assemblies, r_w is its effective rolling radius, and ρ is the density of air. This form of the tractive force is calculated at the wheels of the vehicle and therefore does not consider the components within the vehicle system such as the power train (i.e., rotational inertia of engine components and internal friction).

The tractive energy required to travel an incremental distance dS is $F_{TR}\,Vdt$, and its integral over all portions of a driving schedule in which $F_{TR} > 0$ (i.e., constant-speed driving and accelerations) is the total tractive-energy requirement, E_{TR}. For each of the EPA driving schedules, Sovran and Blaser (2006) calculated tractive energy for a large number of vehicles covering a broad range of parameter sets (r_0, C_D, A, M) representing the spectrum of current vehicles. They then fitted the data with a linear equation of the following form:

$$\frac{E_{TR}}{MS} = \alpha r_0 + \beta\left(\frac{C_D A}{M}\right) + \gamma\left(1 + \frac{4I_w}{Mr_w^2}\right) \tag{2.2}$$

where S is the total distance traveled in a driving schedule, and α, β, and γ are specific but different constants for the UDDS and HWFET schedules. Sovran and Blaser (2006) also identified that a combination of five UDDS and three HWFET schedules very closely reproduces the EPA combined fuel consumption of 55 percent UDDS plus 45 percent HWFET, and provided its values of α, β, and γ.

The same approach was used for those portions of a driving schedule in which $F_{TR} < 0$ (i.e., decelerations), where the power plant is not required to provide energy for propulsion. In this case the rolling resistance and aerodynamic drag retard vehicle motion, but their effect is not sufficient to follow the driving cycle deceleration, and so some form of wheel braking is required. When a vehicle reaches the end of a schedule and becomes stationary, all the kinetic energy of its mass that was acquired when $F_{TR} > 0$ has to have been removed. Consequently the decrease in kinetic energy produced by wheel braking is

$$E_{BR}/MS = \gamma\left(1 + 4I_w/Mrw^2\right) - \alpha' r_0 - \beta'\left(C_D A/M\right). \tag{2.3}$$

The coefficients α' and β' are also specific to the test schedule and are given in the reference. Two observations are of interest: (1) γ is the same for both motoring and braking as it relates to the kinetic energy of the vehicle; (2) since the energy used in rolling resistance is $r_0\, M\, g\, S$, the sum of α and α' is equal to g.

Sovran and Blaser (2006) considered 2,500 vehicles from the EPA database for 2004 and found that their equations fitted the tractive energy for both the UDDS and HWFET schedules with an $r = 0.999$, and the braking energy with an

$r = 0.99$, where r represents the correlation coefficient based on least squares fit of the data.

To illustrate the dependence of tractive and braking energy on vehicle parameters, Sovran and Blaser (2006) used the following three sets of parameters. Fundamentally the energy needed by the vehicle is a function of the rolling resistance, the mass, and the aerodynamic drag times frontal area. By combining the last three into the results shown in Table 2.4, Sovran and Blaser (2006) covered the entire fleet in 2004. The "high" vehicle has a high rolling resistance, and high aerodynamic drag relative to its mass. This would be typical of a truck or an SUV. The "low" vehicle requires low tractive energy and would be typical for a future vehicle. These three vehicles cover the entire spectrum in vehicle design.

The data shown in Table 2.5 were calculated using these values. The low vehicle has a tractive energy requirement that is roughly two-thirds that of the high vehicle. It should also be noted that as the vehicle design becomes more efficient (i.e., the low vehicle), the fraction of energy required to overcome the inertia increases. As expected, for both driving schedules the *normalized* tractive energy, E_{TR}/MS, decreases with reduced rolling and aerodynamic resistances. What is more significant, however, is that at each level, the *actual tractive* energy is strongly dependent on vehicle mass, through its influence on the rolling and inertia components. This gives mass reduction high priority in efforts to reduce vehicle fuel consumption.

TABLE 2.4 Vehicle Characteristics

Vehicle	r_o	C_dA/M
High	0.012	0.00065
Mid	0.009	0.0005
Low	0.006	0.0003

SOURCE: Based on Sovran and Blaser (2006).

TABLE 2.5 Estimated Energy Requirements for the Three Sovran and Blaser (2006) Vehicles in Table 2.4 for the UDDS and HWFET Schedules

	E_{TR}/MS (Normalized)	Rolling Resistance (%)	Aerodynamic Drag (%)	Inertia (%)	Braking/ Tractive (%)
UDDS Vehicle					
High	0.32	28	22	50	36
Mid	0.28	24	19	57	45
Low	0.24	19	14	68	58
HWFET Vehicle					
High	0.34	32	56	13	6
Mid	0.27	30	54	16	10
Low	0.19	29	47	24	18

Effect of Driving Schedule

It is evident from Table 2.5 that inertia is the dominant component on the UDDS schedule, while aerodynamic drag is dominant on the HWFET. The larger any component, the greater the impact of its reduction on tractive energy.

On the UDDS schedule, the magnitude of required braking energy relative to tractive energy is large at all three vehicle levels, increasing as the magnitude of the rolling and aerodynamic resistances decreases. The high values are due to the many decelerations that the schedule contains. The braking energy magnitudes for HWFET are small because of its limited number of decelerations.

In vehicles with conventional power trains, the wheel-braking force is frictional in nature, and so all the vehicle kinetic energy removed is dissipated as heat. However, in hybrid vehicles with regenerative-braking capability, some of the braking energy can be captured and then recycled for propulsion in segments of a schedule where $F_{TR} > 0$. This reduces the *power plant* energy required to provide the E_{TR} necessary for propulsion, thereby reducing fuel consumption. The significant increase in normalized tractive energy (E_{TR}/MS) with decreasing rolling and aerodynamic resistances makes reduction of these resistances even *more* effective in reducing fuel consumption in hybrids with regenerative braking than in conventional vehicles. The relatively small values of braking-to-tractive energy on the HWFET indicate that the fuel consumption reduction capability of regenerative braking is minimal on that schedule. As a result, hybrid power trains only offer significant fuel consumption reductions on the UDDS cycle. However, as pointed out in Chapter 6, hybridization permits engine downsizing and engine operation in more efficient regions, and this applies to the HWFET schedule also.

Effect of Drive Train

Given the tractive energy requirements (plus idling and accessories), the next step is to represent the efficiency of the power train. The power delivered to the output shaft of the engine is termed the *brake output power*, and should not be confused with the *braking energy* mentioned in the previous section. The brake output power, P_b, of an engine is the difference between its indicated power, P_i, and power required for pumping, P_p; friction, P_f; and engine auxiliaries, P_a (e.g., fuel, oil, and water pumps).

$$P_b = P_i - P_p - \left(P_f + P_a\right) \tag{2.4}$$

Brake thermal efficiency is the ratio of brake power output to the energy rate into the system (the mass flow rate of fuel times its energy density).

$$\eta_b = \eta_i - \frac{P_p}{m_f H_f} - \frac{\left(P_f + P_a\right)}{m_f H_f} \tag{2.5}$$

The brake thermal efficiency is η_b, while η_i is the indicated thermal efficiency, and H_f is the lower heating value of the fuel. This equation provides the means for relating pumping losses, engine friction, and auxiliary load to the overall engine efficiency. Equations for fuel use during braking and idling are not shown here but can be found in Sovran and Blaser (2003), as can the equations for average schedule and maximum engine efficiency.

Ultimately the fuel consumption is given by Equation 2.6:

$$g^* = \left\{ \frac{\frac{E_{TR}}{\eta^*_{dr}} + E_{Accessories}}{H_f \eta_{b,\max} \left(\frac{\eta^*_b}{\eta_{b,\max}} \right)} \right\} + g^*_{braking} + g^*_{idling} \qquad (2.6)$$

where in addition to the terms defined earlier, g^* is the fuel consumption over the driving schedule, $g^*_{braking}$ and g^*_{idling} represent the fuel consumed during idling and braking, H_f is the fuel density of fuel, η^*_{dr} is the average drive train efficiency for the schedule, $\eta_{b,max}$ is the maximum engine brake thermal efficiency, η^*_b is the average engine brake thermal efficiency, and $E_{Accessories}$ is the energy to power the accessories. The term $\eta_{b,max}$ is repeated in the denominator to show that to minimize fuel consumption the fraction in the denominator should be as large as possible. Thus things should be arranged so that the average engine efficiency be as close to the maximum.

The principal term in Equation 2.6 is the bracketed term. Clearly fuel consumption can be reduced by reducing E_{TR} and $E_{Accessories}$. It can also be reduced by increasing $\eta^*_b/\eta_{b,max}$. As stated earlier, this can be done by downsizing the engine or by increasing the number of gears in the transmission so that average engine brake thermal efficiency, η^*_b, is increased. Equation 2.6 explains why reducing rolling resistance or aerodynamic drag without changes in engine or transmission may not maximize the benefit, since it may move $\eta^*_b/\eta_{b,max}$ farther from its optimum point. In other words, changing to lower-rolling-resistance tires without modifying the power train will not give the full benefit.

The tractive energy E_{TR} can be precisely determined given just three parameters, rolling resistance r_0, the product of aero coefficient and frontal area C_DA, and vehicle mass M. However, many of the other terms in Equation 2.6 are difficult to evaluate analytically. This is especially true of the engine efficiencies, which require detailed engine maps. Thus converting the tractive energy into fuel consumption is best done using a detailed step-by-step simulation. This simulation is usually carried out by breaking down the test schedule into 1-second intervals, computing the E_{TR} for each interval using detailed engine maps along with transmission characterizations, and adding up the interval values to get the totals for the drive cycle analyzed. Such a simulation is frequently called a full system simulation, FSS.

The discussion above on tractive energy highlights the fact that the effects of the three principal aspects of vehicle design—vehicle mass, rolling resistance, and aerodynamic drag—can be used to calculate precisely the amount of energy needed to propel the vehicle for any kind of drive schedule. Further, the equations developed highlight both the effect of the various parameters involved and at the same time demonstrate the complexity of the problem. Although the equations provide understanding, in the end estimating the fuel consumption of a future vehicle must be determined by FSS modeling and ultimately by constructing a demonstration vehicle.

DETAILED VEHICLE SIMULATION

The committee obtained results of a study by Ricardo, Inc. (2008) for a complete simulation for a 2007 Camry passenger car. This FSS is discussed further in Chapter 8; one set of results is used here for illustration. Table 2.6 gives the specifications of the vehicle in terms of the parameters used in the simulation.

First, the tractive energy and its components for this vehicle were calculated to illustrate how these vary with different test schedules. Although the US06 cycle described in Table 2.2 is not yet used for fuel economy certification, it is interesting to note how it affects the energy distribution. Table 2.7 shows the results. Energy to the wheels and rolling resistance increase from the UDDS to the US06, with the total tractive energy requirement being almost double that of the UDDS. The aero energy requirement increases from the UDDS to the HWFET, but it is not much increased in going to the US06, in spite of the higher peak speed. What is somewhat surprising is the amount of braking energy for the UDDS and the US06 compared to the HWFET. This is where hybrids excel.

For the highway, rolling resistance and aero dominate, and very little energy is dissipated in the brakes. As expected, the aero is dominant for the US06, where it is more than

TABLE 2.6 Specifications of Vehicle Simulated by Ricardo, Inc. (2008)

Mass	1,644 kg
C_D	0.30
A	2.3 m^2

TABLE 2.7 Energy Distribution for Various Schedules (in kilowatt-hours)

	Total Tractive Energy	Total Rolling Resistance	Total Aerodynamic Drag	Braking Energy	Braking/ Tractive (%)
Urban	1.250	0.440	0.310	0.500	40.00
Highway	1.760	0.610	1.000	0.150	8.52
US06	2.390	0.660	1.170	0.560	23.43

half the total tractive energy. Note, though, that the US06 has a significant amount of energy dissipated in the brakes.

As discussed earlier, some people will drive in a UDDS environment and some on the highway. A vehicle optimized for one type of driving will not perform as well for the other, and it is not possible to derive a schedule that fits all driving conditions. Table 2.7 shows the impractically of developing a test that duplicates the actual driving patterns.

Note that the data in Table 2.7 show the actual energy in kilowatt-hours used to drive each schedule. The unit of total energy is used to allow for an easier comparison between the schedules on the basis of energy distribution. Since as shown in Table 2.2, the distances are 7.45 miles for the UDDS, 10.3 miles for the HWFET, and 8 miles for the US06, the energies should be divided by distance to provide the energy required per mile.

An FSS provides a detailed breakdown of where the energy goes, something that is not practical to do with real vehicles during a test schedule. Figure 2.4 illustrates the total energy distribution in the midsize car, visually identifying where the energy goes.

Table 2.8 shows the fuel consumed for this vehicle for the UDDS, HWFET, and US06 schedules. Efficiency is the ratio of tractive energy divided by "fuel energy input." Clearly this gives a more succinct picture of the efficiency of an internal combustion engine power train in converting fuel to propel a vehicle and to power the accessories. Depending on the drive schedule, it varies from 15 to 25 percent (including the energy to power accessories). This range is significantly less than the peak efficiency $\eta_{b,max}$ discussed earlier.

In addition to the specific operating characteristics of the particular components, the computation of engine fuel consumption depends on the following inputs: (1) the transmission gear at each instant during the driving schedule and (2) the engine fuel consumption rate during braking and idling. None of these details is available, so the data in Table 2.8 should be considered as an illustrative example of the energy distribution in 2007 model-year vehicles with conventional SI power trains.

FINDINGS AND RECOMMENDATIONS

Finding 2.1: Fuel consumption has been shown to be the fundamental metric to judge fuel efficiency improvements from both an engineering and a regulatory viewpoint. Fuel economy data cause consumers to undervalue small increases (1-4 mpg) in fuel economy for vehicles in the 15- to 30-mpg range, where large decreases in fuel consumption can be realized with small increases in fuel economy. For example, consider the comparison of increasing the mpg rating from 40 mpg to 50 mpg, where the total fuel saved in driving 10,000 miles is only 50 gallons, compared to 500 gallons for a change from 10 mpg to 20 mpg.

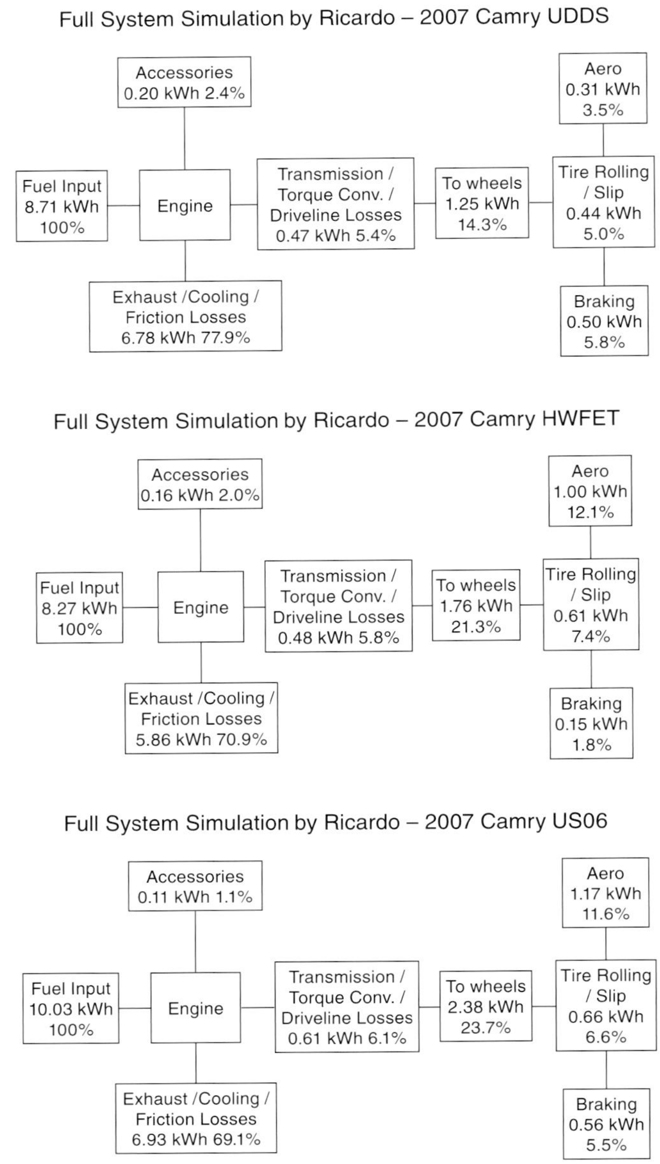

FIGURE 2.4 Energy distribution obtained through full-system simulation for UDDS (top), HWFET (middle), and US06 (bottom). SOURCE: Ricardo, Inc. (2008).

TABLE 2.8 Results of Full System Simulation (energy values in kilowatt-hours)

	Total Tractive Energy	Fuel Input Energy	Power Train Efficiency (%)
Urban	1.250	8.59	14.6
Highway	1.760	8.01	22.0
US06	2.390	9.66	24.7

Recommendation 2.1: Because differences in the fuel consumption of vehicles relate directly to fuel savings, the labeling on new cars and light-duty trucks should include information on the gallons of fuel consumed per 100 miles traveled in addition to the already-supplied data on fuel economy so that consumers can become familiar with fuel consumption as a fundamental metric for calculating fuel savings.

Finding 2.2: Fuel consumption in this report is evaluated by means of the two EPA schedules: UDDS and HWFET. In the opinion of the committee, the schedules used to compute CAFE should be modified so that vehicle test data better reflect actual fuel consumption. Excluding some driving conditions and accessory loads in determining CAFE discourages the introduction of certain technologies into the vehicle fleet. The three additional schedules recently adopted by the EPA for vehicle labeling purposes—ones that capture the effects of higher speed and acceleration, air-conditioner use, and cold weather—represent a positive step forward, but further study is needed to assess to what degree the new test procedures can fully characterize changes in in-use vehicle fuel consumption.

Recommendation 2.2: The NHTSA and the EPA should review and revise fuel economy test procedures so that they better reflect in-use vehicle operating conditions and also better provide the proper incentives to manufacturers to produce vehicles that reduce fuel consumption.

REFERENCES

EPA (U.S. Environmental Protection Agency). 2008. Light-Duty Automotive Technology and Fuel Economy Trends: 1975 Through 2008. EPA420-R-08-015. September. Washington, D.C.

Fischer, C. 2009. Let's turn CAFE regulation on its head. Issue Brief No. 09-06. May. Resources for the Future, Washington, D.C.

Foljambe, E.S. 1916. The automobile fuel situation. SAE Transactions, Vol. 11, Pt. I.

General Motors Company. 2009. Chevy Volt gets 230 mpg city EPA rating. Press release. August 11.

Heywood, J.B., 1988. Internal Combustion Engine Fundamentals. McGraw-Hill, New York.

Kettering, C.F. 1918. Modern aeronautic engines. SAE Transactions, Vol. 13, Pt. II.

Larrick, R., and J. Soll. 2008. The mpg illusion. Science 320(5883):1593-1594.

Ricardo, Inc. 2008. A Study of Potential Effectiveness of Carbon Dioxide Reducing Vehicle Technologies. Prepared for the U.S. Environmental Protection Agency. EPA420-R-08-004. Contract No. EP-C-06-003. Work Assignment No. 1-14. Ann Arbor, Michigan.

Sovran, G., and D. Blaser. 2003. A contribution to understanding automotive fuel economy and its limits. SAE Paper 2003-01-2070. SAE International, Warrendale, Pa.

Sovran, G., and D. Blaser. 2006. Quantifying the potential impacts of regenerative braking on a vehicle's tractive-fuel consumption for the US, European and Japanese driving schedules. SAE Paper 2006-01-0664. SAE International, Warrendale, Pa.

3

Cost Estimation

INTRODUCTION

As a general rule, reduced fuel consumption comes at a cost. The cost may be due to more expensive materials, increased manufacturing complexity, or a tradeoff with other vehicle attributes such as power or size. In addition to increased manufacturing costs, other costs of doing business are likely to be affected to a greater or lesser degree. These indirect costs include research and development (R&D), pensions and health care, warranties, advertising, maintaining a dealer network, and profits. The most appropriate measure of cost for the purpose of evaluating the costs and benefits of fuel economy regulations is the long-run increase in retail price paid by consumers under competitive market conditions.[1] The retail price equivalent (RPE) cost of decreasing fuel consumption includes not only changes in manufacturing costs but also any induced changes in indirect costs and profit.

Most methods for estimating manufacturing costs begin by identifying specific changes in vehicle components or designs, and they then develop individual cost estimates for each affected item. Most changes result in cost increases, but some, such as the downsizing of a V6 engine to an I4, will reduce costs. Component cost estimates can come from a variety of sources, including interviews of original equipment manufacturers (OEMs) and suppliers, prices of optional equipment, and comparisons of models with and without the technology in question. Total costs are obtained by adding up the costs of changes in the individual components.

An alternative method, which has only just begun to be used for estimating fuel economy costs, is to tear down a component into the fundamental materials, labor, and capital required to make it, and then to estimate the cost of every nut and bolt and every step in the manufacturing process (Kolwich, 2009). A potential advantage of this method is that total costs can be directly related to the costs of materials, labor, and capital so that as their prices change, cost estimates can be revised. However, this method is difficult to apply to new technologies that have not yet been implemented in a mass-production vehicle, whose designs are not yet finalized and whose impact on changing related parts is not yet known.

Differences in cost estimates from different sources arise in a number of ways:

- Assumptions about the costs of commodities, labor, and capital;
- Judgments about the changes in other vehicle components required to implement a given technology;
- Definitions of "manufacturing cost" and what items are included in it; and
- Assessments of the impacts of technologies on indirect costs.

This chapter discusses the premises, concepts, and methods used in estimating the costs of fuel economy improvement, highlights areas where differences arise, and presents the committee's judgments on the key issue of RPE markup factors.

Information on costs can be used with assumptions on payback periods, discount rates, price of fuel, and miles driven per year to provide an estimate of the cost-effectiveness of technologies. However, the statement of task given to the committee is to look at the costs and fuel consumption benefits of individual technologies. Performing cost-effectiveness analysis was not included within the committee's task and was not done by the committee. The accurate calculation of benefits of improved fuel efficiency is a complex task that is being undertaken by the National Highway Traffic Safety Administration (NHTSA) and the U.S. Environmental Protection Agency (EPA) as part of their current joint regulatory efforts.

[1] As explained below, this rests on the premise that the global automotive market can be reasonably characterized (in economic jargon) as either a perfectly competitive or a monopolistically competitive market. Under such market conditions, products are sold, in the long run, at their average cost of production, including a normal rate of return to capital but no excess profits. Increased costs of production will therefore be fully passed on to consumers. The total cost of resources plus the consumers' surplus loss due to the price increase is, to a close approximation, equal to the increase in long-run retail price times the volume of sales.

PREMISES

In the committee's judgment, the concept of incremental retail price equivalent cost is most appropriate for the NHTSA's purposes because it best represents the full, long-run economic costs of increasing fuel economy. The NHTSA has used the RPE method in its rulemakings on fuel economy, for example in the final rule for model year 2011 light-duty vehicles (DOT/NHTSA, 2009, pp. 346-352). Incremental RPE estimates are intended to represent the average additional price that consumers would pay for a fuel economy technology implemented in a typical vehicle under average economic conditions and typical manufacturing practices. These estimates are intended to represent long-run, high-volume, industry-average production costs, incorporating rates of profit and overhead expenses including warranties, transport, and retailing. Although learning and technological progress never stop, RPEs are intended to represent costs after an initial period of rapid cost reduction that results from learning by doing.[2] The committee uses the term *substantially learned* as opposed to *fully learned* to convey that cost reductions due to increasing volumes may continue to occur. RPEs are not intended to replicate the market price of a specific vehicle or a specific optional feature at a specific time. The market price of a particular vehicle at a particular time depends on many factors (e.g., market trends, marketing strategies, profit opportunities, business cycles, temporary shortages or surpluses) other than the cost of manufacturing and retailing a vehicle or any given component. It is not appropriate to base a long-term policy such as fuel economy standards on short-run conditions or special circumstances.

The RPE concept, unfortunately, is not easy to apply. It raises a number of difficult questions about appropriate premises and assumptions and reliable sources of data. It frequently relies on the application of markup factors, which could vary depending on the nature of the technology and the basis for the original cost estimate. When an RPE markup factor is used, the definition of the cost to which it applies is critical. Much of the disagreement over RPE multipliers can be traced to inconsistent definition of the cost to be marked up. The following are key premises of the committee's application of the RPE method.

- *Incremental RPE.* The relevant measure of cost is the change in RPE in comparison to an equivalent vehicle without the particular fuel economy technology. More often than not, a fuel economy technology replaces an existing technology. For example, a 6-speed automatic transmission replaces a 5-speed, a compression-ignition (CI) engine replaces a spark-ignition (SI) engine, or a set of low-rolling-resistance tires replaces a set with higher rolling resistance. What matters is the change in RPE rather than the total RPE of the new technology. This requires that an estimate of the RPE of the existing technology be subtracted from that of the new technology.
- *Equivalent vehicle size and performance.* Estimating the cost of decreasing fuel consumption requires one to carefully specify a basis for comparison. The committee considers that to the extent possible, fuel consumption cost comparisons should be made at equivalent acceleration performance and equivalent vehicle size. Other vehicle attributes matter as well, such as reliability, noise, and vibration. Ideally, cost and fuel economy comparisons should be made on the basis of no compromise for the consumer. Often there are differences of opinion about what design and engineering changes may be required to ensure no compromise for the consumer. This, in turn, leads to differing bills of materials to be costed out, which leads to significant differences in incremental RPE estimates.
- *Learning by doing, scale economies, and competition.* When new technologies are first introduced and only one or two suppliers exist, costs are typically higher than they will be in the long run due to lack of scale economies, as-yet-unrealized learning by doing, and limited competition. These transitional costs can be important to manufacturers' bottom lines and should be considered. However, nearly all cost estimates are developed assuming long-run, high-volume, average economic conditions. Typical assumptions include (1) high volume, (2) substantially learned component costs, and (3) competition provided by at least three global suppliers available to each manufacturer (Martec Group, Inc., 2008a, slide 3). Under these assumptions, it is not appropriate to employ traditional learning curves to predict future reductions in cost as production experience increases. However, if cost estimates are for novel technology and do not reflect learning by doing, then the application of learning curves as well as the estimation of scale economies may be appropriate. The use of such methods introduces substantial uncertainty, however, since there are no proven methods for predicting the amount of cost reduction that a new technology will achieve.
- *Normal product cycles.* As a general rule, premises include normal redesign and product turnover schedules. Accelerated rates of implementation can increase costs by decreasing amortization periods and by demanding more engineering and design resources than are available. Product cycles are discussed in Chapter 7.
- *Purchased components versus in-house manufacture.* Costs can be estimated at different stages in the manufacturing process. Manufacturing cost estimates gen-

[2] Learning by doing represents the increase in productivity and decrease in cost that occurs during a technology's lifetime as a result of manufacturers' gaining experience in producing the technology. The impacts of learning on costs can be represented as a volume-based learning where costs reductions occur with increasing production levels or as a time-based learning where cost reductions occur over time.

erally do not include warranty, profit, transportation, and retailing costs, and may not include overhead or research and development. Other estimates are based on the prices that original equipment manufacturers (OEMs) would pay a Tier 1 supplier for a fully manufactured component.[3] These estimates include the supplier's overhead, profit, and R&D costs, but not costs incurred by the OEM. RPEs attempt to estimate the fully marked-up cost to the ultimate vehicle purchaser. A key issue for cost estimates based on Tier 1 supplier costs is the appropriate markup to RPE. This will depend on the degree to which the part requires engineering and design changes to be integrated into the vehicle, and other factors.

- *Allocation of overhead costs.* Specific changes in vehicle technology and design may affect some of an OEM's costs of doing business and not others. A reduction in engine friction, for example, might not affect advertising budgets or transportation costs. To date there is a very limited understanding of how to determine which costs of doing business are affected by each individual technology and how to develop technology-specific markups (e.g., Rogozhin et al., 2009). In theory, this approach has the potential to yield the most accurate results. However, in practice, unambiguous attribution of costs to specific vehicle components is difficult. For example, despite extensive reliability testing, it is not possible to predict with certainty what impact a technology or design change will have on warranty costs. Furthermore, there are significant cost components that cannot logically be allocated to any individual component. Among these are the maintenance of a dealer network and advertising. Yet, these costs must be paid. The RPE method assumes that such costs should be allocated in proportion to the component's cost and that overall overhead costs will increase in proportion to total vehicle cost. This will not necessarily produce the most accurate estimate for each individual item but is consistent with the goal of estimating long-run average costs.

COMPONENTS OF COST

Although different studies describe and group the components of the retail price equivalent (long-run average cost) in different ways, there are four fundamental components: (1) the variable costs of manufacturing components, (2) fixed costs of manufacturing components, (3) variable costs of vehicle assembly, and (4) fixed costs of vehicle assembly and sale. The distinction between variable and fixed costs is not a sharp one, because many "fixed" costs scale to some extent with production volume. In fact, the degree to which fixed or overhead costs scale with variable costs is a key area of uncertainty.

Although many components are manufactured in-house by OEMs, it is useful to distinguish between component and vehicle assembly costs, because many manufacturers purchase 50 percent or more of a vehicle's components from suppliers. Transaction prices and price estimates from Tier 1 and Tier 2 suppliers are a major source of information on the costs of fuel economy technologies.

Variable manufacturing costs of components include materials, labor, and direct labor burden (Table 3.1). Variable manufacturing costs are sometimes referred to as *direct manufacturing costs,* although when this term is used it typically includes the depreciation and amortization of manufacturing equipment. Fixed costs of component manufacturing include tooling and facilities depreciation and amortization associated with capital investments, manufacturing overhead (e.g., R&D, engineering, warranty, etc.), and profit (or return to capital). Unfortunately, terminology frequently differs from one study to another. Total manufacturing costs (variable plus fixed) are equivalent to the price that a Tier 1 supplier would charge an OEM for a finished component, ready for installation.

OEM or assembly costs include the variable costs of materials, labor, and direct labor burden for vehicle assem-

TABLE 3.1 Components of Vehicle Retail Price Equivalent (Long-Run Average Cost)

Component Manufacturing (Subassembly)
Variable component manufacturing costs
Materials
Labor
Direct labor burden
Fixed component manufacturing costs
Tooling and facilities depreciation and amortization
R&D
Engineering
Warranty
Other overhead
Profit
Vehicle Assembly and Marketing
Variable costs
Assembly materials
Assembly labor
Direct labor burden
Fixed costs
Tooling and facilities depreciation and amortization
Warranty
R&D
Engineering
Warranty
Other overhead
Transportation
Marketing and advertising
Dealer costs and profit
Original equipment manufacturer profit

[3]Tier 1 suppliers contract directly with OEMs, whereas Tier 2 suppliers contract with Tier 1 suppliers.

bly. Fixed costs include facilities and tooling depreciation and amortization, warranty, R&D, engineering, advertising, dealer expenses and profit, transportation, and OEM return on investment (profit). The sum total of all costs, divided by the Tier 1 supplier price (or equivalent), is called the RPE markup.

The costs of inputs to the production process can vary over time. Some key components, such as electrical systems, emissions controls, and hybrid vehicle batteries, use relatively expensive metals whose prices can be volatile, significantly impacting manufacturing costs. The prices of many of these metals increased dramatically prior to the global recession beginning in 2008, but have since returned to previous levels. Most publicly available estimates of technology costs do not explicitly reflect uncertainties about future commodity prices.

FACTORS AFFECTING COSTS OVER TIME AND ACROSS MANUFACTURERS

Cost estimates for fuel economy technologies are typically presented as a single point estimate or as a range. In fact, costs will vary over time and even across manufacturers owing to technological progress, experience (learning by doing), prices of commodities, labor and capital, and the nature of the vehicles manufactured.

Economies of Scale

Scale economies describe the tendency for average manufacturing costs to decrease with increasing volume, as fixed costs are distributed over a greater number of units produced. The automobile industry is characterized by large economies of scale. Although sources differ, full scale economies are generally considered to be reached at between 100,000 and 500,000 units per year. Martec Group, Inc. (2008a), for example, asserts that production efficiencies are maximized at 250,000 to 300,000 units. Honda cited a maximum efficiency of 300,000 units in its comments to the DOT/NHTSA (2009, p. 185).

Technological Progress and Learning by Doing

Although cost estimates are generally premised on full scale economies and fully learned technologies, both the EPA and the NHTSA believe that not all Tier 1 supplier or piece cost estimates represent fully learned technology costs. In their view, learning curves should be applied for the more novel technologies not in widespread use today.[4] The EPA listed 16 advanced technologies that, in its judgment, would experience future cost reductions relative to current estimates through learning by doing. Technologies such as cylinder deactivation, camless valve trains, gasoline direct injection with lean burn, turbocharging with engine downsizing, and hybrid systems from stop-start to full hybrids and plug-in hybrids were all assumed to have progress ratios of 0.8 (i.e., a doubling of cumulative production would reduce costs by 20 percent). Diesel emissions control systems were assumed to have smaller progress ratios of 0.9 (EPA, 2008a, Table 4.2-3).

If supplier cost estimates truly represent fully learned costs (at full scale economies), then there is no justification for assuming future learning by doing. The cost estimates made by Martec for the Northeast States Center for a Clean Air Future (NESCCAF), for example, were intended to reflect cost reductions by learning that would occur over the period 2009-2011. In its study for the Alliance of Automobile Manufacturers, Martec intended that its cost estimates reflect full scale economies and full learning: "Martec specified an extremely high annual volume target [500,000 units per year] specifically to drive respondents to report mature, forward costs expected in the future with the impact of learning fully reflected" (Martec Group, Inc., 2008b, p. 7). But Martec identifies two sources of learning: (1) improvement in manufacturing productivity, largely as a result of production volume; and (2) changes in system design. Martec considered the latter to be technological innovations that would change the system architecture and thus the technology itself, requiring new cost estimates. Thus, the learning considered by Martec in its estimates is based on the belief that the Tier 1 and Tier 2 suppliers would implicitly include learning effects of the first type in their high-volume cost estimates, and would exclude learning of the second type.

In its 2011 corporate average fuel economy (CAFE) rulemaking, the NHTSA recognized two types of learning by doing: "volume-based" learning and "time-based" learning. Neither is based on cumulative production, as is much of the literature on learning by doing. DOT/NHTSA (2009, p. 185) judged that a first cycle of volume-based learning would occur at a volume of 300,000 units per year and that costs would be reduced by 20 percent over low-volume estimates. A second learning threshold was set at 600,000 units per year, at which point a second cost reduction of 20 percent was taken. No further volume-based learning was assumed. The NHTSA applied this procedure to only three technologies in its 2011 rule: integrated starter generator, two-mode hybrid, and plug-in hybrid.

DOT/NHTSA (2009, p. 188) also applies time-based or year-over-year learning by doing to widely available, high-volume, mature technologies. Either time-based or volume-based learning, but not both, is applied to a particular technology. Time-based learning is applied at the rate of 3 percent per year in the second and all subsequent years of a technology's application.

The use of learning curves poses a dilemma. On the one hand, there is no rigorous method for determining how much

[4] The EPA generally does not use typical continuous learning curves but instead stepwise learning as a function of time, rather than cumulative production. Usually, costs are assumed to decrease by 10 percent after the first year of production, and by another 10 percent after the second year, and then to remain constant.

and how rapidly a specific technology's costs can be reduced by learning by doing.[5] On the other hand, the phenomenon of learning by doing is widely and generally observed in the manufacturing of new technologies (e.g., Wene, 2000). This does not mean that no learning should be assumed. Rather, learning curves should be applied cautiously and should reflect average rates of learning based on empirical evidence from the motor vehicle industry. Expert judgment should be used to determine the potential for learning, depending on the nature of the technology in question.

Vehicle Type or Class

The costs of fuel economy technology also vary across vehicle classes. To a large extent this is a function of vehicle size and power. For example, an eight-cylinder engine has twice as many valves as a four-cylinder, and so the costs of valve train technologies will be higher. When technologies, such as turbocharging, increase the power output per unit of displacement and thereby enable engine downsizing at constant performance, the starting cylinder count can affect the options for downsizing. In general, an eight-cylinder engine can be replaced by a smaller six-cylinder engine of equivalent performance without additional costs for mitigating vibration. Downsizing a four-cylinder to a three-cylinder would require significant modifications to offset increased vibration, and this might even rule out reducing the cylinder count. Since most of the cost savings from downsizing accrue from reducing the number of cylinders, technologies that enable engine downsizing will be relatively more expensive for four-cylinder engines. Since different vehicle classes have different distributions of cylinder counts, the costs of certain technologies should be class-dependent. As another example, the cost of a 1 percent weight reduction by material substitution will depend on the initial mass of the vehicle.

National Research Council (2002) did not vary technology costs by vehicle class. The NHTSA's Volpe model's algorithm, however, operates at the level of make, model, engine, and transmission configuration. Some technology costs are scaled to the specific attributes of each vehicle. Other costs are class-dependent. In its final rule for 2011, DOT/NHTSA (2009, p. 165) specified eight passenger car classes and four light truck classes (Table 3.2). Passenger cars were divided into size classes on the basis of their footprint. Each class was divided into a standard and high-performance class on the basis of class-specific cut-points determined using expert judgment. This reflects the NHTSA's view that in addition to size, performance is the key factor determining differences in technology applicability and cost. The classification of light trucks was based on structural and design considerations (minivans) and footprint size (sport utility vehicles [SUVs], pickups, and vans).

TABLE 3.2 Vehicle Classification by the National Highway Traffic Safety Administration

Passenger Cars
Subcompact
Subcompact performance
Compact
Compact performance
Midsize
Midsize performance
Large
Large performance
Light Trucks
Minivans
Small SUV/pickup/van
Midsize SUV/pickup/van
Large SUV/pickup/van

Although classification can improve the accuracy of cost estimates, there is no perfect classification system, and there will always be some heterogeneity within a class.

METHODS OF ESTIMATING COSTS

As a generalization, there are two basic methods of cost estimation. The first and most common is to obtain estimates of the selling prices of manufactured components. The second is to tear down a technology into its most basic materials and manufacturing processes and to construct a bottom-up estimate by costing out materials, labor, and capital costs for every step. Both methods ultimately rely heavily on the expertise and the absence of bias on the cost estimator's part.

Estimation Using Supplier Prices for Components, or "Piece Costs"

The supplier price method relies on comparing an estimate of the price that a Tier 1 component manufacturer would charge an OEM for a reference component to an estimate of the price that it would charge for an alternative that delivered reduced fuel consumption. In the past, information on the prices that manufacturers pay to Tier 1 suppliers for components has come from a variety of sources, including the following:

- The NRC (2002) report on the CAFE standards;
- The NESCCAF (2004) study on reducing light-duty vehicle greenhouse gas emissions;
- The California Air Resources Board study in support of its greenhouse gas regulations;
- The study by Energy and Environmental Analysis, Inc. (EEA, 2006) for Transport Canada;

[5] Not only the progress ratio, but also the assumed initial cumulative production (or threshold volume) strongly influences estimated future cost reductions. Numerous after-the-fact estimations of progress ratios are available. However, in general, there is no scientific method for deciding on these parameters ex ante.

- Confidential data submitted by manufacturers to the NHTSA in advance of rulemakings; and
- Confidential data shared by manufacturers in meetings with the NHTSA and the EPA in 2007.

Component cost estimates can be obtained from discussions with suppliers or OEMs, from published reports, or by comparing the prices of vehicles with and without the component in question (Duleep, 2008), bearing in mind that costs and market prices may differ significantly. The NHTSA also receives cost estimates in the form of confidential data submitted by manufacturers. Depending on how fuel economy technologies are defined, estimates for more than one component may be involved. Given a supplier price estimate, a markup factor is applied to estimate the RPE. A single markup factor is often used for all components, but different markups may be used according to the nature of the component. The key issues are, therefore, the accuracy of the supplier price estimates and the accuracy of the markup factor(s).

First at the request of NESCCAF (2004) and later at the request of the Alliance of Automobile Manufacturers, Martec Group, Inc. (2008b) estimated the variable (or manufacturing) costs of fuel economy technologies based on the bill of materials (BOM) required. The term *materials* as used in the Martec studies refers to manufactured components supplied by Tier 1 and Tier 2 suppliers. The direct and indirect changes in vehicle components associated with a particular technology were determined in discussions with engineering consultants and OEM engineers. The Tier 1 and Tier 2 suppliers were the primary sources of information on the costs of manufactured components required to implement the fuel economy increases (Martec Group, Inc., 2008b, p. 7).

Teardown or Bottom-Up Estimation

A change in the design and content of a vehicle induces changes in the materials of which it is made, the quantity and types of labor required to construct it, and changes in the capital equipment needed to manufacture it. Such estimates not only are time-consuming but also require analysts with a thorough knowledge of and experience with automotive manufacturing processes.

Bottom-up cost estimation methods have been used by the NHTSA for assessing the impacts of safety regulations. For example, in a study of air bag costs, an NHTSA contractor used a teardown method to identify all components of 13 existing air bag systems. This study (Ludtke and Associates, 2004) is described in Appendix F. The contractor analyzed each part or assembly and identified each manufacturing process required for fabrication, from raw material to finished product. The analysis identified parts purchased from suppliers as well as parts made in-house. Process engineers and cost estimators then carried out a process and cost analysis for each part and assembly. Two costs were developed: (1) variable costs associated with the actual manufacturing and (2) fixed or burden costs. Estimating costs to the consumer (analogous to the retail price equivalent) requires additionally estimating the OEM's amortized costs, as well as other costs and profit. Dealers' costs are added to the manufacturer's cost plus profit to obtain the consumer's cost (Figure 3.1). As the NHTSA report is careful to point out, estimating costs "is not an exact science" but rather one strongly dependent on the expertise and judgment of the estimators at every step.

The teardown method was applied by Kolwich (2009) to estimate the incremental manufacturing cost of a downsized 1.6-liter, four-cylinder, stoichiometric direct injection, turbocharged engine versus a 2.4-liter, four-cylinder, naturally aspirated base engine. The study did not attempt to estimate the markup from manufacturing costs to RPE. Rather, the cost estimated is equivalent to the price that a Tier 1 supplier would charge an OEM for the fully manufactured engine. Unit costs are composed of direct manufacturing costs (material + labor + fixed manufacturing costs) + "markup costs" (scrap + overhead + profit) + packaging costs (Figure 3.2).

Manufacturing costs are estimated in a series of highly detailed steps based on what is learned in disassembling the technology. Both the new and the base technologies must be torn down and costed in order to estimate the difference in cost. First, the technology to be evaluated is identified and defined. Next, candidate vehicles for teardown are identified (this limits the analysis to technologies already in production). A pre-teardown, high-level bill of materials (consisting of subsystems and components) is then created, subject to amendment, as discoveries might be made during the teardown process. At that point, the actual teardown process begins. During the teardown, all of the processes necessary for assembly are identified and recorded, and every component and the material of which it is made are identified. The data generated in the disassembly are then reviewed by a team of experts. Following the review, the components are torn down and assembly processes are identified, as is each and every piece of each component. A worksheet is then constructed for all parts, containing all cost elements. Parts with high or unexpected cost results are double-checked, and then entered into a final spreadsheet in which they are totaled and formatted.

Once manufacturing costs have been estimated, a markup reflecting all other costs of doing business is typically applied to estimate the long-run cost that consumers will have to pay. Applying this markup was outside the scope of the FEV (2009) study but was included in the Ludtke and Associates (2004) study. Estimates of the consumer's cost of curtain air bag systems installed in five different vehicles from the Ludtke and Associates study are shown in Table 3.3. Although costs vary, it is clear that Ludtke and Associates used the same markup factors for Tier 1 manufacturers' markups over their direct costs (24 percent), OEM markups (36 percent), and dealer markups (11 percent). These markups result in multipliers for the consumer's cost over the Tier 1 supplier's cost of

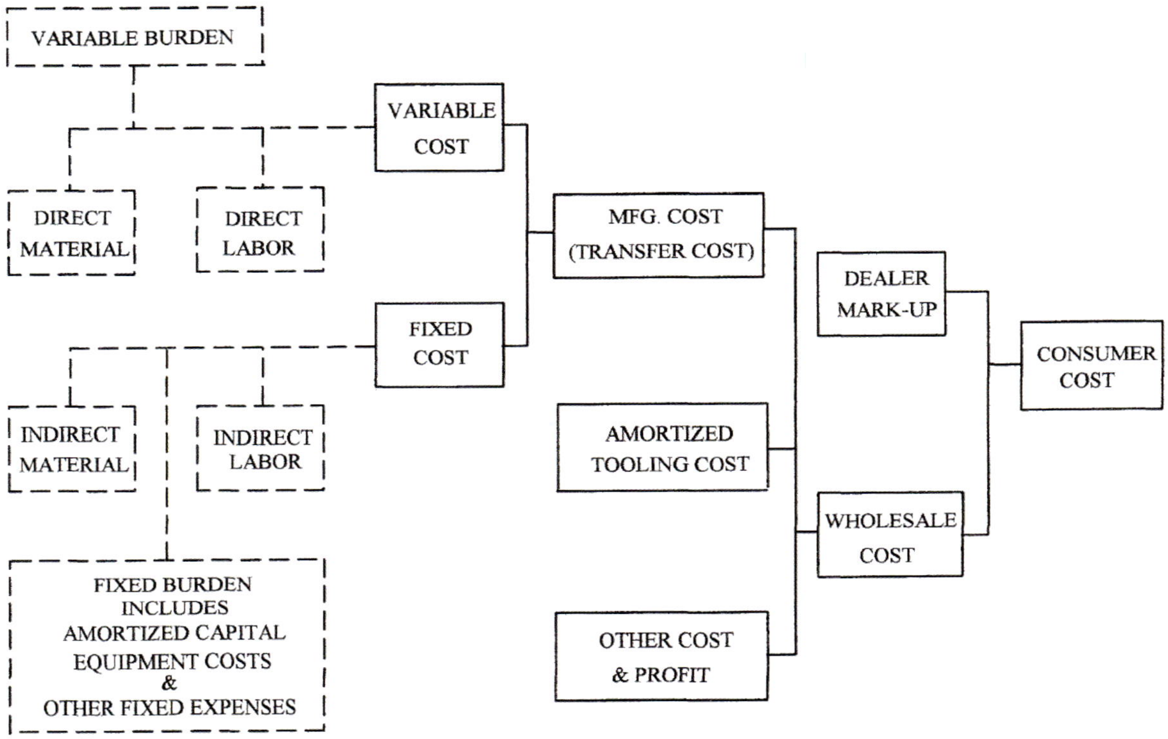

FIGURE 3.1 Determination of manufacturing and consumer cost. SOURCE: Ludtke and Associates (2004), p. B-10.

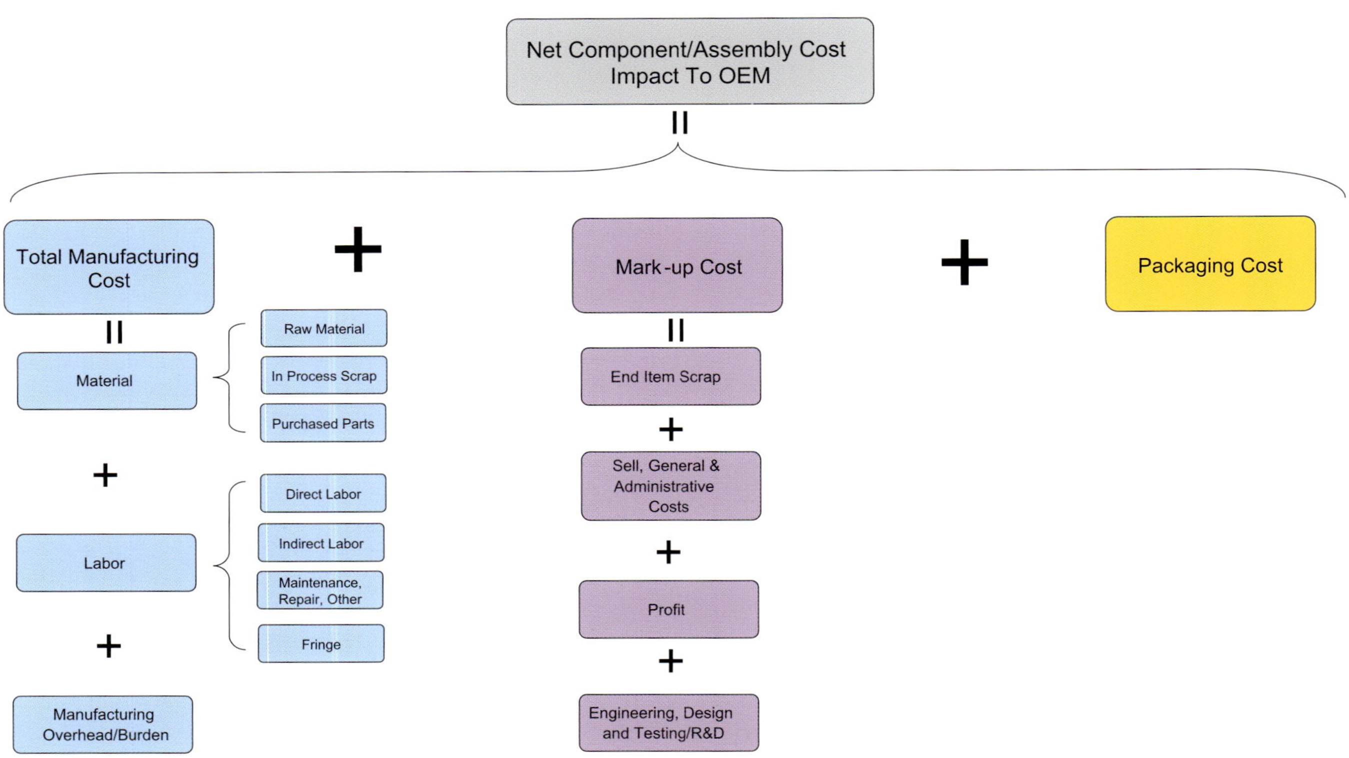

FIGURE 3.2 Unit cost model. SOURCE: FEV, Inc. (2009) (FEV.com), Figure 5.

TABLE 3.3 Estimated Consumer Cost (2003 dollars) for Installed Air Bag Systems and Markups

Item	VW Jetta	Toyota Camry	Cadillac CTS	Mercury Monterey[a]	Jeep Grand Cherokee
Material	$30.04	$27.45	$48.46	$69.88	$54.43
Direct labor	$11.11	$20.54	$16.54	$37.62	$17.68
Direct labor burden	$22.59	$34.40	$24.61	$55.91	$23.93
Tier 1 markup	$15.40	$19.89	$21.93	$39.66	$23.21
Manufacturer markup	$28.49	$36.82	$40.15	$73.11	$42.93
Dealer markup	$11.84	$15.30	$16.69	$30.38	$17.84
Consumer's cost	$119.47	$154.40	$168.38	$306.55	$180.02
Variable cost	$63.74	$82.39	$89.61	$163.41	$96.04
Variable manufacturing cost	$79.14	$102.28	$111.54	$203.07	$119.25
Markup Tier 1 cost	1.51	1.51	1.51	1.51	1.51
Markup variable manufacturing cost	1.87	1.87	1.88	1.88	1.87
Tier 1 markup	24.2%	24.1%	24.5%	24.3%	24.2%
OEM markup	36.0%	36.0%	36.0%	36.0%	36.0%
Dealer markup	11.0%	11.0%	11.0%	11.0%	11.0%

NOTE: Original equipment manufacturer (OEM) manufacturing costs (2003$) per vehicle—head protection air bag systems (curtain-type system without a torso airbag already installed in vehicle).

[a]Cost estimates for the Mercury Monterey are substantially higher than those for the other vehicles. Ludtke and Associates (2004) do not offer an explanation for the design differences that account for the higher cost.

SOURCE: Ludtke and Associates (2004).

1.51 ($1.36 \times 1.11 = 1.51$), and for the consumer's cost over the direct variable costs of manufacturing ("Total Manufacturing Costs" minus "Manufacturing Overhead Burden" in the FEV [2009] study; see Figure 3.2 above) the component of 1.87 ($1.24 \times 1.36 \times 1.11 = 1.87$). The costs shown in Table 3.3 are in 2003 dollars and assume a manufacturing scale of 250,000 units per year for the air bags.

While Ludtke and Associates (2004) use a markup factor of 1.24 for direct manufacturing costs, Kolwich (2009) uses markup factors ranging from 10.3 percent to 17.7 percent, depending on the complexity of the component (Table 3.4). Note that the Kolwich rates do not include manufacturing overhead whereas the Ludtke rates do, and thus the former should be higher.

The FEV teardown study (FEV, 2009; Kolwich, 2009) allows total manufacturing costs to be broken down by engine subsystem as well as cost component. Figure 3.3 shows the incremental manufacturing costs by cost component. The largest single component of the $537.70 total is material ($218.82), followed by manufacturing burden ($154.24), labor ($72.58), corporate overhead ($33.96), profit ($33.12), engineering and R&D ($12.36), and scrap ($11.72). The total markup on manufacturing costs is just over 20 percent. Figure 3.4 shows the same total cost broken down by engine subsystem. By far the largest components are the induction air charging system ($258.89) and the fuel induction system ($107.32). Cost savings occur in counterbalance ($35.95) and intake systems ($12.73).

TABLE 3.4 Total Manufacturing Cost Markup Rates for Tier 1 and Tier 2/3 Suppliers

Primary Manufacturing Equipment Group	End Item Scrap Markup (%)	SG&A Markup (%)	Profit Markup (%)	ED&T Markup (%)	Total Markup (%)
Tier 2/3—large size, high complexity	0.7	7.0	8.0	2.0	17.7
Tier 2/3—medium size, moderate complexity	0.5	6.5	6.0	1.0	14.0
Tier 2/3—small size, low complexity	0.3	6.0	4.0	0.0	10.3
Tier 1 complete system/subsystem supplier (system/subsystem integrator)	0.7	7.0	8.0	6.0	21.7
Tier 1 high-complexity-component supplier	0.7	7.0	8.0	4.0	19.7
Tier 1 moderate-complexity-component supplier	0.5	6.5	6.0	2.5	15.5
Tier 1 low-complexity-component supplier	0.3	6.0	4.0	1.0	11.3

SOURCE: Kolwich (2009), Table 2.

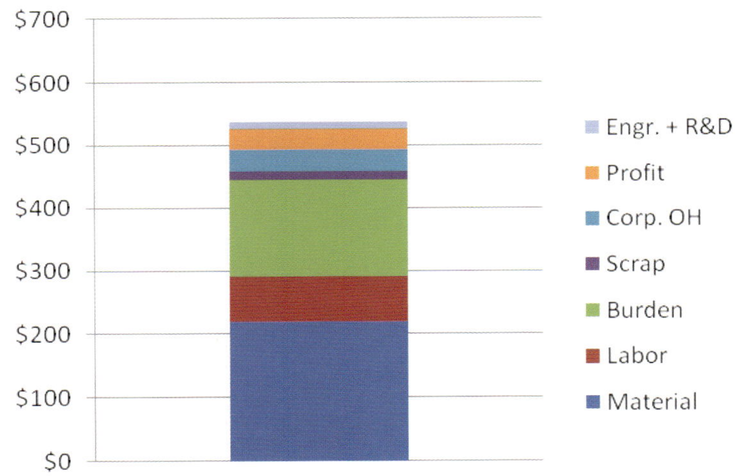

FIGURE 3.3 Incremental cost of turbocharged, downsized, gasoline direct-injection I4 engine broken down by cost category. SOURCE: Kolwich (2009), Figure 19.

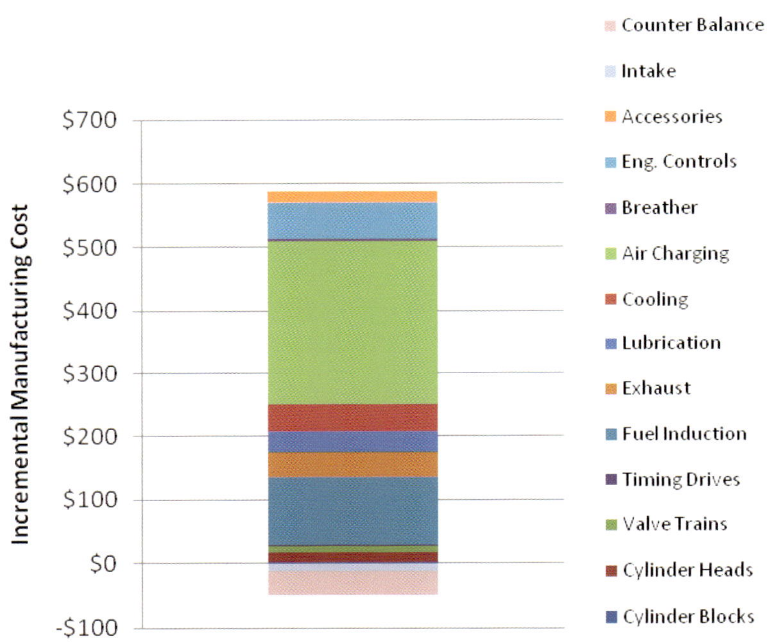

FIGURE 3.4 Incremental cost of turbocharged, downsized, gasoline direct-injection I4 engine broken down by engine subsystem. SOURCE: Kolwich (2009), Figure 19.

RETAIL PRICE EQUIVALENT MARKUP FACTORS

Markup factors relating component costs to RPE add significantly to the estimated costs of automotive technologies and are the subject of continuing controversy. The cost of making and selling light-duty vehicles is not limited to the manufacture of components and their assembly. Even for a single technological or design change, cost impacts are generally not limited to the component that is changed. Engineering expertise must be supplied to design these changes, which may or may not induce other changes in the cost of manufacturing. These integration costs can be substantial for major components, such as engines, or when, as is more often the case than not, many changes are made simultaneously. There are also indirect costs for research and development, administrative overhead, warranties, and marketing and advertising. Vehicles must be transported to dealers who have their own labor, material, and capital costs. All of these additional costs are represented by RPE markup factors.

Existing RPE Markup Factors

For the automobile industry, there is a reasonable consensus on the ratio of total costs of doing business to the cost of fully manufactured components (the price that a Tier 1 supplier would charge an OEM). This average RPE markup factor is approximately 1.5, according to the available evidence, reviewed in detail in Appendix F of this report. Part of the disagreement over the size of the RPE markup factor arises from the difference between the variable costs versus the variable plus fixed costs of a manufactured component. An appropriate RPE markup over the variable (or direct) costs of a component is approximately 2.0 (Bussmann and Whinihan, 2009). Part of the disagreement arises over the difficulty of attributing indirect and other fixed costs to a particular vehicle component.

Every fuel economy technology does not affect fixed or indirect costs in the same way. Some costs may be affected by engineering and design changes to decrease fuel consumption; others may not. This can have a very large impact on the appropriate RPE of a given fuel economy technology. Some studies use a single, average RPE markup factor (e.g., NRC, 2002; Albu, 2008; DOT/NHTSA, 2009), while others attempt to tailor the markup to the nature of the technology (Rogozhin et al., 2009; Duleep, 2008). The problem of how best to attribute indirect and fixed costs to a specific change in vehicle technology remains unresolved.

Existing estimates of the RPE markup factor are similar when interpreted consistently. Vyas et al. (2000) compared their own markup factors to estimates developed by EEA, Inc., and Chrysler. Unfortunately, differences in the definitions of categories of costs preclude precise comparisons. Vyas et al. concluded that an appropriate markup factor over the variable costs of manufacturing a motor vehicle was 2.0. The Vyas et al. (2000) report also summarized the cost methodology used by EEA, Inc., in a study for the Office of Technology Assessment (OTA, 1995). Vyas et al. (2000) concluded that the markup over variable manufacturing costs used in that study was 2.14, while the markup over outsourced parts (e.g., purchased from a Tier 1 supplier) was 1.56 (Table 3.5).

A markup factor of 1.5 was also used by the NHTSA (2009, p. 173) in its final fuel economy rule for 2011. A somewhat lower RPE markup factor of 1.4 was used by the NRC (2002) and Albu (2008), while the EPA has used a markup of 1.26 (EPA, 2008a).

The use of a markup of approximately 2 over the direct manufacturing costs of parts manufactured in-house by an OEM was also supported by Bussmann (2008), who cited a 2003 study of the global automotive industry by McKinsey Global Institute that produced a markup factor of 2.08, and his own analysis of Chrysler data for 2003-2004 that produced factors of 1.96 to 1.97. Information supplied by EEA, Inc., to the committee (Duleep, 2008) implies higher markup factors: 2.22 to 2.51 for the markup over variable costs and 1.65 to 1.73 for the markup over Tier 1 supplier costs.

Average RPE factors can be inferred by costing out all the components of a vehicle, summing those costs to obtain an estimate of OEM Tier 1 costs or fully burdened in-house manufacturing costs, and then dividing the sum into the selling price of a vehicle. The committee contracted with IBIS Associates (2008) to conduct such an analysis for two high-selling model-year 2009 vehicles: the Honda Accord sedan and the Ford F-150 pickup truck. For the Honda, the RPE multipliers were 1.39 to market transaction price and 1.49 to manufacturer's suggested retail price (MSRP). The multiplier to dealer invoice cost is 1.35, implying that dealer costs, including profit, amount to about 4 percent of manufacturing costs, not considering any dealer incentives provided by OEMs. For the Ford F-150, the RPE multipliers were 1.52 for market price and 1.54 for MSRP. The markup factor for dealer invoice is 1.43, implying that dealer costs and profit amount to about 9 percent of total manufacturing costs, not including any possible OEM incentives to dealers.

TABLE 3.5 Comparison of Markup Factors

Markup Factor for	ANL	Borroni-Bird	EEA
In-house components	2.00	2.05	2.14
Outsourced components	1.50	1.56	1.56

SOURCE: Vyas et al. (2000).

The EPA Study on RPE Factors and Indirect Cost Multipliers

Concerns with the Existing RPE Method

Objections have been raised with respect to the use of a single RPE markup factor for components manufactured by Tier 1 suppliers and sold to OEMs. The EPA has pointed out that not all technologies will affect indirect costs equally, and it has proposed to investigate technology-specific markups, by attempting to identify only those indirect costs actually affected by each technology (EPA, 2008b). In a similar vein, the importance of "integration costs" has been cited as a factor that would justify different markup factors for different technologies (Duleep, 2008).[6] Because a vehicle is a system, it is almost always the case that the design of one part affects others. Manufacturers cannot simply buy a list of parts and

[6] Duleep (2008) recommends using different markup factors for different kinds of components to account for differences in the cost of integrating components into the overall vehicle design. For parts purchased from Tier 1 suppliers, Duleep recommends a range of markup factors from 1.45 to 1.7, depending chiefly on integration costs. As an example, Duleep presented to the committee an estimated markup factor of 1.72 for injector, pump, and rail costs for a stoichiometric GDI engine. This is at the high end of his markup range, reflecting the greater integration costs for engine technologies. Duleep (2008) proposed using judgment to divide technologies into three groups. He recommended a markup factor of 1.7 for technologies requiring extensive integration engineering, 1.56 for those having average integration costs, and 1.4 for those with little or no integration costs.

bolt them together to produce a vehicle that meets customers' expectations and satisfies all regulatory requirements.[7] Integrating a new engine or transmission to decrease fuel consumption will have much greater ramifications for vehicle design and is likely to generate greater integration costs than simpler components.

In a presentation to the committee, the EPA raised concerns that markup factors on piece or supplier costs tended to overestimate the costs of most fuel economy technologies: "Our first preference is to make an explicit estimate of all indirect costs rather than rely on general markup factors" (EPA, 2008b, slide 4). Nonetheless, in its assessment of the costs of greenhouse gas mitigation technologies for light-duty vehicles, the EPA staff assumed a uniform markup of 50 percent over supplier costs (i.e., a markup factor of 1.5). Still, the EPA maintains that such a markup is too large: "We believe that this indirect cost markup overstates the incremental indirect costs because it is based on studies that include cost elements—such as funding of pensions—which we believe are unlikely to change as a result of the introduction of new technology" (EPA, 2008a, p. 47).

Following up on this assertion, the EPA commissioned a study of RPE factors and indirect cost (IC) multipliers (Rogozhin et al., 2009). The IC multiplier attempts to improve on the RPE by including only those specific elements of indirect costs that are likely to be affected by vehicle modifications associated with environmental regulation. In particular, fixed depreciation costs, health care costs for retired workers, and pensions may not be affected by many vehicle modifications caused by environmental regulations.

The EPA study (Rogozhin et al., 2009) also criticizes the RPE method on the grounds that an increase in the total cost of producing a vehicle will not be fully reflected in the increased price of the vehicle due to elasticities of supply and demand. For this reason, the report argues that manufacturer profits should not be included in the RPE multiplier. The committee disagrees with this assertion for two reasons. First, as noted earlier, the global automotive industry approximates what economists term a monopolistically competitive market, that is, a market in which there is product differentiation but a high degree of competition among many firms. In a monopolistically competitive market, in the long run the full costs of production will be passed on to consumers. In the long run, monopolistically competitive market supply is perfectly elastic at the long-run average cost of production (this includes a normal rate of return on capital). Since cost estimates by convention assume long-run conditions (full scale economies and learning), long-run supply assumptions should be used to ensure consistency. The increase in RPE is a reasonable estimate of the change in welfare associated with the increased vehicle cost especially, as noted above, in the long run.

[7] For some parts, the effort required for integration may be small. Tires are often cited as an example. Still, even tires have implications for a vehicle's suspension and braking systems.

The EPA study (Rogozhin et al., 2009) estimated RPEs for the largest manufacturers for the year 2007 using publicly available data in manufacturers' annual reports. Several assumptions were required to infer components not reported, or reported in different ways by different manufacturers. The method is similar to that used by Bussman (2008) and produced similar results. One notable difference is that the estimates shown in Table 3.6 attempt to exclude legacy health care costs, estimated at 45 percent of total health care costs, which in turn were estimated to be 3 percent of fully burdened manufacturing costs. This would lower the estimated RPEs by 1 to 2 percent relative to estimates in other reports, all else being equal. The estimated RPE multipliers were remarkably consistent across manufacturers (Table 3.6) and very comparable to the studies cited above. Estimated RPE multipliers ranged from 1.42 for Hyundai to 1.49 for Nissan, with an industry average of 1.46. Adding 1 to 2 percent for health care costs would bring the average multiplier even closer to 1.5.

Estimating Technology-Specific Markup Factors and IC Multipliers

The assertion that different technologies will induce different changes in indirect costs seems evident. The question is how to identify and measure the differences. At the present time a rigorous and robust method for estimating these differential impacts does not exist (Bussmann and Whinihan, 2009). Therefore, it is not clear that the accuracy of fuel consumption cost assessment would be increased by the use of technology-specific, as opposed to an industry-average, markup factor. The EPA (Rogozhin et al., 2009), however, has taken the first steps in attempting to analyze this problem in a way that could lead to a practical method of estimating technology-specific markup factors.

The EPA-sponsored study (Rogozhin et al., 2009) went on to estimate IC multipliers as a function of the complexity or scope of the innovation in an automaker's products caused by the adoption of the technology. A four-class typology of innovation was used:

- *Incremental innovation* describes technologies that require only minor changes to an existing product and permit the continued use of an established design. Low-rolling-resistance tires were given as an example of incremental innovation.
- *Modular innovation* is that which does not change the architecture of how components of a vehicle interact but does change the core concept of the component replaced. No example was given for modular innovation.
- *Architectural innovation* was defined as innovation that requires changes in the way that vehicle components are linked together but does not change the core design concepts. The dual-clutch transmission was offered as an example, in that it replaces the function of an

TABLE 3.6 Individual Manufacturer and Industry Average Retail Price Equivalent (RPE) Multipliers: 2007

	Relative to Cost of Sales								
RPE Multiplier Contributor	Industry Average	Daimler Chrysler	Ford	GM	Honda	Hyundai	Nissan	Toyota	VW
Vehicle Manufacturing									
Cost of sales	1.00	1.00	1.00	1.00	1.00	1.00	1.00	1.00	1.00
Production Overhead									
Warranty	0.03	0.04	0.03	0.03	0.01	0.02	0.03	0.04	0.02
R&D product development	0.05	0.04	0.02	0.06	0.07	0.04	0.06	0.05	0.06
Depreciation and amortization	0.07	0.11	0.05	0.06	0.05	0.06	0.09	0.08	0.09
Maintenance, repair, operations cost	0.03	0.03	0.03	0.03	0.03	0.03	0.03	0.03	0.03
Total production overhead	0.18	0.22	0.13	0.17	0.16	0.15	0.21	0.19	0.20
Corporate Overhead									
General and administrative	0.07	0.05	0.12	0.07	0.11	0.08	0.03	0.06	0.03
Retirement	<0.01	0.01	0.00	0.01	<0.01	<0.01	<0.01	<0.01	<0.01
Health	0.01	<0.01	<0.01	0.01	0.01	0.01	0.01	0.01	0.01
Total corporate overhead	0.08	0.06	0.13	0.08	0.14	0.09	0.04	0.07	0.04
Selling									
Transportation	0.04	0.04	0.04	0.04	0.04	0.04	0.04	0.04	0.10
Marketing	0.04	0.02	0.04	0.05	0.03	0.05	0.08	0.03	0.02
Dealers									
Dealer new vehicle net profit	<0.01	<0.01	<0.01	<0.01	<0.01	<0.01	<0.01	<0.01	<0.01
Dealer new vehicle selling cost	0.06	0.06	0.06	0.06	0.06	0.06	0.06	0.06	0.06
Total selling and dealer contributors	0.14	0.12	0.14	0.14	0.13	0.15	0.18	0.12	0.17
Sum of Indirect Costs	0.40	0.40	0.39	0.40	0.44	0.39	0.43	0.38	0.41
Net income	0.06	0.07	0.05	0.05	0.04	0.03	0.06	0.09	0.02
Other costs (not included as contributors)	0.04	0.04	0.11	0.06	0.02	0.01	0.01	<0.01	0.05
RPE multiplier	1.46	1.47	1.45	1.45	1.47	1.42	1.49	1.48	1.43

SOURCE: Rogozhin et al. (2009), Table 3-3.

existing transmission but does require redesign and reintegration with other components.

- *Differential innovation* involves significant changes in the core concepts of vehicle components, as well as their integration. Hybrid vehicle technology was cited as an example because it changes the functions of such key components as the engine, brakes, and battery.

An industry average was computed for each component of the RPE, omitting profit, or net income. As stated above, the committee considers this omission to be in error. The resulting components are shown in Table 3.7. Next, based on the judgment of an expert panel, short- and long-term effects on the RPE components were estimated for the four categories of technology innovation (Rogozhin et al., 2009). A value of zero for the effect of a technology innovation on an RPE component implies that the application of that technology has no impact on the cost of that particular RPE component. There will be no increase in expenditure on that RPE component as a result of the adoption of the technology. A value of 1 implies that the cost of the component will increase directly with the increased cost of the component. Values greater than 1 imply a greater-than-proportional increase. Each RPE component is multiplied by its respective short- or long-term effect, and the results are summed and

TABLE 3.7 Weighted Industry Average RPE Components Omitting Return on Capital

Cost Contributor	Light Car Industry Average
Production Overhead	
Warranty	0.03
R&D (product development)	0.05
Depreciation and amortization	0.07
Maintenance, repair, operations cost	0.03
Total production overhead	0.18
Corporate Overhead	
General and administrative	0.07
Retirement	0.00
Health care	0.01
Total corporate overhead	0.08
Selling	
Transportation	0.04
Marketing	0.04
Dealers	
Dealer new vehicle selling cost	0.06
Total selling and dealer costs	0.14
Sum of Indirect Costs	0.40

SOURCE: Rogozhin et al. (2009), Table 4-1.

TABLE 3.8 Short- and Long-Term Indirect Cost Multipliers

	Low Complexity	Medium Complexity	High Complexity	Industry Average RPE
Short term	1.05	1.20	1.45	1.46
Long term	1.02	1.05	1.26	1.46

SOURCE: Rogozhin et al. (2009), Table 4-5.

added to 1.0 to produce the IC multipliers. The multipliers range from 1.05 to 1.45 in the short run and 1.02 to 1.26 in the long run (Table 3.8). This implies that none of the fuel economy technologies considered, no matter how complex, could cause an increase in indirect costs as large as the industry average indirect costs, especially in the long run. This result would imply that the more that regulatory requirements increase the cost of automobile manufacturing, the lower the overall industry RPE would be.

FINDINGS

Large differences in technology cost estimates can result from differing assumptions. Carefully specifying premises and assumptions can greatly reduce these differences. These include the following:

- Whether the total cost of a technology or its incremental cost over the technology that it will replace is estimated;
- Whether long-run costs at large-scale production are assumed or short-run, low-volume costs are estimated;
- Whether learning by doing is included or not;
- Whether the cost estimate represents only direct in-house manufacturing costs or the cost of the purchase of a component from a Tier 1 supplier;
- Whether the RPE multiplier is based on industry average markups or is specific to the nature of the technology; and
- What other changes in vehicle design, required to maintain vehicle quality (e.g., emissions, towing, gradability, launch acceleration, noise, vibration, harshness, manufacturability), have been included in the cost estimate.

Finding 3.1: For fully manufactured components purchased from a Tier 1 supplier, a reasonable *average* RPE markup factor is 1.5. For in-house direct (variable) manufacturing costs, including only labor, materials, energy, and equipment amortization, a reasonable *average* RPE markup factor is 2.0. In applying such markup factors, it is essential that the cost basis be appropriately defined and that the *incremental* cost of fuel economy technology is the basis for the markup. The factors given above are averages; markups for specific technologies in specific circumstances will vary.

Finding 3.2: RPE factors certainly do vary depending on the complexity of the task of integrating a component into a vehicle system, the extent of the required changes to other components, the novelty of the technology, and other factors. However, until empirical data derived by means of rigorous estimation methods are available, the committee prefers to use average markup factors.

Finding 3.3: Available cost estimates are based on a variety of sources: component cost estimates obtained from suppliers, discussions with experts at OEMs and suppliers, comparisons of actual transaction prices when publicly available, and comparisons of the prices of similar vehicles with and without a particular technology. There is a need for cost estimates based on a teardown of all the elements of a technology and a detailed costing of material costs, accounting for labor time and capital costs for all fabrication and assembly processes. Such studies are more costly than the current approaches listed above and are not feasible for advanced technologies whose designs are not yet finalized and/or whose system integration impacts are not yet fully understood. Nonetheless, estimates based on the more rigorous method of teardown analysis are needed to increase confidence in the accuracy of the costs of reducing fuel consumption.

Technology cost estimates are provided in the following chapters for each fuel economy technology discussed. Except as indicated, the cost estimates represent the price that an OEM would pay a supplier for a finished component. Thus, on average, the RPE multiplier of 1.5 would apply.

REFERENCES

Albu, S. 2008. ARB perspective on vehicle technology costs for reducing greenhouse gas emissions. Presentation by Assistant Chief, Mobile Source Division, California Air Resources Board, to the Committee on Technologies for Improving Light-Duty Vehicle Fuel Economy, January 24, Detroit, Mich.

Bussmann, W.V. 2008. Study of industry-average markup factors used to estimate retail price equivalents (RPE). Presentation to the Committee on Assessment of Technologies for Improving Light-Duty Vehicle Fuel Economy, January 24, Detroit, Mich.

Bussmann, W.V., and M.J. Whinihan. 2009. The estimation of impacts on retail prices of regulations: A critique of "Automobile Industry Retail Price Equivalent and Indirect Cost Multipliers." Prepared for Alliance of Automobile Manufacturers, May 6, Southfield, Mich.

DOT/NHTSA (Department of Transportation/National Highway Traffic Safety Administration). 2009. Average fuel economy standards, passenger cars and light trucks, model year 2001: Final rule. 49 CFR Parts 523, 531, 533, 534, 536, and 537, Docket No. NHTSA-2009-0062, RIN 2127-AK29. March 23. Washington D.C.

Duleep, K.G. 2008. Analysis of technology cost and retail price. Presentation to Committee on Assessment of Technologies for Improving Light-Duty Vehicle Fuel Economy, January 25, Detroit, Mich.

EEA (Energy and Environmental Analysis, Inc.). 2006. Technologies to Reduce Greenhouse Gas Emissions from Light-Duty Vehicles. Report to Transport Canada, Ottawa, Ontario, March. Arlington, Va.

EPA (U.S. Environmental Protection Agency). 2008a. EPA Staff Technical Report: Cost and Effectiveness Estimates of Technologies Used to Reduce Light-Duty Vehicle Carbon Dioxide Emissions. EPA420-R-08-008. Washington, D.C.

EPA. 2008b. The use of markup factors to estimate indirect costs. Presentation from National Vehicle and Fuel Emissions Laboratory, Office of Transportation and Air Quality, to the Committee on Assessment of Technologies for Improving Light-Duty Vehicle Fuel Economy, National Research Council, March 31, Washington, D.C.

FEV, Inc. 2009. Light-duty technology cost analysis overview. Presentation to the Committee on the Assessment of Technologies for Improving Light-Duty Vehicle Fuel Economy, March 16, Washington, D.C.

IBIS Associates, Inc. 2008. Data Collection and Analysis: Vehicle Systems Costs. Report to the Committee on Assessment of Technologies for Improving Light-Duty Vehicle Fuel Economy, National Research Council, December, 2008, Waltham, Mass.

Kolwich, G. 2009. Light-Duty Technology Cost Analysis Pilot Study. Report FEV 07-069-103F. Prepared for the U.S. Environmental Protection Agency, Ann Arbor, Michigan, September 3. FEV, Inc., Auburn Hills, Mich.

Ludtke and Associates. 2004. Perform Cost and Weight Analysis, Head Protection Air Bag Systems, FMVSS 201. DOT HS 809 842. December. U.S. Department of Transportation, National Highway Traffic Safety Administration, Washington, D.C.

Martec Group, Inc. 2008a. Variable costs of fuel economy technologies. Presentation to the Committee on Assessment of Technologies for Improving Light-Duty Vehicle Fuel Economy, National Research Council, January 24, Detroit, Mich.

Martec Group, Inc. 2008b. Variable Costs of Fuel Economy Technologies. Report prepared for the Alliance of Automobile Manufacturers. June 1. Washington, D.C.

NESCCAF (Northeast States Center for a Clean Air Future). 2004. Reducing Greenhouse Gas Emissions from Light-Duty Motor Vehicles. Available at http://www.nesccaf.org/documents/rpt040923ghglightduty.pdf. Accessed July 6, 2009.

NRC (National Research Council). 2002. Effectiveness and Impact of Corporate Average Fuel Economy (CAFE) Standards. National Academy Press, Washington, D.C.

OTA (Office of Technology Assessment). 1995. Advanced Automotive Technology: Visions of a Super-Efficient Family Car. OTA-ETI-638. Washington, D.C.

Rogozhin, A., M. Gallaher, and W. McManus. 2009. Automobile Industry Retail Price Equivalent and Indirect Cost Multipliers. Report by RTI International to Office of Transportation and Air Quality. U.S. Environmental Protection Agency. RTI Project Number 0211577.002.004. February. Research Triangle Park, North Carolina.

Vyas, A., D. Santini, and R. Cuenca. 2000. Comparison of Indirect Cost Multipliers for Vehicle Manufacturing. Center for Transportation Research, Argonne National Laboratory. April. Argonne, Ill.

Wene, C. 2000. Experience Curves for Energy Technology Policy. International Energy Agency, Paris.

4

Spark-Ignition Gasoline Engines

INTRODUCTION

A large majority of light-duty vehicles in the United States are powered with spark-ignition (SI) engines fueled with gasoline. Several technologies have been developed to improve the efficiency of SI engines. This chapter updates the status of various SI engine technologies described in the National Research Council report that focused on reduction of fuel consumption (NRC, 2002). As stated in Chapter 2 of the present report, the objective is to evaluate technologies that reduce fuel consumption without significantly reducing customer satisfaction—therefore, power and acceleration performance are not to be degraded. The primary focus is on technologies that can be feasibly implemented over the period to 2025.

The present study examines these SI engine technologies in the context of their incremental improvements in reducing fuel consumption, as well as the associated costs of their implementation. It also discusses the mechanisms by which fuel consumption benefits are realized along with the interactions that these technologies have with the base-engine architecture. As with the other vehicle technologies examined in this report, the committee's estimates of incremental reduction of fuel consumption and the costs of doing so for the SI technologies presented in this chapter are based on published data from technical journals and analyses conducted by Northeast States Center for a Clean Air Future (NESCCAF), Energy and Environmental Analysis, Inc. (EEA), U.S. National Highway Traffic Safety Administration (NHTSA), U.S. Environmental Protection Agency (EPA), and other organizations. In addition, the expert judgment of committee members whose careers have focused on vehicle and power train design, development, and analysis, as well as the results of consultation with individual original equipment manufacturers (OEMs) and suppliers, were also incorporated in the estimates.

SI ENGINE EFFICIENCY FUNDAMENTALS

It is common practice to group engine-efficiency-related factors with their respective process fundamentals (i.e., thermodynamic factors, friction losses, etc.). For example, consider the basic stages of the SI engine cycle that contribute to positive work: heat released during fuel combustion, volumetric expansion, and associated heat transfer. The factors related to this process can be grouped together as the thermodynamic component. In addition, there are several processes within the engine that mitigate the positive work produced; these can be grouped as either gas exchange losses (pumping losses) or frictional losses within the engine. Furthermore, the engine architecture and the use of accessory/operational components (i.e., power steering, coolant, oil and fuel pumps) can be the source of additional parasitic losses. The fundamental aspects of each category of engine efficiency factors are discussed further in the following sections.

Thermodynamic Components

Thermodynamic factors include combustion interval, effective expansion ratio, and working fluid properties. In consideration of these factors there are some fundamental methods that can be used to improve efficiency, including:

- *Short combustion intervals*—allow for more of the heat of combustion to undergo more expansion and thus yield an increase in positive work.
- *High compression ratios and late exhaust-valve-opening event*—can be used to influence the expansion ratio in order to improve efficiency. However, these factors are constrained by other considerations.
- *High specific heat ratio of working fluid* (i.e., c_p/c_v.)—working-fluid property of significance related to the specific heat ratio. Atmospheric air is preferred over exhaust gas as a combustion diluent thermodynamically, but exhaust emis-

sions after-treatment challenges limit this as an option for reducing fuel consumption.

• *Optimize timing of spark event*—an important factor since this affects the countervailing variables of in-cylinder heat loss and thermodynamic losses. This is discussed in more detail below.

Maximum efficiency occurs when the two countervailing variables, heat loss and thermodynamic losses, sum to a minimum. The optimum spark timing is often referred to as minimum advance for best torque or maximum brake torque (MBT). At low to moderate speeds and medium to high loads, SI engines tend to be knock-prone, and spark-timing retardation is used to suppress the knock tendency. Spark-timing adjustments are also made to enable rapid-response idle load control to compensate for such things as AC compressor engagement. For this to be effective, idle spark timing must be substantially retarded from MBT. Retardation from MBT for either of the aforementioned reasons compromises fuel consumption.

Gas Exchange or Pumping Losses

Gas exchange or pumping losses, in the simplest terms, refer to the pressure-gradient-induced forces across the piston crown that oppose normal piston travel during the exhaust and intake strokes. The pumping loss that principally affects fuel consumption is that which occurs during the intake stroke when the cylinder pressure and the intake manifold are approximately equal. The pumping loss component that occurs during the exhaust stroke mainly affects peak power. Both of these oppose the desired work production of the engine cycle and thus are seen as internal parasitic losses, which compromise fuel efficiency.

Frictional Losses

The main source of friction losses within an SI engine are the piston and crankshaft-bearing assemblies. The majority of the piston-assembly friction comes from the ring-cylinder interface. The oil-control ring applies force against the cylinder liner during all four strokes while the compression rings only apply minor spring force but are gas-pressure loaded. Piston-assembly friction is rather complex as it constantly undergoes transitions from hydrodynamic to boundary-layer friction. Hydrodynamic piston-assembly friction predominates in the mid-stroke region while boundary-layer friction is common near the top center. Avoidance of cylinder out-of-roundness can contribute to the minimization of piston-ring-related friction. Crankshaft-bearing friction, while significant, is predominately hydrodynamic and is relatively predictable.

Engine Architecture

Engine architecture refers to the overall design of the engine, generally in terms of number of cylinders and cylinder displacement. The engine architecture can affect efficiency mainly through bore-stroke ratio effects and balance-shaft requirements.

Trends in power train packaging and power-to-weight ratios have led in-line engines to have under-square bore-stroke ratios (i.e., less than unity) while most V-configuration engines have over-square ratios. Under-square ratios tend to be favored for their high thermodynamic efficiency. This is due to the surface-area-to-volume ratio of the combustion chamber; under-square designs tend to exhibit less heat transfer and have shorter burn intervals. Over-square designs enable larger valve flow areas normalized to displacement and therefore favor power density. These interactive factors play a role in determining overall vehicle fuel efficiency.

Balance-shafts are used to satisfy vibration concerns. These balance shafts add parasitic losses, weight, and rotational inertia, and therefore have an effect on vehicle fuel efficiency. I4 engines having displacement of roughly 1.8 L or more require balance shafts to cancel the second-order shake forces. These are two counter-rotating balance shafts running at twice crankshaft speed. The 90° V6 engines typically require a single, first-order balance shaft to cancel a rotating couple. The 60° V6 and 90° V8 engines need no balance shafts. Small-displacement I3 engines have received development attention from many vehicle manufacturers. These require a single first-order balance shaft to negate a rotating couple. While low-speed high-load operation of small displacement I3 engines tends to be objectionable from a noise, vibration, and harshness (NVH) perspective, they could be seen as candidate engines for vehicles such as hybrid-electric vehicles (HEVs) where some of the objectionable operating modes could be avoided.

Parasitic Losses

Parasitic losses in and around the engine typically involve oil and coolant pumps, power steering, alternator, and balance shafts. These impose power demands and therefore affect fuel consumption. Many vehicle manufacturers have given much attention to replacing the mechanical drives for the first three of these with electric drives. Most agree that electrification of the power steering provides a measurable fuel consumption benefit under typical driving conditions. Fuel consumption benefit associated with the electrification of oil or coolant pumps is much less clear. Electrification of these functions provides control flexibility but at a lower efficiency. Claims have been made that the coolant pump can be inactive during the cold-start and warm-up period; however, consideration must be given to such things as gasket failure, bore or valve seat distortion, etc. These factors result from local hot spots in the cooling system since much of the waste heat enters the cooling system via the exhaust ports.

Further discussion on the parasitic losses associated with these types of engine components is provided in Chapter 7 of this report.

THERMODYNAMIC FACTORS

Fast-Burn Combustion Systems

Fast-burn combustion systems are used to increase the thermodynamic efficiency of an SI engine by reducing the burn interval. This is generally achieved either by inducing increased turbulent flow in the combustion chamber or by adding multiple spark plugs to achieve rapid combustion.

Fluid-mechanical manipulation is used to increase turbulence through the creation of large-scale in-cylinder flows (swirl or tumble) during the intake stroke. The in-cylinder flows are then forced to undergo fluid-motion length-scale reduction near the end of the compression stroke due to the reduced clearance between the piston and the cylinder head. This reduction cascades the large-scale fluid motion into smaller scale motions, which increases turbulence. Increased turbulence increases the turbulent flame speed, which thereby increases the thermodynamic efficiency by allowing for reduced burn intervals and by enabling an increase in knock-limited compression ratio by 0.5 to 1.0. This decrease in burn interval increases dilution tolerance of the combustion system. Dilution tolerance is a measure of the ability of the combustion system to absorb gaseous diluents like exhaust gas. Exhaust gas is introduced by means of an exhaust-gas-recirculation (EGR) system or by a variable-valve-timing scheme that modulates exhaust-gas retention without incurring unacceptable increases in combustion variability on a cycle-by-cycle basis. Combustion variability must be controlled to yield acceptable drivability and exhaust emissions performance.

Multiple spark plugs are sometimes used to achieve rapid combustion where fluid-mechanical means are impractical. Here, multiple flame fronts shorten the flame propagation distance and thus reduce the burn interval. High dilution-tolerant combustion systems can accept large dosages of EGR, thereby reducing pumping losses while maintaining thermodynamic efficiency at acceptable levels.

Fuel Consumption Benefit and Cost of Fast-Burn Combustion Systems

Combining fast-burn and strategic EGR usage typically decreases fuel consumption by 2 to 3 percent, based on manufacturer's input. The implementation of this technology is essentially cost neutral. Variable mixture-motion devices, which may throttle one inlet port in a four-valve engine to increase inlet swirl and in-cylinder mixture momentum, may add another 1 to 2 percent benefit at a cost of $50, $80, and $100 for I4, V6, and V8 engines, respectively, based on manufacturer's input. As of 2007 the implementation of this technology has become common; therefore, fast burn and strategic EGR is considered to be included in the baseline of this analysis.

Variable Compression Ratio

If an engine's compression ratio could be adjusted to near the knock-limited value over the operating range, significant fuel economy gains could be realized. Many mechanisms to realize variable compression ratios have been proposed in the literature and many have been tested. However, to date all these attempts add too much weight, friction, and parasitic load as well as significant cost and have therefore not been implemented into production designs (Wirbeleit et al., 1990; Pischinger et al., 2001; Tanaka et al., 2007). It should be recalled that alterations to the effective compression ratio via intake-valve closing (IVC) timing adjustments with higher-than-normal geometric compression ratios achieves some of this benefit.

VALVE-EVENT MODULATION OF GAS-EXCHANGE PROCESSES

Alteration of valve timing can have a major impact on volumetric efficiency over an engine's speed range, and thus peak torque and power are affected by this. IVC timing is the main determinant of this effect (Tuttle, 1980). Early IVC (compression stroke) favors torque, and later IVC favors power. Implementations of valve-event modulation (VEM) typically are referred to as specific technologies such as variable valve timing, variable valve timing and lift, two-step cam phasing, three-step cam phasing, and intake-valve throttling. VEM aids fuel consumption reduction by means of reducing pumping loss. Pumping loss is reduced by either allowing a portion of the fresh charge to be pushed back into the intake system (late IVC during the compression stroke) or by allowing only a small amount of the mixture to enter the cylinder (early IVC during the intake stroke).

It should be noted that any of the VEM schemes that reduce or eliminate the pumping loss also reduce or eliminate intake-manifold vacuum. Alternative means to operate power brakes, fuel vapor canister purge, and positive crankcase ventilation (PCV) systems, normally driven by intake-manifold vacuum, must then be considered. To overcome this issue, an electrically operated pump may need to be added. It should also be noted that while the implementation of VEM techniques can boost torque output of a given engine, this report assumes that constant torque will be maintained, leading to engine downsizing. The fuel consumption benefits listed in the following section consider a constant-torque engine.

VEM History

The first modern successful production implementation of a varying valve-event setup was Honda's VTEC in the late 1980s. Honda's system allowed a stepped increase in the duration and lift of the intake valves. Prior to the development of a multi-step cam profile system, a cam profile was chosen based on performance compromises. Engineers were confronted with a tradeoff, as it is difficult to satisfy the needs of both good low-speed torque and high-speed torque with a single cam profile. The cam profiles and timings necessary to maximize these needs are completely different in their characteristics.

Honda's technology was one of the first discrete variable valve lift (DVVL)-type systems. Over the years, many other companies have developed various implementations of DVVL-type setups, as well as other innovative VEM technologies. Some newer developments in VEM technology include systems that offer continuously variable lift and duration. Nissan's VEL, BMW's Valvetronic, and Fiat's Multi-Air are all examples of continuously variable lift systems that also incorporate adjustable valve timing (Takemura et al., 2001; Flierl and Kluting, 2000; Bernard et al., 2002). These systems attempt to operate throttle-less and rely on varying lift and timing to throttle the incoming air. Throttle-less operation allows a reduction in pumping losses at part load, and thus reduces fuel consumption. However, these throttle-less approaches also generally result in slight variations in the very small valve lifts necessary for idle operation even with well-controlled manufacturing tolerances. These small variations result in a slightly different charge mass from cylinder to cylinder, causing somewhat rougher idle engine operation, which is detrimental to customer satisfaction.

The cam phaser, used to vary the valve timing, is another technology that has been in constant development by the OEMs. Early cam phasers featured only two-step phasing, allowing two possible cam positions relative to the crankshaft. Today, cam phasing is fully variable, offering a wide range of positions. Due to the system's relative simplicity and long evolution, many production vehicles now utilize cam phasing technology. Until recently, cam phasing had only been applied to overhead cam style setups due to ease of integration. This recently has changed with GM's development and production of an in-block cam phaser applied to its overhead valve (OHV) 6.2-L engine.

Intake-Valve Closing Timing

Intake-valve closing timing, also known as intake cam phasing (ICP), is a form of VEM. At moderate speeds and light loads, late intake valve closing (i.e., during the compression stroke) can reduce the pumping loss; however, it also slows combustion. Typically this configuration yields effective compression ratios that are lower than the effective expansion ratio. To achieve a lower effective compression ratio, the intake valve closing is delayed until later on the compression stroke at light loads. By closing the valve later on the compression stroke, a larger portion of the air that was drawn in on the intake stroke is pushed back out through the valve. This phenomenon allows a decrease in pumping losses by relying on the timing of the intake valve to regulate engine load. From the reduction in pumping losses, a reduction in fuel consumption will occur. Some refer to late IVC as the Atkinson cycle (Boggs et al., 1995), and most engines have some of this character. For boosted engines, late IVC is termed by some as the Miller cycle (Hitomi et al., 1995).

A diagram of a typical oil-actuated variable cam phaser system installed on the intake cam (exhaust cam timing for this engine is fixed), Figure 4.1 shows the complexity of integrating a variable cam phaser into the standard engine architecture with fixed timing. As indicated in the figure, two separate oil passages are fed to the phaser. A solenoid controls the direction of the fluid to the two different passages. These passages are used to control whether the cam will be advanced or retarded relative to the crankshaft. In order for the engine control unit (ECU) to sense the relative position of the camshaft, a position sensor is installed that provides feedback information to the ECU. It is important to note that, like many of the vehicle technologies discussed in this chapter, implementing a variable cam phaser involves a complete system integration as illustrated in Figure 4.1 and is not as simple as bolting on a component.

Fuel Consumption Benefit and Cost of IVC Timing

OEM input suggests intake cam phasing results in roughly a 1 to 2 percent fuel consumption reduction. Both the EPA and NESCCAF also estimate approximately 1 to 2 percent fuel consumption reduction (EPA, 2008; NESCCAF, 2004). EEA claims a fuel consumption improvement of 1.1 to 1.7 percent can occur with the addition of an ICP (EEA, 2007). In agreement with most sources, the committee has also estimated a 1 to 2 percent reduction in fuel consumption using ICP. However, as with the other VEM technologies that are listed in the chapter, a generalized statement can be made that smaller-cylinder-count engines (i.e., four cylinders) will be closer to the low end of this improvement range, and higher-cylinder-count engines will be closer to the high end of the fuel consumption reduction ranges that are listed.

OEM input suggests that fixed-duration intake systems add a cost of about \$35/phaser. OEM input does not reflect a retail price equivalent (RPE) factor. The EPA estimates an RPE cost increase of \$59/phaser (EPA, 2008). NESCCAF quoted a literature RPE of \$18 to \$70 (NESCCAF, 2004) and EEA estimates an RPE of \$52/phaser (EEA, 2007). A 1.5 RPE factor was used to develop the committee estimate of \$52.50 for an in-line engine and \$105 for a V-configuration that requires two phasers.

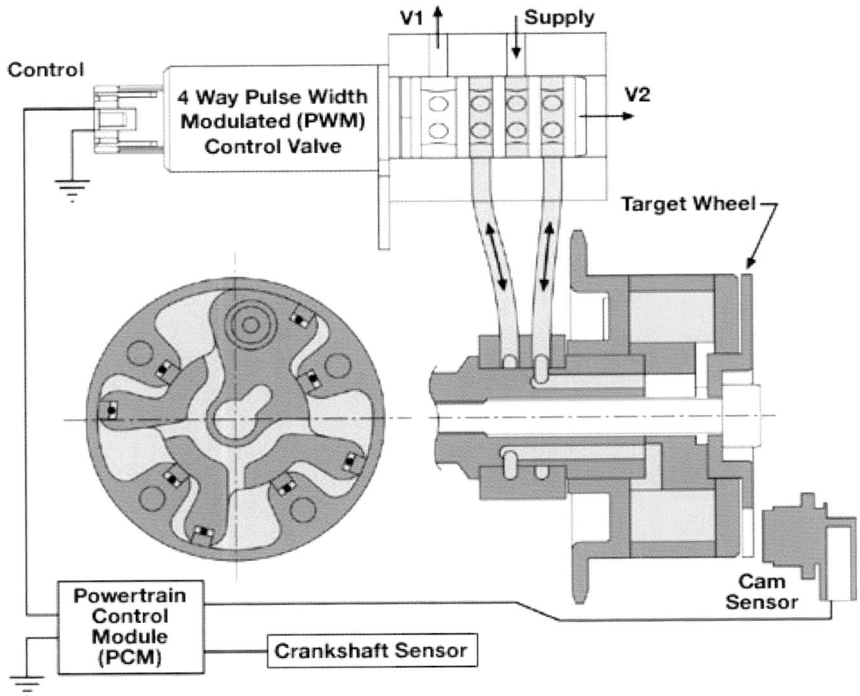

FIGURE 4.1 System-level mechanization of the variable cam phaser, oil control valve, control module, crank sensor, and cam sensor to the engine. SOURCE: Delphi (2009). Reprinted by permission from Delphi Corporation.

Valve Overlap Control

Valve overlap control, also known as dual cam phasing (DCP), is another form of VEM. Valve overlap (i.e., the interval between intake-valve opening [IVO] and exhaust-valve closing [EVC]) can affect residual-gas retention at low loads and can reduce pumping loss in a manner similar to that with EGR (exhaust gas recirculation). Valve overlap control can also be utilized to tune performance at high engine speeds, resulting in increased torque, which, in principle, can allow for minor engine downsizing. Valve overlap can be modulated by changing the phasing of either the intake or exhaust cam. Typically it is done with the exhaust cam because exhaust-cam phasing for increased overlap also delays exhaust-valve opening (EVO) timing. Thus both EVO and EVC move in ways favorable to low-speed and light-load fuel consumption reduction. Modulating valve overlap with an intake cam yields countervailing effects, i.e., increased valve overlap in this manner tends to reduce pumping loss while the corresponding IVC event will occur earlier, thus offsetting some of the increased-overlap benefit. At idle, too much valve overlap will destabilize combustion. When variable phasing, fixed-duration intake and exhaust cams are implemented, valve-overlap control may eliminate the need for an external EGR system.

Fuel Consumption Benefit and Cost of Valve Overlap Control

The fuel consumption reduction from valve overlap control/DCP is expected to be slightly greater than just controlling the IVC timing at about 2 percent over intake phasing alone, based on manufacturer input. The EPA and NESCCAF both estimate a reduction in consumption of 2 to 4 percent (EPA, 2008; NESCCAF, 2004). EEA estimates a 1.8 to 2.6 percent improvement in fuel economy (EEA, 2007). The committee concluded that adding variable exhaust cam phasing to ICP will yield an incremental 1.5 to 3 percent reduction in fuel consumption. This would mean the total estimated effect of adding DCP would be about 2.5 to 5 percent over an engine without any variable valve timing technology. The high end of 5 percent has been verified by OEMs and Ricardo, Inc.'s full-vehicle system simulation (FSS) (Ricardo, Inc., 2008).

Dual overhead cam (DOHC) V-engines with variable intake and exhaust would require four cam phasers, adding roughly $140 of manufacturer cost based on manufacturer input, but a portion of this is offset by the elimination of the external EGR system. EEA estimates an RPE of $76 to $84 for an I4, and $178 to $190 for V6 and V8 engines (EEA, 2007). The EPA estimates an incremental cost increase of $89 for an I4 and $209 for V6 and V8 engines (EPA, 2008).

NESCCAF quotes a literature RPE of $35 to $140 for dual cam phasers (NESCCAF, 2004). Discussion with OEMs also verified that by simply doubling the cost of ICP, a reasonably accurate DCP cost can be attained. The committee has estimated an RPE cost of $52.50 for an in-line engine and $105 for a V-configuration, incremental to the cost of ICP technology.

Intake-Valve Throttling

Using very short duration and low-lift intake-valve-opening events during the intake stroke can reduce (or eliminate) the pumping loss. This VEM, also known as intake-valve throttling, also tends to slow combustion, mainly at low engine speeds. (Small-scale turbulence generated by this approach dissipates rapidly, well before the start of combustion, and thus this does not generally contribute to rapid combustion). Note that low valve lift is simply a consequence of short-duration cam design. Manufacturing tolerance control is of extreme importance with intake valve throttling if cylinder-to-cylinder variability at idle is to be acceptable. BMW and Nissan currently offer this technology on some of their engine models, which use varying lift and timing to throttle the engine. Other manufacturers have announced plans to introduce engines with throttle-less operation within the next few years.

The above options (DCP and ICP) are focused mainly on pumping-loss reduction by means of late IVC timing and exhaust-gas recycling via variable valve overlap. Very early IVC (i.e., during the intake stroke) is another effective means of reducing pumping losses, but it involves much more complex and costly means of implementation. Two types of intake-valve-opening techniques are considered: discrete variable valve lift and continuously variable valve lift.

Discrete Variable Valve Lift

A discrete variable valve lift (DVVL) system is one which typically uses two or three different cam profiles over the range of engine speeds and loads. This system attempts to reduce pumping losses by varying the lift profile of the camshaft. By varying the lift of the valves, it is possible to limit the use of the throttle and significantly reduce the pumping losses.

As described earlier, Honda has been using a DVVL-type setup on its vehicles known as VTEC. To engage the different cam profile on Honda's system, there is a third cam lobe and follower, located in between the two main lobes, which is hydraulically activated by an internal solenoid controlled oil passage. During low-speed and low-load operation, the engine runs using the base cam profile(s). Once a certain load point is reached, the ECU activates a control valve to direct oil pressure from the main gallery to an oil passage that engages the third follower. Once the third follower engages, it is then locked into place by a locking pin. Honda's VTEC system is more cost-effective on its single overhead cam (SOHC) engines, due simply to the fact that a DOHC engine would require more hardware. This is an example of one manufacturer's method of DVVL implementation. It should be noted that other manufacturers have developed different designs to accomplish the same goal, and as a result the different systems have differing amounts of pumping loss reduction and friction increase. This situation reinforces the point that advanced VEM technologies are not simply "bolt-on" parts that provide a uniform fuel consumption reduction to all OEMs.

Delphi performed testing on a GM 4.2-L I6 equipped with a two-step variable valve actuation system and a camshaft phaser on the intake (Sellnau et al., 2006). The engine was already outfitted with an exhaust cam phaser. Delphi's two-step valve actuation system consisted of oil-actuated switchable rocker arms. Testing on the engine revealed a 4.3 percent fuel consumption reduction during the EPA city drive cycle, compared to the base engine with no variable lift and timing. These results were obtained with no other modifications besides the VVL, a phaser, and control system reconfiguration. Delphi claimed that "mixture motion is nearly absent for low lift profiles, so an enhanced combustion system, with higher tumble for low-lift profiles, would likely yield significant improvements in fuel economy." In the second portion of the test Delphi modified the cylinder head and added flow restriction that generates turbulence in an attempt to speed up combustion, thereby furthering the fuel economy gain. Chamber masks were used to increase the tumble motion. The lift profile on the exhaust cam and the port were also modified. For the second phase of testing with the altered cylinder head and calibration, the fuel consumption reduction was estimated to be 6.5 percent in comparison to the original engine. These values were estimated from data taken at multiple load points rather than over a driving cycle (Sellnau et al., 2006).

Fuel Consumption Reduction and Cost of DVVL

Two (or three)-step cams that yield short intake durations using DVVL can yield fuel consumption reductions in the 4 to 5 percent range based on vehicle OEM input. A reduction of 3 to 4 percent in fuel consumption (FC) is estimated from the EPA (EPA, 2008). FEV has developed a two-stage switch of the intake valve lift that is claimed to offer up to a 6 to 8 percent reduction in consumption when combined with variable valve timing, during the New European Drive Cycle (Ademes et al., 2005). NESCCAF and EEA estimate that a 3 to 4 percent reduction is possible (NESCCAF, 2004; EEA, 2007) on the U.S. driving cycles. EEA also estimates a fuel economy improvement of 7.4 to 8.8 percent when DVVL is combined with DCP and the engine is downsized to maintain constant torque. Simulation work by Sierra Research indicated a 6.3 to 6.8 percent benefit when combined with variable valve timing, which accounts for up to 5 percent of that amount (Sierra Research, 2008). The committee concluded that a 1.5 to 4.0 percent drive-cycle-based FC reduction is possible, incremental to an OHC engine with DCP or an OHV engine with CCP.

Vehicle OEM input suggests a $35 to $40/cylinder cost for implementing DVVL. The Martec Group estimates an OEM cost of $320 to implement a two-step VVL on a V6 DOHC engine (Martec Group, Inc., 2008). The EPA estimates an incremental cost increase of $169 for an I4, $246 for a V6, and $322 for a V8 (EPA, 2008). EEA estimates RPEs for an OHC-4V; $142 to $158 (equivalent to $95 to $105 assuming an RPE multiplier of 1.5) for an I4, $188 to $212 (equivalent to $125 to $141 assuming an RPE multiplier of 1.5) for a V6, and $255 to $285 (equivalent to $170 to $190 assuming an RPE multiplier of 1.5) for a V8 (EEA, 2007). The committee estimates the manufacturing cost of implementing DVVL to be about $30 to $40/cylinder.

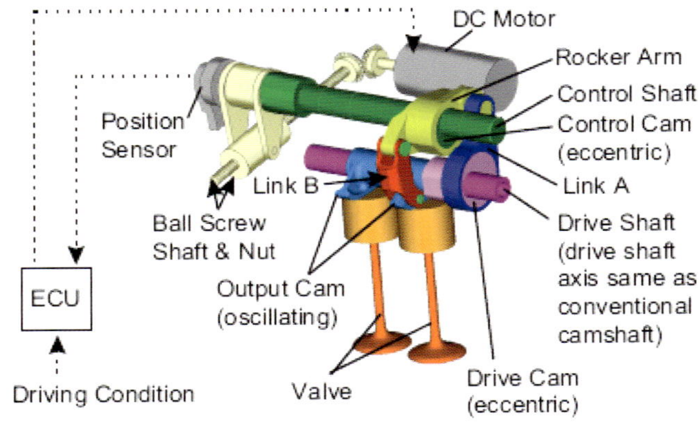

FIGURE 4.3 Nissan valve event and lift design. SOURCE: Takemura et al. (2001). Reprinted with permission from SAE Paper 2001-01-0243, Copyright 2001 SAE International.

Continuously Variable Valve Lift

The continuously variable valve lift (CVVL) system allows a wide control range of the camshaft profile (see Figures 4.2 and 4.3 for schematics). A continuous system allows for calibration of the optimal valve lift for various load conditions, versus the discrete system, which will only offer two or three different profiles. The combination of a continuous VVL system and an intake cam phaser has the potential to allow the engine to operate throttle-less. In the following, greater detail of this particular VEM technology is given due to its relative novelty to the mass production environment and the large fuel consumption benefits it offers. Two approaches to CVVL have been considered, electromechanical and electrohydraulic.

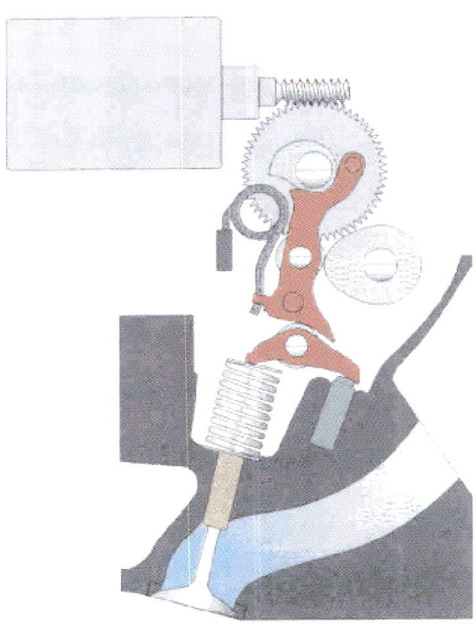

FIGURE 4.2 BMW Valvetronic. SOURCE: Flierl et al. (2006). Reprinted with permission from SAE Paper 2006-01-0223, Copyright 2006 SAE International.

Electromechanical CVVL Systems

BMW was the first to offer a mass production fully variable valve train incorporating CVVL in 2001, which it calls Valvetronic, Figure 4.2. This system is an electromechanical system that when combined with variable intake and exhaust cam phasers provides a fully throttle-less induction system. To vary the lift of the valve, an intermediate lever was added along with an eccentric shaft. The eccentric shaft is operated by an electric motor that adjusts the positioning of the lever over the camshaft. The lever contains a profile with one side being relatively flat and the other side being relatively steep. Adjusting the relative positioning of the lever controls the valve lift. BMW claims that up to a 10 percent reduction in fuel consumption is possible with this system (Sycomoreen). Figure 4.2 shows the many added components needed for the Valvetronic system.

Nissan Motor Company has also developed a continuous variable valve event and lift (VEL) system (Figure 4.3). The electromechanical system allows continuous variation of valve timing and lift events similar to the BMW system, but achieves this using a different architecture. Nissan estimates a 10 percent reduction in fuel consumption over the Japanese 10-15 drive cycle (Takemura et al., 2001) for its VEL system. The 10-15 drive cycle is intended to simulate a typical urban drive cycle, and an EPA combined FTP cycle rating would be somewhat lower. Nissan attributes the reduction in consumption to "lower friction loss due to the use of extremely small valve lift-timing events and reduction of pumping loss resulting from effective use of internal gas recirculation." Nissan evaluated the consumption benefits distribution at a fixed speed and load of 1,600 rpm and 78 N-m. The distribution of effects was the following: (1) pumping loss decrease yielded a consumption reduction of 5.2 percent, (2) friction reduction yielded a consumption benefit of 1.1 percent, and (3) an improvement in combustion caused a reduction in consumption of 3 percent.

Figure 4.3 shows the layout of Nissan's VEL system. The electromechanical system uses an oscillating cam to open and close the valve. An oscillating cam (output cam) looks like half of a camshaft, but it is hinged on one end to allow full opening and closing of the valve on the same cam face. To change the valve lift and duration of the cam, the control shaft is adjusted by a motor to change the distance between the control cam and the oscillating cam. An increase in distance is caused by the lobe on the control shaft turning and pushing the rocker arm assembly out. This changes which portion of the output cam contacts the valve to control the amount of lift.

Toyota Motor Company has recently developed its own type of a CVVL timing system. The new system will first be applied to their newly developed 2.0-L engine. Toyota's system features separate cam phasers on the intake and exhaust camshafts to vary the camshaft timing, along with a continuously variable valve lift system. Toyota claims that the system "improves fuel efficiency by 5 to 10 percent (depending on driving conditions), boosts output by at least 10 percent and enhances acceleration." Toyota did not state what features the base engine already had in order to generate fuel efficiency improvement percentages (Toyota Motor Co., 2007).

The Technical University of Kaiserslautern performed testing on a 2.0-L four-cylinder gasoline engine that was outfitted with a fully variable lift and timing system (VVTL) called Univalve, Figure 4.4. The Univalve system allows for either the use of standard throttle or unthrottled operation. At a load point of 2000 rpm and a BMEP of 2 bar, a 13 percent reduction in fuel consumption occurred compared to the base engine with a nonvariable valve train. This reduction is due to the reduction in the pumping work and an improvement in the formation of the mixture. The Univalve system varies the lift and duration of the valve by adjusting the eccentric contour (see Figure 4.4). Adjusting the eccentric shaft changes the rocker arm pivot point (Flierl et al., 2006).

The Univalve system in Figure 4.4 operates similar to BMW's version of a CVVL system. In Figure 4.4 the image to the left demonstrates a fixed pivot ratio on the rocker with constant valve lift. The image to the right features variable valve lift. To vary the lift the rocker arm is no longer fixed to a single pivot point. An eccentric shaft creates a varying pivot point by adjustment of the shaft's contour contact point on the rocker.

Honda has also patented its new Advanced-VTEC system, which turns its current DVVL VTEC system into a throttleless CVVL setup. While initial claims are up to a 10.5 percent reduction in fuel consumption, this system is not currently in

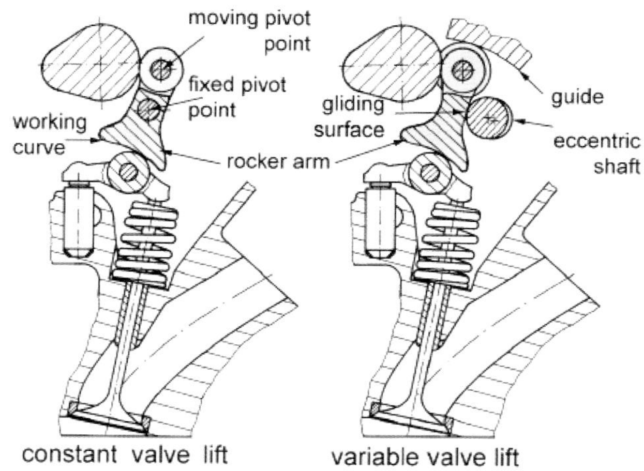

FIGURE 4.4 Univalve. SOURCE: Flierl et al. (2006). Reprinted with permission from SAE Paper 2006-01-0223, Copyright 2006 SAE International.

production and the testing cycle used to produce this estimate is unclear. Therefore, Advanced-VTEC is only mentioned to demonstrate an example of emerging CVVL technology.

Electrohydraulic CVVL Systems

The electrohydraulic approach to CVVL has been under development for over a decade. One of the organizations which has been active in this development is Fiat Central Research (CRF). The major focus of the work by CRF is a system that it calls Uniair (Bernard et al., 2002). Fiat recently announced a system it calls Multiair that is derived from Uniair. Multiair is a joint development between Fiat and valve train component supplier INA that promises a 10 percent reduction in fuel consumption. Other organizations have also been active in the development of systems using similar principles (Misovec et al., 1999). The Uniair/Multiair system has been described as a lost-motion system wherein the camshaft lobe drives the piston of a small pumping chamber, one for each cylinder intake and one for each exhaust. Multiair utilizes the system only for the intake valves.

The output from the pump is controlled by a solenoid-actuated flow control valve that directs the hydraulic output of the pump directly to the hydraulic actuator on the valve(s) or to the accumulator. If the control valve directs the hydraulic pressure to the valve actuator(s), the valve(s) open normally following the camshaft profile. In principle a lost-motion system allows opening the valve(s) at any fraction of the normal valve lift profile by directing part of the hydraulic pressure to the accumulator rather than to the valve actuator. By appropriately controlling the application of the hydraulic pressure to the valve actuators or to the accumulator, a wide range of valve lift profiles can be achieved, including multiple small lifts during one valve event. This latter capability is not achievable with mechanical CVVL systems. However, electrohydraulic CVVL systems tend to be less efficient considering the energy lost by the hydraulic pump and the increased friction losses from the additional number of components. The committee believes that the large increase in parasitic losses that will offset the perceived fuel consumption reduction benefit, combined with the high component cost, will limit the market penetration of this technology. In addition, achieving consistent and uniform valve lifts under idle conditions to maintain a smooth idle may be more challenging than with mechanical CVVL systems.

Fuel Consumption Benefit and Cost of CVVL

The above discussion reviewed the technology of VEM approaches and various FC benefits ascribed to each system. As noted in Chapter 2, the fuel consumption reduction benefits for the technology approaches considered are based on the combined city and highway driving cycles, while some of the benefits described earlier are not necessarily based on these driving cycles. CVVL is expected to be in the 5 to 7 percent range based on manufacturer input. The EPA and NESCCAF both estimate a 4 to 6 percent reduction in fuel consumption (EPA, 2008; NESCCAF, 2004), while EEA estimates a 6.5 to 8.3 percent reduction in fuel consumption at constant engine size and 8.1 to 10.1 percent with an engine downsize to maintain constant performance (EEA, 2007). Sierra Research's simulation work resulted in a 10.2 to 11.0 percent benefit when combined with variable valve timing (Sierra Research, 2008). The committee has estimated that CVVL will have an additional 3.5 to 6.5 percent reduction in fuel consumption over an engine already equipped with DCP. Going from a base DOHC engine to one with continuously variable lift and timing could provide a 6 to 11 percent fuel consumption reduction assuming engine size adjustments for constant acceleration performance.

Vehicle OEM input suggests that the cost of a continuously variable intake-valve is two to three times that of the two-step system plus the cost of the actuation system ($40 to $80) plus the cost of the intake and exhaust cam-phasing system. Vehicle integration could add another cost in the range of $140. The EPA estimates an RPE incremental cost of $254 (or $169 cost assuming an RPE multiplier of 1.5) for I4, $466 (or $311 cost) for V6, and $508 (or $339 cost) for V8 engines (EPA, 2008). The Martec Group estimates a manufacturing cost of $285 for an I4, $450 for a V6, and $550 for a V8 (Martec Group, Inc., 2008). For a CVVL system, EEA (2007) estimates RPEs of $314 to $346 (or $209 to $231 cost) for an I4, $440 to $480 (or $293 to $320 cost) for a V6, and $575 to $625 (or $383 to $417 cost) for a V8 (EEA, 2007), all assuming an RPE multiplier of 1.5. The committee estimates the manufacturing cost of CVVL to be $159 to $205 for I4 engines, $290 to $310 for V6 engines, and $350 to $390 for V8 engines, not including an RPE factor.

VEM Implementation Techniques

Many of the above-mentioned VEM systems are often implemented as a package combining varying valve lift and timing events. The combination of these technologies will provide further reduction in the use of the throttle.

General Motors Research and Development (Kuwahara et al., 2000; Cleary and Silvas, 2007) performed testing on a single-cylinder model of their 3.4-L DOHC engine. The model made use of varying intake valve cam timing, duration, and intake valve lift. A combination of the varying parameters allowed for the engine to operate without a throttle. From the study by General Motors, an approximate reduction in fuel consumption of up to 7 percent occurred at part load conditions. By unthrottling the engine, a large reduction in throttling losses occurs and the engine was able to operate at higher intake manifold pressures. It is important to note that the cost and fuel consumption reductions of the various VEM approaches are highly variable and dependent upon the basic engine architecture to which they are applied.

Cylinder Deactivation

Cylinder deactivation is utilized during part load situations to reduce thermal and throttling losses. During constant speed operation, the power demand is relatively low. By shutting off multiple cylinders, a higher load is placed on the remaining operating cylinders. The higher load requires the throttle to be open further and therefore reduces the throttling losses. The decrease in losses reduces the overall fuel consumption. Cylinder deactivation via valve deactivation has been applied to four-, six-, and eight-cylinder engines, in some cases rather successfully. Most commonly, cylinder deactivation is applied to engines that have at least six cylinders; four-cylinder engines typically are not equipped with deactivation due to additional noise, vibration, and harshness concerns that are deemed unsatisfactory for consumers. Even current production V6 offerings have NVH levels that are very noticeable to customers. Increased NVH can be perceived as a low-quality characteristic that deters potential customers from purchasing vehicles with this technology.

History of Cylinder Deactivation

Cylinder deactivation was first implemented on a production vehicle in 1981 on the Cadillac V8-6-4. The engine could operate in four-, six-, and eight-cylinder mode depending on power demand. To deactivate the cylinders, a solenoid mounted on top of the rocker arm assembly would disconnect the pivot point for the rocker and the rocker would then pivot against a soft spring. The valves would remain closed and the cylinder would not fire, but rather act as a compressed air spring. This system helped to reduce fuel consumption at cruising type conditions. However, drivability and the need for quick re-engagement of the cylinders caused customer dissatisfaction, and the technology was soon taken out of production. Since then, engine control systems and programming ability have diminished the drivability concerns with modern day deactivation systems. New solutions have been developed to address the NVH concerns that arise when cylinders become deactivated. The NVH is a concern during deactivation due to the "lower frequency, higher amplitude torque pulsations at the crankshaft" (Leone and Pozar, 2001). With the addition of active engine mounts, any vibrations which would normally transfer to the passenger compartment of the vehicle, causing customer dissatisfaction, are nearly eliminated. However, active engine mounts add cost. Today's trend toward overhead cam (OHC) valve trains has an added a level of cost and complexity to integrate cylinder deactivation.

Implementation of Cylinder Deactivation

The integration of a cylinder deactivation system varies depending on the engine layout. For overhead valve V8 and V6 engines, this can be accomplished fairly simply by modifications to the passages that supply oil to the valve lifters along with different valve lifters (Falkowski et al., 2006). Implementation of a deactivation system on an OHC engine is slightly different than on an OHV engine. One of the methods utilized for cylinder deactivation in an OHC roller finger follower system involves the use of a switchable roller finger follower. In the follower's normal mode, the valve will operate as usual and maximum lift will still be achieved. To deactivate the cylinder, a locking mechanism must be released on the follower by oil pressure (Rebbert et al., 2008), collapsing the follower and rendering the valve inactive.

Fuel Consumption Benefit and Cost of Cylinder Deactivation

Vehicle OEMs estimate cylinder deactivation typically yields fuel consumption reductions in the 6 to 10 percent range on V8 configurations. Testing done by FEV on a V8 engine found that a decrease in fuel consumption of 7 percent occurred on the New European Drive Cycle (NEDC). According to FEV, these reductions would be "even higher for the US driving cycle, because of the US cycle's higher proportion of part load operating conditions" (Rebbert et al., 2008). NESCCAF estimates a 4 to 6 percent reduction in fuel consumption (NESCCAF, 2004). The EPA estimates a 6 percent reduction in fuel consumption (EPA, 2008). Sierra Research's simulation estimated a reduction in consumption of 7.5 to 8.8 percent (Sierra Research, 2008). EEA estimates a 5.3 to 7.1 percent reduction in fuel consumption (EEA, 2007). For OHV engines, the committee estimates a 4 to 6 percent drive-cycle fuel consumption reduction on a V6, and a 5 to 10 percent reduction on a V8. For OHC engines, the committee assumes manufacturers would have already implemented DCP and VVL based on the cost/benefit ratio. This means that there is less pumping loss left to reduce, resulting in an incremental 1 to 2.5 percent reduction for a V6 and a 1.5 to 4 percent reduction for V8 configurations. The lower cost-benefit ratio for cylinder deactivation makes the technology far less attractive on DOHC engines. Despite the existence of prototype four-cylinder engines with cylinder deactivation, the committee believes the cost and customer dissatisfaction issues related to NVH outweigh the benefits of implementing this technology on four-cylinder engines.

Vehicle OEMs estimate the cost for deactivation is approximately \$115. Vehicle integration items that mitigate NVH issues may incur additional costs in the \$140 range. The cost of applying cylinder deactivation to OHC engines is much higher, i.e., \$340 to \$400 because more complex and costly valve train elements must be changed. The EPA estimates the incremental RPE cost to be \$203 (or \$135 cost) for six cylinders and \$229 (or \$153 cost) for eight cylinders (EPA, 2008) (both assuming an RPE multiplier of 1.5). NESCCAF quotes a literature RPE of \$112 to \$746 (NESCCAF, 2004) (or \$75 to \$497 cost). Martec estimates a manufacturing cost increase of \$220 for a V6 DOHC engine (Martec Group, Inc., 2008). Sierra Research estimates an incremental cost of \$360 to \$440 (Sierra Research, 2008). EEA (2007) estimates for six-cylinder engines an RPE of \$162 to \$178 (or cost of \$108 to \$119) with an additional cost of \$140 for NVH. For eight-cylinder engines, EEA estimates an RPE of \$205 to \$225 (EEA, 2007) (or cost of \$137 to \$150 assuming an RPE of 1.5). The committee estimates that the manufacturing cost of implementing cylinder deactivation for OHV would be \$220 to \$255 and \$340 to \$420 for engines with SOHC (not including RPE).

Camless Valve Trains

A fully camless valve train eliminates the need for camshafts, as well as various other supporting hardware, and operates the valves individually by means of actuators. This would allow for VEM fuel consumption saving technologies, such as cylinder deactivation and continuously variable valve lift and timing, to be applied all in one package. However, the complexity of the controls required makes for a difficult integration. Camless valve trains are electromagnetic, hydraulic, pneumatic, or combinations of these that all face fundamental obstacles. By replacing the valve train, BMW claims the frictional saving from just the roller-bearing valve train achieves a further 2 percent reduction in fuel consumption. BMW also claims an overall reduction of up to 10 percent from camless operation (Hofmann et al., 2000). However, none of these has been shown to offer advantages not observed with the aforementioned cam-based systems. The very high valve-timing precision associated with most cam-driven systems is subject to compromise with camless approaches. The ballistic character of the valve assembly with any camless system presents many control challenges. In addition, the power demand for camless systems is generally higher than that of their cam-driven counterparts.

Camless systems are perceived to have significant durability risk, and as a result, no production implementations of camless systems have been announced. It is the judgment of the committee that camless systems need further development and are not expected on the market before 2015.

GASOLINE DIRECT INJECTION

The most recent development of direct injection spark ignition (DISI) (also known as GDI) systems (Wurms et al., 2002) have focused on early-injection, homogeneous-charge implementations using stoichiometric mixture ratios under most operating conditions. These conditions allow for the use of highly effective and well-proven closed-loop fuel control and three-way catalyst exhaust aftertreatment systems. Fuel consumption benefits of these homogeneous versions are derived mainly from a knock-limited compression ratio increase (typically +1.0) enabled by forcing all of the fuel to vaporize in the cylinder. This yields a charge-cooling effect that suppresses the knocking tendency. Another added benefit of charge-cooling is an increase in the volumetric efficiency from the increase in density of the incoming charge. In contrast, with port fuel injection (PFI) systems some of the fuel vaporizes in the intake port, and this conveys heat from outside of the cylinder, i.e., from the intake port, to the in-cylinder charge. While heating of the intake charge is a negative (relative to the knock-limited compression ratio and performance) it does provide a measure of "thermal throttling" at typical road loads, which reduces negative pumping work. Thermal throttling, like common pressure throttling, lowers the mass of inducted fuel-air mixture thus reducing power, which is the objective of throttling. It does this, however, with less pumping loss than the conventional throttling used with homogeneous DISI.

In terms of additional losses, DISI relies on fuel pressures that are higher than those typically used with PFI systems (e.g., 150-200 bar versus 3-5 bar for PFI), and the increase in required fuel pump work increases parasitic loss. Finally, these homogeneous, stoichiometric DISI systems cannot exploit the thermodynamic expansion efficiency gains possible with lean overall mixtures.

History of Direct Injection

Early (1960s and 1970s) versions focused on late-injection, lean overall stratified-charge implementations as exemplified by the Texaco TCCS (Alperstein et al., 1974) and Ford PROCO (Simko et al., 1972) systems, neither of which entered volume production. These systems were attempts to utilize gasoline and other fuels in spark-ignited engines designed to take advantage of two of the three thermodynamic advantages of diesels, namely lack of throttling to eliminate pumping losses, and lean overall mixture ratios to achieve more thermodynamically efficient expansion processes. However, the TCCS and PROCO systems suffered from injector fouling, high exhaust emissions and low power density. Nonetheless, the goals of these engine systems remained valid and interest returned to DISI following progress in fuel-injection systems and engine controls during the 1980s and early 1990s. Mitsubishi introduced the first production implementation of DISI (which they called GDI) in Europe in 1996 (Iwamoto et al., 1997) in a 1.8-L four-cylinder engine, followed shortly after by a 3.5-L V6 in 1997. These GDI systems utilized lean-overall stratified-charge combustion but with some inlet throttling. It was soon found that typical in-use fuel consumption was significantly higher than European emissions-test-schedule results suggested.

Following an initial burst of interest, Mitsubishi GDI sales were lower than expected. Hence, this system was withdrawn from the market, and there was a return to conventional PFI systems. It was believed that this withdrawal stemmed not only from disappointing sales but also because meeting upcoming NO_x emissions standards in Europe and especially the United States using only combustion system control was more difficult than anticipated, and lean NO_x aftertreatment systems were seen as very costly and of questionable reliability for volume production.

Implementation of Direct Injection

A concern today (as in the past) with DISI systems is the matter of fuel-based carbonaceous deposits forming from residual fuel in the injector nozzle upon hot engine shutdown. Carbonaceous deposits can restrict fuel flow and also modify fuel-spray geometry in some unfavorable manner (Lindgren et al., 2003). Locating the injector in a relatively cool part of the cylinder head is one approach to alleviating this problem. Fuel variability in the United States is of some concern relative to this issue based largely upon the olefin content of the fuel, which typically is higher than that found in European gasoline. While some concerns with deposits remain, they are being alleviated mainly by injector design improvements.

DISI researchers often make reference to wall-guided, flow-guided, or spray-guided injection (Kuwahara et al., 2000), and in general these terms refer to different geometric arrangements of the fuel injection and mixture preparation processes. For example, wall-guided usually refers to placement of the fuel injector to the side of the cylinder near the corner of the cylinder head with the cylinder wall. The spray is then aimed across the cylinder toward the top of the piston when the piston is near the top of the cylinder. In this case the piston crown shape is the "wall" which guides the spray (Kuwahara et al., 2000). In spray-guided engines, the injector is located in the cylinder head near the center of the cylinder with the spray aimed down the cylinder axis (Schwarz et al., 2006). Injection in this case would be timed later during the induction process. The fuel-spray trajectory is then guided mainly by the direction of the spray and its interaction with the cylinder gas motion rather than by directly impinging on a surface such as the piston.

BMW performed a fuel consumption comparison study using a four-valve port fuel injection engine with fixed timing and lift as the base engine for comparison. For the study a direct injection system operating at stoichiometric was applied to an engine, and a fuel consumption benefit of 5 percent resulted. BMW claimed that if a spray-guided system were adapted, the engine could operate with lean mixtures, which would allow for up to a 20 percent fuel consumption reduction (EEA, 2007).

Fuel Consumption Benefit and Cost of Direct Injection

The increase in knock-limited compression ratio possible for DISI configurations would be expected to yield a fuel consumption reduction in the 2 percent range based on vehicle OEM input, but the countervailing effect of pumping and parasitic loss increases may reduce this benefit somewhat to about 1.8 percent. Based on modeling by EPA, consumption reduction estimates for converting from a PFI to a DISI system are in the range of 1 to 2 percent for four-, six-, and eight-cylinder engines (EPA, 2008). Sierra Research estimates a reduction in consumption of 5.9 to 6.2 percent (Sierra Research, 2008). EEA estimates a 2.9 to 3.8 percent reduction in fuel consumption (EEA, 2007). Ricardo Inc.'s simulation work (Ricardo, Inc., 2008) attributes a 2 to 3 percent benefit to DISI. The committee believes that a 1.5 to 3 percent fuel consumption reduction can be realized from stoichiometric direct injection.

Vehicle OEMs estimate that the variable cost of DISI for parts is in the range of $60 per cylinder plus about $136 for vehicle noise abatement features, excluding the cylinder-head design and retooling costs. This input does not reflect an RPE factor. The EPA estimates the incremental cost for converting from a PFI to a DISI system on a four-cylinder engine to be from $122 to $420, on a six-cylinder from $204 to $525, and on an eight-cylinder from $228 to $525 (EPA, 2008). Martec Group estimates incremental costs of $293 for a four-cylinder, $372 for a six-cylinder, and $497 for an eight-cylinder engine (Martec Group, Inc., 2008). The estimates from Martec were based on converting to a homogeneous, side-mount direct injection from a port injection system. The committee estimates that the manufacturing cost of implementing a stoichiometric direction injection system would be $117 to $351 depending on the cylinder count (not including RPE). The cost range for noise abatement-related items causes the most uncertainty in the estimates, as the various manufacturers have different standards for acceptable noise levels. Luxury vehicles, for example, require more money to be spent to reduce noise to levels that customers expect. See Table 4.A.1 in the annex at the end of this chapter for a complete breakdown of cost and fuel consumption benefits for each engine size, including ranges for costs.

DOWNSIZED ENGINES WITH TURBOCHARGING

Turbocharging and downsizing engines (Petitjean et al., 2004) reduces engine mass and pumping losses, but the fuel consumption benefit is based somewhat on the measures taken to avoid knock and pre-ignition. Some engines in this category are developed and calibrated in such a way that premium fuel is required in order to avoid knock without decreasing the compression ratio. If this is the case, any fuel consumption benefit cannot be solely attributed to turbocharging and downsizing. Based on vehicle OEM input, a compression-ratio reduction of 1 to 2 from non-turbocharged versions is typically required if this system is to be regular-fuel compatible. Furthermore, reduction in the number of cylinders, e.g., V6 to I4, may require countermeasures necessary to satisfy NVH expectations.

Implementation of Downsizing and Turbocharging

Several conditions must be addressed in implementing downsizing and turbocharging. Piston oil squirters aimed at the underside of the piston and oil coolers are employed to mitigate knock and pre-ignition conditions. An increase in intake air temperature is a natural by-product of compressing the air. To counter this effect, charge-air coolers are frequently employed to reduce charge temperature prior to its entry into the cylinder. In order to maximize the power output of the engine, the charge cooler acts as a heat exchanger and typically uses ambient air for cooling. The addition of a charge cooler creates packaging concerns since a location must be chosen where the cooler will experience a large amount of cross flow in order to avoid becoming heat soaked during prolonged high load conditions.

Additional parasitic loads are often imposed by the use of increased oil and coolant pump capacities relative to their non-turbocharged counterparts. The increase in capacities results from the increase in power and heat rejection with the same physical displacement.

As mentioned above, a port fuel-injected engine typically requires a decrease in compression ratio, which decreases the thermal efficiency and the part load response of a turbocharged engine. Direct injection alleviates some of the knock tendencies associated with turbocharging through the charge cooling effect created by the high atomization of the fuel that results from high injection pressure. This cooling effect allows for a less significant reduction in compression ratio compared to a port fuel-injected engine. A concern with direct injection is the injector nozzle fouling upon hot engine shutdown, as noted previously. However, a positive synergism is possible by combining DISI, turbocharging, and dual cam phasers, because under some operating conditions the intake manifold pressure is higher than that of the exhaust manifold. This positive pressure difference enables improved exhaust scavenging and thus improved volumetric efficiency. This condition is sometimes referred to as blow-through because it occurs during valve overlap. This synergism of turbocharging, DISI, and blow-through can enable further engine downsizing, and an additional fuel consumption benefit may thus result. Unfortunately, this engine performance opportunity occurs in the knock-sensitive operating range. As a result, establishing acceptable vehicle launch performance with turbocharged and downsized engines is challenging.

The distinction between research octane number (RON) and motor octane number (MON) is particularly noteworthy when fuels other than traditional gasoline are considered. The test methodology on which RON is based reflects resistance to thermal auto-ignition resulting from both chemical and heat-of-vaporization (evaporative cooling) properties, whereas MON is relatively insensitive to the latter of these. The difference between these two metrics is termed *sensitivity* (RON – MON = sensitivity). When fuels like ethanol are considered, the aforementioned distinction should be emphasized as this fuel has a very high RON, but its MON is moderate. Hence, the sensitivity of ethanol is 18, whereas that of a typical gasoline is considerably lower, e.g., 10. The consequence of high-sensitivity fuels when aggressive boosting and high compression ratios are pursued is an increased vulnerability to pre-ignition problems. This typically results from engine operation in the peak-power range where all surface temperatures to which the fuel is exposed are very high. This tends to reduce the heat-of-vaporization benefit associated with ethanol. It has been widely recognized for most of the history of the SI engine that water induction along with the fuel and air can reduce the thermal auto-ignition tendency and thus can increase the torque and power output. While this has been widely used in racing communities, there are some practical limitations to the general applicability of this, e.g., water can find its way into the crankcase and form an emulsion with the oil and therefore compromise the lubrication system.

The evaporative characteristic of any liquid largely depends upon intermolecular affinity, and in the cases cited above the so-called hydrogen bonding is a major component. This involves the polarized bonds between hydrogen and oxygen atoms where there is a slight positive charge on the hydrogen atom that is bound to an adjacent oxygen atom, which carries a slight negative charge. Hence, the positive charge on the hydrogen atom of the –OH group applies an attractive force acting on the negative charge on the oxygen atom of a nearby molecule. This grouping of –OH-containing molecules, be they ethanol or water, is responsible for their relatively high evaporative-cooling characteristic. This evaporative cooling characteristic can be utilized to prevent knock at certain engine operating conditions by implementing a system that can selectively inject the charge cooling liquid. This system is discussed below in this chapter in the section "Ethanol Direct Injection."

Exhaust-gas recirculation (EGR) is well known as a means to reduce pumping losses and thereby increase fuel efficiency. With downsized turbocharged engines (including those with direct fuel injection) it has been found that cooled EGR can be seen as an alternative means for controlling knock at moderate engine speeds and medium to high loads. Under certain operating and base-engine conditions, passing the EGR through a heat exchanger to reduce its temperature can be a more fuel-efficient means of controlling knock compared to spark-timing retardation and fuel-air ratio enrichment. The fuel consumption benefits of this feature are highly dependent upon the base engine to which it is applied and the engine's operating map in a particular vehicle. As the heat exchanger must be equipped with a diverter valve to accommodate heat-exchanger bypass for lighter-load operation, the sequences of carbonaceous deposit formation in the heat exchanger, in the diverter and control valves, and in the turbine are among the real-world factors that can compromise the overall performance of this feature. This feature is in production for CI engines for which the exhaust particulate level is much higher than for downsized and boosted SI engines; however, packaging the system into certain vehicles can make implementation difficult.

Variable geometry turbochargers (VGTs), commonly used on CI diesel engines, have not reached mainstream use on SI engines. The concern with using VGTs on gasoline-engine exhaust has been the ability of the adjustable blades and their adjustment mechanism to withstand the higher temperatures of the gasoline exhaust gases. A diesel engine typically has lower exhaust gas temperatures, and material selection for the adjustable blades has been successful in production. Recently, Porsche and Borg Warner have developed a variable geometry turbo to be used on the Porsche 911. This turbocharger required the development of new material specifications that could withstand the higher temperatures of the exhaust gases. Due to the high cost of material to withstand the heat and ensure long-term functionality of the vane guides, VGTs are currently seen only for use in high-end vehicles. Alternatively, a downsized, fixed-geometry turbocharger may be used, but this approach will compromise power output because the fixed exhaust turbine geometry will restrict airflow through the engine in order to provide acceptable low-speed turbocharger transient response. Extra-slippery torque converters (e.g., those with higher stall speed) can help to alleviate turbo lag issues, but they will also impose a fuel consumption penalty from increased slippage.

General Motors performed simulation testing on its 2.4-L port fuel-injected four-cylinder engine in the Chevrolet Equinox. The port fuel-injected 2.4-L engine was compared to an engine of the same displacement equipped with direct injection, turbocharger, and dual VVT. GM claims that this approach "can improve fuel consumption on the FTP cycle by up to 10 percent relative to an engine with VVT" but without DI and turbocharging (EEA, 2007).

Ford Motor Company has been developing downsized and turbocharged engines equipped with direct injection. The company plans to offer these engines in nearly all its upcoming models in the future. One of the engines is 3.5 L in displacement and features twin turbochargers with direct injection. From testing, Ford has claimed that this engine will reduce fuel consumption by 13 percent when compared to a V8 with similar performance (EEA, 2007).

Fuel Consumption Benefit and Cost of Downsizing and Turbocharging

The EPA estimates that a fuel consumption reduction of 5 to 7 percent can occur with downsizing and turbocharging (EPA, 2008). This estimate assumes that the vehicle is currently equipped with a DISI fuel system. NESCCAF estimates a 6 to 8 percent reduction in fuel consumption (NESCCAF, 2004). A study performed by Honeywell Turbo Technologies estimates that a 20 percent reduction in fuel consumption is possible from downsizing by 40 percent (Shahed and Bauer, 2009). FEV claims by downsizing and turbocharging a consumption reduction of 15 percent can occur in the New European Drive Cycle. An additional 5 to 6 percent is possible with the addition of a DI fuel system (Ademes et al., 2005). The expected consumption reductions are highly load dependent. The highest benefits will occur at low load conditions. Reduction in consumption is due to higher engine loads and lower friction loss. Sierra Research estimates midsize sedans will increase fuel consumption by 0.3 percent and pickup trucks will decrease consumption by 0.3 percent (Sierra Research, 2008). Sierra's values are lower than others since Sierra did not increase the octane requirement for the engine or combine it with direct injection. Sierra was therefore forced to lower the compression ratio in order to reduce the knocking tendencies while avoiding an octane requirement increase. Sierra claims that "turbocharging and downsizing without the use of gasoline direct injection does not yield benefits on a constant performance basis, based on a statistical analysis of available CAFE data done in 2004" (Sierra Research, 2008). The committee concluded that for the purposes of this report, turbocharging and downsizing will always be applied following DI in order to minimize the need to reduce compression ratio. This order of implementation is in agreement with recent industry trends. The committee estimates that a 2 to 6 percent reduction in fuel consumption is possible when downsizing and turbocharging is added to an engine with DI.

There is a large variation in the cost estimates from the various sources, which arises from a couple of key items. One item is whether or not there is a credit included in the cost from decreasing the engine cylinder count (e.g., going from V6 to I4) and the amount of the credit. Another source of difference is from the use of a split scroll turbine housing or a standard housing on the turbocharger. The split scroll adds cost compared to the standard-type housing.

Vehicle OEM input indicates that basic, fixed-geometry turbochargers add roughly $500 system cost, and dual-scroll turbocharger systems can add about $1,000 (not considering an RPE factor). Currently no pricing information is available for gasoline VGTs. System detail choices depend largely on vehicle performance targets. Martec estimates that the manufacturing cost of downsizing a six-cylinder to a turbocharged four-cylinder engine is $570, and a downsize from an eight-cylinder to a six-cylinder turbo adds a manufacturer cost of $859 (Martec Group, Inc., 2008). For the six-cylinder to a four-cylinder case, Martec is including a $310 downsizing credit and a $270 credit for eight cylinders to six cylinders. Martec's system price includes a water-cooled charge air cooler, split scroll turbo, and upgraded engine internals (not including "modifications to cylinder heads, con-rods, and piston geometry or coatings") (Martec Group, Inc., 2008). It should be noted that most manufacturers tend to use air-cooled charge air coolers. Sierra research estimates an incremental RPE adjusted cost increase of $380 to $996 (Note: values have been adjusted from Sierra's 1.61 RPE factor to 1.5) (Sierra Research, 2008). Sierra's price estimate is based on a "relatively simple turbocharger system that would not be able to match the launch performance of the larger, naturally aspirated engine." The value provided by Sierra is "not including the catalyst plus $650 in additional variable cost for a turbo system marked up to RPE using a factor of 1.61" (Sierra Research, 2008). The EPA provided incremental costs for large cars, minivans, and small trucks at $120. This cost included a downsizing credit. For the small car classification, the EPA has estimated an incremental cost of $690. The higher cost for the small car is due to the lack of significant engine downsizing possibilities (EPA, 2008). EEA estimates a V6 approximately 3 liters in displacement to have an RPE adjusted cost of $540 (or $360 cost assuming an RPE factor of 1.5) (EEA, 2007). Pricing for the EEA study was based on a standard turbo, air-to-air intercooler, engine upgrades, additional sensors and controls, and intake and exhaust modifications.

The committee estimated that the manufacturing costs for integrating downsizing and turbocharging would be in the range of a $144 cost savings to a $790 additional cost, depending on the engine size and configuration. See Table 4.A.1 in the annex at the end of the chapter for a complete breakdown of cost benefits for each engine size. The teardown studies currently being performed for the EPA by FEV (Kolwich, 2009, 2010) have been deemed the most accurate source of cost information by the committee, and therefore these studies were the primary source used for these cost estimates. As with other sources, the committee encourages the reader to view the original document to gain a better understanding of how the costs were derived. The cost increase for an I4 is somewhat obvious, due to the cost of additional components and a lack of significant downsizing credit. The downsizing credit is small because the cylinder count remains the same and generally the same number of valve train, fuel system, and other supporting components are still required. The very low cost of converting from a DOHC V6 to a turbocharged DOHC I4 is due to the very large downsizing credit from removing two cylinders and the supporting hardware for a whole bank of the engine, such as moving from four camshafts to two. In this report, the conversion from a Vee-type engine to an in-line is used only when moving from a V6 to an I4, as an I6 (from a V8) is far less common in the market. When converting from a V8 to a V6, the downsizing credit is much smaller, as you lose two cylinders but still have a Vee engine with two banks requiring two cam drive systems, four camshafts, etc. Also, turbocharging a V6 usually requires a more expensive twin-turbo system, versus the single turbo on the I4. To summarize, the downsizing credit is much smaller and the turbocharging cost is much higher for going from a V8 to a V6 than for going from a V6 to an I4.

ENGINE FRICTION REDUCTION EFFORTS

Engine friction can account for up to 10 percent of the fuel consumption in an IC-powered vehicle (Fenske et al., 2009). Therefore, reducing friction is a constant aim of engine development for improved fuel economy. A large majority of the friction in an IC engine is experienced by three components: piston-assembly, bearings (i.e., crankshaft journal bearings), and the valve train. Within these components friction comes in two general forms: hydrodynamic viscous shear of the lubricant (mainly in journal bearings) and surface contact interactions, depending on the operating conditions and the component.

There are several approaches to reduce frictional losses in an SI engine, mainly through the design of the engine and lubricant. A common trend has been to utilize low-viscosity lubricants (LVL) to reduce energy loss through lowered viscous shear (Nakada, 1994); significant fuel economy improvements have been demonstrated through this adaptation (Taylor and Coy, 1999; Fontaras et al., 2009). However, lowering viscosity also effectively reduces the lubricant thickness between interacting component surfaces, which can increase the occurrence of surface contact. Increased surface contact can have the detrimental effect of increased wear and heat generation, which can in turn affect engine durability. In addition to lowered lubricant viscosity, other SI technology trends (in particular turbo charging and downsizing) lead to increased power density, which can cause increased surface interaction (Priest and Taylor, 2000). In order to maintain engine durability, improving mixed lubrication performance in vulnerable components should be considered. Improvements in lubricant additives (low friction modifiers) and surface engineering (surface coatings and surface topography design) are methods that have been used to improve performance in these surface contact conditions (Erdemir, 2005; Etsion, 2005; Sorab et al., 1996; Priest and Taylor, 2000).

The following sections discuss in more detail specific engine design considerations for reducing friction, and also provide further discussion of low-viscosity lubricants.

Piston-Assembly Friction

Piston-assembly friction is a major component of overall engine friction, and of this the oil-control ring is the biggest contributor. Efforts have been underway for several decades to minimize the radial dimension of the rails to render them more conformable, with minimum spring force, to bores that may not be perfectly circular. Unlike oil-control rings, which are forced against the cylinder liner surface only by their expander spring, the forces pushing the compression rings against the cylinder are gas-pressure forces in the ring groove behind the rings. This gas pressure comes from the cylinder gases that pass down into the ring groove by way of the ring end gap, and little can be done to reduce the frictional contribution of compression rings. It should be noted that it is only during the high-pressure portions of the cycle that their frictional contribution is significant. It is noteworthy that bore distortion either due to thermal distortion of the cylinder block when the engine heats up to operating temperature or to mechanical distortion caused by the forces resulting from torquing the cylinder-head attachment bolts must be minimized if ring friction is to be minimized (Abe and Suzuki, 1995; Rosenberg, 1982).

Crankshaft Offset

Crankshaft offset from the cylinder centerlines will alter connecting-rod angularity. If this is done in a manner that reduces the piston side loading during the high-pressure portion of the engine cycle (i.e., the expansion stroke), a piston-skirt friction reduction is theoretically possible. Some early 20th-century engines employed this concept, and some relatively recent claims have been made on this design strategy. Recent efforts to document any friction reduction have failed to show any benefit (Shin et al., 2004). It is likely that the tribological state at the piston-skirt-to-cylinder-wall interface will affect this, i.e., presence or absence of a hydrodynamic oil film in the critical area under typical operating conditions.

Valve Train Friction

Valve train friction underwent a major reduction in the mid-1980s with near-universal adoption of roller cam followers. Valve-spring tension reduction may also reduce valve train friction, but reduction down to the valve-motion dynamic-stability limit have been found to yield susceptibility to compression loss under circumstances where carbonaceous deposits become detached from chamber surfaces and become trapped between the valve seat and valve face and thus cause major valve leakage.

Crankshaft Journal Bearing Friction

Energy loss due to crankshaft journal bearing friction tends to scale as the cube of the diameter times the length, or (diameter)^3 × (length). Efforts are always made to minimize this source of friction, but adequate crankshaft stiffness at the pin-to-main joints and overall length constrain this option. In V6 engines adequate pin-to-pin joint strength integrity must also be maintained.

Low-Viscosity Lubricants

As discussed previously, lowering lubricant viscosity reduces viscous shear. Therefore moving to advanced low-viscosity lubricants has the potential to improve fuel economy; however, there is debate about the range of effectiveness. Several studies have examined the effectiveness of LVL in lowering friction and reducing fuel consumption (Sorab et al., 1996; Taylor and Coy, 1999; Fontaras et al., 2009). Variations in test methodologies, i.e., vehicle fuel consumption measurement versus engine-dynamometer motoring tests, have led to some confusion in this area. Sorab tested the effectiveness of low-viscosity lubricants on one component of an IC engine, the connecting rod journal bearing. Experimental testing showed significant friction reduction; however, it is difficult to extend these results to an overall fuel consumption benefit. Taylor and Coy (1999) reviewed several modeling techniques that analyzed the fuel consumption benefit of designed lubricants. It was shown that lubricants with designed low-viscosity properties can reduce FC by up to 1 percent. Fontaras et al. (2009) tested the fuel consumption benefit of LVL in different drive cycles. The benefit ranged from 3.6 percent down to negligible depending on the driving cycle. For a cycle that includes a cold start, the LVL effectiveness is higher since the low-temperature viscous behavior prevails in this cycle. In a fully warmed-up engine the FC benefits are not as noticeable and can even be negligible.

Fuel Consumption Benefit and Cost of Reducing Engine Friction

The effectiveness of low-viscosity lubricants has limited drive cycle testing. Fontaras et al. (2009) performed several tests of LVL over different drive cycles, with the conclusion that a benefit of 1 to 1.5 percent can be achieved without affecting the overall engine performance. It was noted that the actual consumption reduction will vary by the amount of time spent in transient operation and if the drive cycle is one in which the engine must be started cold (Fontaras et al., 2009). The EPA estimated that a reduction in consumption of 0.5 percent can occur with the use of LVL at a cost of $3 per vehicle (EPA, 2008). Considering the more relevant U.S. drive cycle and the current widespread use of 5W30, the committee estimates that an additional 0.5 percent FC benefit can be realized with more advanced synthetic LVL at a cost of $3 to $5 per vehicle.

Improved engine friction reduction is a constant aim, yet there is still opportunity for additional FC benefit. Additional friction reduction can be achieved through engine component design and through improvements of surface engineering (surface coatings, material substitutions, selective surface hardening and surface topography control). The EPA estimated potential FC benefit at a range of 1 to 3 percent with a cost of $7 per cylinder (EPA, 2008). Given recent advancements in engine friction reduction, the committee estimates that the potential FC benefit is 0.5 to 2.0 percent at a manufacturing cost of $8 to $13 per cylinder.

ENGINE HEAT MANAGEMENT

As there is never a shortage of waste heat in and around IC engines, efforts to utilize this in productive ways have been ongoing for decades. Following are some methods of improving heat management; however, these techniques are not assigned a fuel consumption benefit or cost for this analysis.

Piston-Crown Design

Piston-crown design can affect its temperature. In some cases moving the piston-ring pack upward motivated by hydrocarbon-emissions reduction efforts to reduce crevice volume also tended to reduce piston-crown temperatures and thus reduced the knock tendency in some cases. To the extent that this enabled a small increase in compression ratio, a small fuel consumption benefit may result along with a significant reduction in hydrocarbon emissions. In some cases this piston modification shortened the heat-conduction pathway by which heat in the piston crown is transferred through the second piston land and then into the top ring and to the cylinder and into the coolant.

Cylinder-Temperature Profile

Cylinder-temperature profile has been found to have subtle effects on efficiency. If the upper portion of the cylinder can be made to run cooler and the lower portion hotter, then both friction and hydrocarbon emissions may benefit. This result can readily be achieved by shortening the coolant jacket such that only about 75 percent of the piston stroke equivalent is cooled by the coolant. At a fixed coolant pump capacity, higher coolant flow velocities are available at the top of the cylinder. This can enable an overall friction reduction by reducing the extent of boundary-layer piston ring friction at the top and a lubricant viscosity reduction at the bottom of the stroke. In addition, the higher temperature of the lower portion of the cylinder promotes post-flame oxidation of the fuel-air mixture that leaves the piston top-land crevice late in the expansion stroke.

Exhaust Port Surface Area

Exhaust port surface area can affect the heat input to the cooling system, and this has subtle efficiency and exhaust emissions consequences. A significant portion (~50 percent) of the heat that enters the cooling system does so by way of the exhaust port. Typically, the high temperature of the exhaust that leaves the cylinder at the beginning of the exhaust-valve open period is also characterized by its highly turbulent state. The associated high rates of heat transfer can affect both the heat load on the cooling system as well as the time required for the catalyst system to achieve operating temperatures following cold start. It is noteworthy that at peak power the highest exhaust flows occur during the blowdown process when the valve flow area is a limiting factor, and when the valve is fully open near mid-exhaust stroke, the so-called displacement flow is somewhat lower.

Typically if the exhaust-port cross-sectional area is reduced until there is evidence of incremental exhaust pumping work under peak power operating conditions, no power loss is to be expected. Efforts to reduce exhaust-port surface area may reduce the heat load on the cooling and also cause the exhaust temperatures to be somewhat higher. This can yield a fuel consumption benefit if ignition-timing retardation, which is often used to facilitate rapid catalyst light-off, can be minimized. A downsized coolant pump, cooling fan, and radiator core may also be beneficial.

Electrically Driven Coolant Pumps

Electrically driven coolant pumps are also frequently mentioned as fuel consumption enablers. While these tend to decrease parasitic loads during warm-up, local hot spots may cause bore and valve-seat distortion or gasket failures. Fuel consumption reduction derived from the above items depends on the details of the initial engine design. A more detailed discussion of the electrification of water pumps can be found in Chapter 5 of this report.

HOMOGENEOUS-CHARGE COMPRESSION IGNITION

While homogeneous-charge compression ignition (HCCI) has received much attention in the recent past, some fundamental control-related challenges remain. The absence of a discrete triggering event in close temporal proximity to the desired time of combustion is the basis for these challenges. In this type of combustion system, temperature is all important; many real-world factors can come into play that will yield unexpected outcomes, e.g., previous-cycle effects and piston and valve temperature swings. As HCCI combustion is essentially instantaneous, it produces very high rates of pressure rise and high peak pressures. Engine structural attributes must take this into account.

Unthrottled HCCI combustion at light loads may produce very high hydrocarbon emissions when the exhaust-gas temperature is relatively low, and this may challenge exhaust aftertreatment processes. Nonetheless, advanced prototype vehicles using HCCI over a portion of the operating range were shown to the public (Alt et al., 2008) suggesting that controls-related progress has been made. As system definition, fuel consumption benefits, and costs are uncertain at this time, HCCI is believed to be beyond the 15-year time horizon of this study.

COMBUSTION RESTART

Combustion restart can be seen as an enabler for idle-off operation, which has the potential to reduce fuel consumption under drive conditions that have significant idle time. The principle challenge relates to the crankshaft position when the engine comes to rest. One cylinder must be in the early phase of the expansion stroke such that fuel can be injected via DISI and spark(s) delivered to initiate combustion and expansion with sufficient potency to initiate sustained engine rotation. Overcoming the aforementioned challenge is highly dependent upon many real-world conditions over which there are limited opportunities without the addition of some form of electro-machine to properly position the crankshaft prior to restart. Given this challenge, it is believed that this approach will not attain significant market penetration during the time horizon of this study.

ETHANOL DIRECT INJECTION

An approach to cooling the charge to control knock and detonation ties in with both the octane ratings of fuels and their heats of vaporization. This approach is to inject into the intake charge or into the cylinder a fluid with a larger heat of vaporization than the fuel itself. This fluid would then vaporize drawing the heat of vaporization from the intake or cylinder gases thus lowering their temperature. Direct-injected (DI) E85 (i.e., a mixture of ~85 percent ethanol and ~15 percent gasoline) has recently been proposed for use both as an anti-knock additive and as a way to reduce petroleum consumption (Cohn et al., 2005) for boosted SI engines. A recent in-depth study of this concept was carried out at Ford (Stein et al., 2009) where engine dynamometer studies were carried out with a turbocharged 3.5-L V6 engine using gasoline PFI combined with DI E85. The promise of this approach is to enable three benefits, namely, allowing increasing the compression ratio of the boosted engine; allowing increasing the level of boost usable without knock and pre-ignition limitations; and enabling operation closer to MBT, timing. These three benefits provide greater thermal efficiency as well as increased power, which allows further downsizing and downspeeding, thus adding potential fuel consumption reductions. The Stein et al. study (2009) used a prototype V6 DI turbocharged engine (termed *Ecoboost* by Ford) with a PFI gasoline injection system added to the original direct-injection fuel system. The DI fuel system was separated from the PFI system and supplied only with E85 from a separate tank and pump. The engine was operated at both the base 9.8:1 compression ratio and a high value of 12:1. E85 injection quantities and spark advance were optimized, and measured results were then extrapolated to application with a 5.0-L engine in a pickup truck by means of full system simulation. The anticipated benefits were observed. Namely, MBT spark timing was achievable up to higher loads than were possible without the E85 injection, leading to a reduction in both gasoline and overall (combined gasoline and E85) fuel consumption. One of the conclusions reached by Stein et al. (2009) was the following:

> By enabling increased CR [compression ratio], engine downsizing, and downspeeding, E85 DI + gasoline PFI makes the engine more efficient in its use of gasoline, thereby leveraging the constrained supply of ethanol in an optimal manner to reduce petroleum consumption and CO_2 emissions. For a hypothetical 5.0 L E85 DI + gasoline PFI engine in a Ford F-series pickup, the leveraging due to 12:1 CR is approximately 5:1 on the EPA M/H drive cycle. That is, 5 gallons of gasoline are replaced by 1 gallon of E85. This leveraging effect will be significantly reduced for more aggressive drive cycles.

Since the focus of the present report is reducing petroleum consumption, the implications of the Stein et al. work on optimizing ethanol utilization will not be considered. However, the combination of increased compression ratio as well as downsizing and increased boosting possible with the ethanol injection enables reducing fuel consumption compared with operation on gasoline alone.

Any approach to inject an anti-knock fluid such as E85 would require an additional tank on the vehicle to provide the anti-knock fluid for injection and would require a willingness on the part of the vehicle driver to fill the anti-knock fluid tank. In the study by Stein et al. (2009), the authors estimated based on vehicle simulations for a full-size pickup truck that E85 usage on the FTP urban/highway schedule would be only about 1 percent of the total fuel used, thus providing an E85 refill driving range of ~20,000 miles with a 26-gallon gasoline fuel tank and a 10-gallon E85 tank. For the higher-load US06 driving cycle, E85 would constitute 16 percent of the fuel used for an E85 refill range of ~900 miles. For towing a trailer up the Davis Dam slope (~6 percent grade for over 10 miles), E85 usage would be 48 percent of the fuel used with an E85 tank refill range of ~100 miles. Once all the anti-knock fluid has been consumed, spark timing would have to be retarded and turbocharger boost reduced to prevent knock if a high compression ratio were chosen for the engine (e.g., 12 versus 9.8) based on reliance on injection of an anti-knock fluid to control knock. Operating with retarded spark timing and reduced boost would not harm the engine but may impact available power.

Based on the costs for the urea dosing systems used for CI engine selective catalytic reduction aftertreatment that has similar componentry (see Chapter 5), the cost of converting a boosted DI engine to PFI gasoline with DI E85 injection is estimated to be $300 to $350.

FINDINGS

SI engines are widely accepted as the primary source of propulsion for light-duty vehicles in the United States. There have been significant improvements in the fuel consumption reduction of SI engines in response to past trends of rising fuel prices. These improvements are in large part due to past advancements in fast-burn combustion systems with strategic exhaust-gas recirculation (EGR), multi-point fuel injection, and reduced engine friction. Newly available SI technologies are assessed with respect to fuel consumption benefit and cost measured against the aforementioned technologies as the baseline. These current technologies address improvements in the areas of continuing friction reduction, reduced pumping losses through advanced VEM, thermal efficiency improvements, and improved overall engine architecture, including downsizing using turbocharging and GDI. The significant finds are as follows:

Finding 4.1: SI technologies offer a means of reducing fuel consumption in relatively small, incremental steps. OEMs can thus create packages of technologies that can be tailored to meet specific cost and effectiveness targets. It is the combination of numerous, affordable SI technologies in a package that makes them appealing when compared to diesel or full hybrid alternatives—which offer a single large benefit at a large cost. Because of this capability, and considering the wide acceptance of SI engine applications, the committee believes that the implementation of SI engine technologies will continue to play a large role in achieving reduced levels of fuel consumption. Table 4.A.1 at the end of this chapter summarizes the fuel consumption reductions and costs for these technologies.

Finding 4.2: Cylinder deactivation is most cost-effective when applied to OHV V6 and V8 engines; it typically affords 4 to 10 percent fuel consumption reduction. The higher cost of applying cylinder deactivation to DOHC V6 and V8 engines, combined with the reduced fuel consumption benefit when cylinder deactivation is added to an engine with VVT, has caused most OEMs to avoid its application to DOHC engines. For this reason, the committee believes that cylinder deactivation will be applied only to OHV engines in most cases.

Finding 4.3: Stoichiometric gasoline direct injection (SGDI) applied to naturally aspirated engines typically affords a knock-limited compression ratio increase of 1.0 to 1.5 and a reduction in fuel consumption of 1.5 to 3.0 percent at a cost of $117 to $351, depending on cylinder count and including noise-abatement items. Versions of direct injection that provide some measure of charge stratification can further reduce fuel consumption, but emissions and implementation issues have inhibited high-volume applications.

Finding 4.4: Turbocharging and downsizing, while maintaining vehicle performance, can yield fuel consumption reductions ranging from 2 to 6 percent, depending on many implementation details such as changes in cylinder count. Industry trends and input from OEMs show that this technology is usually added in combination with direct injection when the goal is improved efficiency. SGDI will help negate the need to reduce compression ratio when turbocharging, giving the combination a positive synergistic effect. If the cylinder count is reduced, NVH-related issues will reduce the benefit level. Cost estimates were based primarily on the 2009 EPA teardown study and range from around zero additional cost when converting from a V6 to an I4, to about $658 when converting from a V8 to a V6.

Finding 4.5: The VEM over the speed-load range of an SI engine can further reduce the pumping loss over that of the previously described configurations and can also cause a slight increase in engine performance, which will offer a potential downsizing opportunity. There are many different implementations of this, and the cost-benefit relationship for these implementations depends on the engine architecture to which they are applied. Fuel consumption reduction can range from 1.0 percent with only intake cam phasing, to about 11 percent with a continuously variable valve lift and timing setup. The total cost range is $35 to $530, depending on the implementation technique and engine architecture.

Finding 4.6: It is important to note that, according to industry trends and input from OEMs, the major OEMs are either pursuing advanced VEM technologies, such as CVVL (Nissan, Toyota, and Honda), or turbocharging and downsizing with SGDI (Ford, GM, and VW), but usually not both (aside from BMW, which has both on its new N55 engine). However, there would still be a benefit, diminished somewhat by synergistic effects, to be gained by adding VVL to a turbo/SGDI engine with VVT. The committee concluded that thus far the industry has deemed the cost-benefit ratio too small to implement both technologies on one engine for mainstream vehicles. Adding continuously variable valve lift and timing to a baseline DOHC engine with intake cam phasing can result in a 5 to 9 percent reduction in fuel consumption. Implementing dual cam phasing, SGDI, and turbocharging and downsizing to a baseline DOHC engine with intake cam phasing can provide a 6 to 11 percent fuel consumption reduction.

Finding 4.7: Variable compression ratio, camless valve trains, homogeneous-charge compression ignition, and cooled EGR were all given careful consideration during the course of this study. Because of either questionable benefits or major implementation issues, it is highly uncertain whether any of these technologies will have any significant market penetration even in the 10- to 15-year time horizon.

BIBLIOGRAPHY

Abe, S., and M. Suzuki. 1995. Analysis of cylinder bore-distortion during engine operation. SAE Paper 950541. SAE International, Warrendale, Pa.

Ademes, N., O. Lang, S. Lauer, W. Salber, and H. Jene. 2005. Valve train variability for advanced gasoline engines. MTZ 66(12; December).

Alperstein, M., G. Shafer, and F. Villforth. 1974. Texaco's stratified-charge-engine—Multifuel, efficient, clean, and practical. SAE Paper 740563. SAE International, Warrendale, Pa.

Alt, M., U.D. Grebe, J.R. Dulzo, V.A. Ramappan, P. Kafarnik, and P.M. Najt. 2008. HCCI—From lab to the road. 29th Vienna Motor Symposium.

Automobile Magazine. 2002. Fiat multiair technology increases power, lowers emissions. Available at http://rumors.automobilemag.com/6476771/green/fiat-multiair-technology-increases-power-lowers-emissions/index.html.

Bernard, L., A. Ferrari, R. Rinolfi, and C. Vafidis. 2002. Fuel economy improvement potential of UNIAIR throttleless technology. SAE Paper 2002-33-0012. SAE International, Warrendale, Pa.

Boggs, D.L., H.S. Hilbert, and M.M. Schecter. 1995. The Otto-Atkinson-Cycle engine: Fuel economy and emissions results and hardware design. SAE Paper 950089. SAE International, Warrendale, Pa.

Cleary, D., and G. Silvas. 2007. Unthrottled engine operation with variable intake valve lift, duration, and timing. SAE Paper 2007-01-1282. SAE International, Warrendale, Pa.

Cohn, D.R., L. Bromberg, and J.B. Heywood. 2005. Direct Injection Ethanol Boosted Gasoline Engines: Biofuel Leveraging for Cost-Effective Reduction of Oil Dependence and CO_2 Emissions. MIT Laboratory for the Environment Report LFEE 2005-01 RP. Massachusetts Institute of Technology, Cambridge, Mass.

Delphi. 2009. Gasoline engine management systems. Available at http://delphi.com/manufacturers/auto/powertrain/gas/valvetrain/vcp/.

EEA (Energy and Environmental Analysis, Inc.). 2007. Update for Advanced Technologies to Improve Fuel Economy of Light Duty Vehicles. Prepared for U.S. Department of Energy. August. Arlington, Va.

EPA (U.S. Environmental Protection Agency). 2008. EPA Staff Technical Report: Cost and Effectiveness Estimates of Technologies Used to Reduce Light-Duty Vehicle Carbon Dioxide Emissions. EPA420-R-08-008. March. Washington, D.C.

Erdemir, A. 2005. Review of engineered tribological interfaces for improved boundary lubrication. Tribology International 38:249.

Etsion, I. 2005. State of the art in laser surface texturing. Journal of Tribology 127(January):248.

Falkowski, A., M. McElwee, and M. Bonne. 2006. Design and development of the DaimlerChrysler 5.7 L Hemi® engine multi-displacement cylinder-deactivation-system. SAE Paper 2004-01-2106. SAE International, Warrendale, Pa.

Fenske, G., R. Erck, O. Ajayi, A. Masoner, and A. Comfort. 2009. Impact of friction reduction technologies on fuel economy for ground vehicles. Proceedings of the 2009 Ground Vehicle Systems Engineering and Technology Symposium (GVSETS).

Flierl, R., and M. Kluting. 2000. The third generation of valvetrains—New fully variable valvetrains for throttle-free load control. SAE Paper 2000-01-1227. SAE International, Warrendale, Pa.

Flierl, R., D. Gollasch, A. Knecht, and W. Hannibal. 2006. Improvements to a four cylinder gasoline engine through the fully variable valve lift and timing system univalve. SAE 2006-01-0223. SAE International, Warrendale, Pa.

Fontaras, G., E. Vouitsis, and Z. Samaras. 2009. Experimental evaluation of the fuel consumption and emissions reduction potential of low viscosity lubricants. SAE Paper 2009-01-1803. SAE International, Warrendale, Pa.

Hitomi, M., J. Sasaki, K. Hatamura, and Y. Yano. 1995. Mechanism of improving fuel efficiency by Miller-cycle and its future prospect. SAE Paper 950974. SAE International, Warrendale, Pa.

Hofmann, R., J. Liebl, M. Kluting, and R. Flierl. 2000. The new BMW 4-cylinder gasoline engine—Reduction in fuel consumption without compromise. SAE Paper 2001-08-0043. SAE International, Warrendale, Pa.

Iwamoto, Y., K. Noma, O. Nakayama, and T. Yamauchi. 1997. Development of gasoline direct injection engine. JSAE Technical Paper No. 9732009.

Kojima, S. 2008. Development of high-performance and low-emission gasoline engine. SAE Paper 2008-01-0608. SAE International, Warrendale, Pa.

Kolwich, G. 2009. Light-Duty Technology Cost Analysis Pilot Study. Report FEV 07-069-103F. Prepared for U.S. Environmental Protection Agency, Ann Arbor, Mich. September 3. FEV, Inc., Auburn Hills, Mich.

Kolwich, G. 2010. Light-Duty Technology Cost Analysis—Report on Additional Case Studies. Prepared for U.S. Environmental Protection Agency. FEV, Inc., Auburn Hills, Mich.

Kuwahara, K., T. Watanabe, H. Tanda, K. Ueda, and H. Ando. 2000. Intake-port design for Mitsubishi GDI-Engine to realize distinctive in-cylinder flow and high charge coefficient. SAE Paper 2000-01-2801. SAE International, Warrendale, Pa.

Leone, T.G., and M. Pozar. 2001. Fuel economy benefit of cylinder deactivation sensitivity to vehicle application and operating constraints. SAE Paper 2001-01-3591. SAE International, Warrendale, Pa.

Lindgren, R., M. Skogsberg, H. Sandquist, and I. Denbratt. 2003. The influence of injector deposits on mixture formation in a DISC SI engine. JSAE Technical Paper No. 20030110.

Martec Group, Inc. 2008. Variable Costs of Fuel Economy Technologies. Prepared for Alliance of Automobile Manufacturers. June 1; amended September 26 and December 10.

Misovec, K., B. Johnson, G. Mansouri, O. Sturman, and S. Massey. 1999. Digital valve technology applied to the control of an hydraulic valve actuator. SAE Paper 1999-01-0825. SAE International, Warrendale, Pa.

Nakada, M. 1994. Trends in engine technology and tribology. Triblogy International 27(1), February.

NESCCAF (Northeast States Center for a Clean Air Future). 2004. Reducing Greenhouse Gas Emissions from Light-Duty Motor Vehicles. March.

NRC (National Research Council). 2002. Effectiveness and Impact of Corporate Average Fuel Economy (CAFE) Standards. National Academy Press, Washington, D.C.

Petitjean, D., L. Bernardini, C. Middlemass, and S.M. Shahad. 2004. Advanced gasoline engine turbocharging technology for fuel economy improvements. SAE Paper 2004-01-0988. SAE International, Warrendale, Pa.

Pischinger, S., K. Yapici, M. Schwanderlapp, and K. Habermann. 2001. Variable compression in SI engines. SAE Paper 2001-24-0050. SAE International, Warrendale, Pa.

Priest, M., and C.M. Taylor. 2000. Automobile engine tribology—Approaching the surface. Wear 241:193-203. SAE International, Warrendale, Pa.

Rebbert, M., G. Kreusen, and S. Lauer. 2008. A new cylinder deactivation by FEV and Mahle. SAE Paper 2008-01-1354. SAE International, Warrendale, Pa.

Ricardo, Inc. 2008. A Study of Potential Effectiveness of Carbon Dioxide Reducing Vehicle Technologies. Prepared for U.S. Environmental Protection Agency. EPA420-R-08-004. Contract No. EP-C-06-003. Work Assignment No. 1-14. Ann Arbor, Mich.

Rosenberg, R.C. 1982. General friction considerations for engine design. SAE Paper 821576. SAE International, Warrendale, Pa.

Schwarz, C., E. Schünemann, B. Durst, J. Fischer, and A. Witt. 2006. Potentials of the spray-guided BMW DI combustion system. SAE Paper 2006-01-1265. SAE International, Warrendale, Pa.

Sellnau, M., T. Kunz, J. Sinnamon, and J. Burkhard. 2006. 2-step variable valve actuation: System optimization and integration on an SI engine. SAE Paper 2006-01-0040. SAE International, Warrendale, Pa.

Shahed, S.M., and Karl-Heinz Bauer. 2009. Parametric studies of the impact of turbocharging on gasoline engine downsizing. SAE Paper 2009-01-1472. SAE International, Warrendale, Pa.

Shin, S., A. Cusenza, and F. Shi. 2004. Offset crankshaft effects on SI engine combustion and friction performance. SAE Paper 2004-01-0606. SAE International, Warrendale, Pa.

Sierra Research. 2008. Basic Analysis of the Cost and Long-Term Impact of the Energy Independence and Security Act Fuel Economy Standards. Sacramento, Calif., April 24.

Simko, A., M. Choma, and L. Repko. 1972. Exhaust emission control by the Ford-programmed combustion process—P ROCO. SAE Paper 720052. SAE International, Warrendale, Pa.

Sorab, J., S. Korcek, C. Brower, and W. Hammer. 1996. Friction reducing potential of low-viscosity-engine oils in bearings. SAE Paper 962033. SAE International, Warrendale, Pa.

Stein, R.A., C.J. House, and T.G. Leone. 2009. Optimal use of E85 in a turbocharged direct-injection engine. SAE Paper 2009-01-1490. SAE International, Warrendale, Pa.

Sycomoreen. BMW Valvetronic. See http://sycomoreen.free.fr/imgs/IMAGES/autoconcept/ BMW_valvetronic.jpg.

Takemura, S., S. Aoyama, T. Sugiyama, T. Nohara, K. Moteki, M. Nakamura, and S. Hara. 2001. A study of a continuous variable valve event and lift (VEL) system. SAE Paper 2001-01-0243. SAE International, Warrendale, Pa.

Tanaka, H., T. Nagashima, T. Sato, and S. Kawauchi. 1999. The effect of 0W-20 low viscosity engine oil on fuel economy. SAE Paper 1999-01-3468. SAE International, Warrendale, Pa.

Tanaka, Y., R. Hiyoshi, S. Takemura,Y. Ikeda, and M. Sugawara. 2007. A study of a compression-ratio-control mechanism for a multiple-link variable-compression-ratio engine. SAE Paper 2007-01-3547. SAE International, Warrendale, Pa.

Taylor, C.M. 1998. Automobile engine tribology—Design considerations for efficiency and durability. Wear 221:1-8.

Taylor, R.I., and C.R. Coy. 1999. Improved fuel efficiency by lubricant design: A review. Proceedings Institution of Mechanical Engineers 214, Part J.

Toyota Motor Company. 2007. Toyota develops next generation engine valve mechanism. June 12, available at http://www2.toyota.co.jp/en/news/07/0612.html.

Tuttle, J.H. 1980. Controlling engine load by means of late intake-valve closing. SAE Paper 800794. SAE International, Warrendale, Pa.

Verhelst, S., J. De Landtsheere, F. De Smet, C. Billiouw, A. Trenson, and R. Sierens. 2008. Effects of supercharging, EGR and variable valve timing on power and emissions of hydrogen internal combustion engines. SAE Paper 2008-01-1033. SAE International, Warrendale, Pa.

Wirbeleit, F., K. Binder, and D. Gwinner. 1990. Development of pistons with variable-compression-height for increasing efficiency and specific power output of combustion engines. SAE Paper 900229. SAE International, Warrendale, Pa.

Wurms, R., M. Grigo, and W. Hatz. 2002. Audi FSI technology—Improved performance and reduced fuel consumption. SAE Paper 2002-33-0002. SAE International, Warrendale, Pa.

ANNEX

TABLE 4.A.1 Summary Table for Fuel Consumption Reduction Techniques for SI Engines: Incremental Percentage Reduction of Fuel Consumption with Associated Incremental Total Cost (with 1.5 RPE). See Figures 9.1 through 9.5 in Chapter 9 to understand the intended order for the incremental values.

Technologies		Consumption Benefit			Incremental Cost $						Comments
		I4	V6	V8	I4		V6		V8		
SI Techniques		(%) Range	(%) Range	(%) Range	Low	High	Low	High	Low	High	
Low-viscosity lubricants	LUB	0.5	0.5	0.5	4.5	7.5	4.5	7.5	4.5	7.5	• Small consumption benefit • Dependent on drive cycle
Engine friction reduction	EFR	0.5-2.0	0.5-2.0	1.0-2.0	48	78	72	117	96	156	• Roller follower valve trains and piston kit friction reduction measures were nearly universally implemented in the mid-1980s
VVT—coupled cam phasing (CCP), SOHC	CCP	1.5-3.0	1.5-3.5	2.0-4.0	52.5		105		105		• On SOHC setup cam phaser adjusts both exhaust and intake valve timing events • Manufacturer cost estimate of $35/phaser
Discrete variable valve lift (DVVL), SOHC/DOHC	DVVL	1.5-3.0	1.5-3.0	2.0-3.0	195	240	270	315	420	480	• Short durations may reduce pumping loss, and the reduced lift is a consequence of this • As intake manifold vacuum vanishes, alternate means must be found to implement power brakes and PCV • DVVL features two to three separate fixed profiles • Manufacturer cost estimate of $40/cylinder + $35/phaser

Technologies		Consumption Benefit			Incremental Cost						Comments
		I4	V6	V8	I4		V6		V8		
SI Techniques		(%) Range	(%) Range	(%) Range	Low	High	Low	High	Low	High	
Cylinder deactivation, SOHC	DEAC	NA	4.0-6.0	5.0-10.0	NA		510	600	536	630	• Effectiveness depends on power to weight ratio, previously added technologies, NVH, and drivability issues • Reduction in pumping losses from higher cylinder loading • Higher cost when applied to OHC engines • Manufacturer cost estimate for OHC engines of $340 to $400 • Additional manufacturer cost of $140 for NVH issues
Cylinder deactivation, OHV	DEAC	NA	4.0-6.0	5.0-10.0	NA		330	375	383		• Effectiveness depends on power to weight ratio, previously added technologies, NVH, and drivability issues • Reduction in pumping losses from higher cylinder loading • OHV has a lower cost when compared to OHC setups
VVT—intake cam phasing (ICP)	ICP	1.0-2.0	1.0-2.0	1.5-2.0	52.5		105		105		• Implementations include intake cam phaser (ICP) • Timing is important, and lift is merely a consequence of duration change • Some of this can be achieved with variable geometry intake manifolds • Manufacturer cost estimate of $35/phaser

continued

TABLE 4.A.1 Continued

Technologies		Consumption Benefit			Incremental Cost						Comments
		I4	V6	V8	I4		V6		V8		
SI Techniques		(%) Range	(%) Range	(%) Range	Low	High	Low	High	Low	High	
VVT—dual cam phasing (DCP)	DCP	1.5-2.5	1.5-3.0	1.5-3.0	52.5		105		105		• Implementations include exhaust only and dual-cam phaser (DCP) • Manufacturer cost estimate of $35/phaser
Continuously variable valve lift (CVVL)	CVVL	3.5-6.0	3.5-6.5	4.0-6.5	239	308	435	465	525	585	• Short durations may reduce pumping loss, and the reduced lift is a consequence of this • As intake manifold vacuum vanishes, alternate means must be found to implement power brakes and PCV • CVVL features wide range of cam profiles • Manufacturer cost estimate of $300 for an I4, and $600 for a V-8
VVT—coupled cam phasing (CCP), OHV	CCP	1.5-3.0	1.5-3.5	2.0-4.0	52.5		52.5		52.5		• Requires in block cam phaser • Manufacturer cost estimate of $35/phaser
Stoichiometric gasoline direct injection (GDI)	SGDI	1.5-3.0	1.5-3.0	1.5-3.0	176	293	254	384	443	527	• Enables about +1.0 knock limited compression ratio • High pressure fuel pump increases parasitic loss • Increased volumetric efficiency increases pumping loss • Injector deposits formed upon hot shut down has been a traditional concern • Manufacturer cost estimates $80/cylinder and $136 for injector noise abatement items

Technologies		Consumption Benefit			Incremental Cost						Comments
		I4	V6	V8	I4		V6		V8		
SI Techniques		(%) Range	(%) Range	(%) Range	Low	High	Low	High	Low	High	
Turbocharging and downsizing	TRBDS	2.0-5.0	4.0-6.0	4.0-6.0	555	735	-50	308	788	1185	• Vehicle launch performance will likely be compromised • Piston underside oil squirters, an oil cooler, and an intercooler may contribute to system merits • Dual scroll and VNT units will improve vehicle launch performance • Manufacturer estimates $550-$920 for a fixed geometry system

5

Compression-Ignition Diesel Engines

INTRODUCTION

Light-duty compression-ignition (CI) engines operating on diesel fuels have the highest thermodynamic cycle efficiency of all light-duty engine types. The CI diesel thermodynamic cycle efficiency advantage over the more common SI gasoline engine stems from three major factors: the CI's use of lean mixtures, its lack of throttling of the intake charge, and its higher compression ratios. In a CI diesel engine-equipped vehicle, there is an additional benefit of reduced volumetric fuel consumption (e.g., gal/100 miles) because diesel fuel provides more energy per gallon than gasoline, as is discussed later in this chapter.

Lean mixtures, whose expansions are thermodynamically more efficient because of their higher ratio of specific heats, are enabled by the CI diesel combustion process. In this process, diesel fuel, which has chemical and physical properties such that it self-ignites readily, is injected into the cylinder late in the compression stroke. Ignition occurs following atomization of the fuel jet into small droplets that vaporize and mix, creating pockets of heterogeneous combustible mixtures. These heterogeneous mixtures burn with localized diffusion flames even though the overall fuel-to-air ratio may be too lean to support turbulent flame propagation such as occurs in an SI gasoline engine. This ability to successfully burn overall lean mixtures allows CI diesel engine power output to be controlled through limiting the amount of fuel injected without resorting to throttling the amount of air inducted. This attribute leads to the second major factor enabling the higher efficiency of CI diesel engines, namely the absence of throttling during the intake process, which otherwise leads to negative pumping work. SI gasoline engines must be throttled to control their power output while still keeping the fuel-air ratio at the stoichiometric ratio necessary for proper functioning of their three-way exhaust catalyst. Finally, the diesel combustion process needs higher compression ratios to ensure ignition of the heterogeneous mixture without a spark. The higher CI diesel compression ratios (e.g., 16-18 versus 9-11 for SI gasoline) improve thermodynamic expansion efficiency, although some of the theoretical gain is lost due to increased ring-to-bore wall friction from the associated higher cylinder pressures.

Fuel economy technologies considered in the NRC's (2002) earlier report on fuel economy did not include diesel-powered CI engines because the costs and emission control systems to meet upcoming nitrogen oxides (NO_x) and particulate emission standards were not developed at that time. The motivation for including light-duty CI engine technology in this report stems from two factors. Light-duty CI engine vehicles are now in widespread use in Europe because a high fuel tax on diesel and gasoline fuel allowed diesel retail prices to be substantially lower than gasoline prices. This differential is disappearing in some countries but still persists in others. European buyers have accepted initial higher CI vehicle purchase prices in return for their lower fuel consumption as well as excellent performance and driving dynamics resulting from their high torque. CI diesel vehicles constitute around 50 percent of the new light-duty vehicle market in Europe (DieselNet, 2008). However, in the 2007 U.S. light-duty market, CI diesel vehicles accounted for only about 1.7 percent of the new light-duty vehicles sales (EIA, 2009a). Recent demonstrations of diesel combustion and exhaust aftertreatment systems have shown the capability to meet U.S. 2010 Tier 2, Bin 5 and LEV II emissions regulations for light-duty vehicles. As a result of the emissions control capability achieved by original equipment manufacturers (OEMs) with their internal development projects, at the 2008 Detroit auto show 12 vehicle manufacturers announced the introduction of 13 new CI diesel powered vehicles for the 50-state 2009 U.S. market (Diesel Forum, 2008). However, due to the large fuel price increases of early 2008 and the resulting reduction in vehicle sales of larger vehicles, many OEMs canceled CI vehicle introductions announced for 2009. Nonetheless, four OEMs have offered 12 2009 CI vehicle models.

TECHNOLOGIES AFFECTING FUEL CONSUMPTION

The fuel consumption of engine systems is driven by two major elements, the base engine (i.e., combustion subsystem, friction, accessories, etc.) and the exhaust aftertreatment subsystem. As a result, the fuel consumption of an engine system depends on both the base engine and the aftertreatment. Technologies affecting engine system fuel consumption through changes to the base engine and to the aftertreatment system are discussed below.

Base Engine Fuel Efficiency Technologies

The strategies being pursued to improve base engine efficiency are the following:

- Downsizing the engine while maintaining equal power,
- Improving thermodynamic cycle efficiency (e.g., improved combustion),
- Reducing engine friction (e.g., reduced piston skirt friction), and
- Reducing accessory loads (e.g., electric water pump, reduced fuel pump loads by avoiding fuel recirculation, modulated oil pump).

Note that all these strategies apply as well to SI engines, although the gains may have different magnitudes due to process differences between CI and SI engines.

Downsizing the Engine

The most significant of these strategies is engine downsizing, which consists of using a smaller displacement engine for a given vehicle mass while still maintaining the same power to give equal vehicle performance.[1] This approach requires higher cylinder pressures (i.e., higher engine brake mean effective pressure [BMEP], which is equivalent to torque) at any given point on the vehicle drive cycle, which reduces engine brake specific fuel consumption (BSFC). To downsize an engine while still maintaining the same vehicle performance, the torque and hence BMEP of the downsized engine must be raised at all speeds including the maximum-power speed. One of the key enablers to raising the BMEP is increasing the intake boost provided by the turbocharger system. The emerging approach to increase intake boost is two-stage turbocharging (Figure 5.1). Increased boosting is also used for downsizing SI engines.

Most current light-duty CI diesel engines use a single-stage, variable-geometry turbocharger (VGT). Two-stage turbocharger (turbo) systems are being actively developed for two reasons. First, they are a key enabler for engine downsizing. Second, they enable increased exhaust gas recirculation (EGR) rates. Cooled EGR is the principal method to reduce engine-out NO_x emissions, as discussed later. With a two-stage turbo system, two separate turbos are combined with additional flow-control valves. The first-stage turbo is usually sized smaller than the normal single-stage VGT used currently, and the second-stage turbo is usually sized larger than the current single-stage VGT. Electronic flow control valves triggered by the engine controller are used to direct exhaust flows to the small turbo and/or to the large one. At lower engine speeds only the smaller turbo is used and a relatively high inlet pressure is generated, even for the low inlet air flow characteristic of operation at high EGR rates.

At higher engine speeds, when the air flow rates have increased and the smaller turbo does not have sufficient flow capacity, air flow rates are sufficient to generate high intake pressures when the exhaust flow is directed through the larger turbo. Therefore, with the use of a two-stage turbo system, the problem of insufficient inlet boost pressure at low speeds with high EGR flow rates is solved without losing engine power at high speeds. The ability of two-stage turbo systems to generate higher boost pressures at low engine speeds is the key characteristic of two-stage systems that makes them enablers for engine downsizing. By providing higher intake boost, two-stage systems provide more air in the cylinder, thus allowing increased BMEP and torque to compensate for the smaller engine displacement. Naturally, two-stage turbo systems are more expensive than single-stage systems.

To utilize the increased charge mass in the cylinder resulting from the higher boost, more fuel must be injected per unit of engine displacement. The resulting increased power output per unit of engine displacement then compensates for the downsized engine displacement. Increasing the fuel flow is generally accomplished by increasing the maximum injection pressure, which enables higher injection-pressures at all loads. To support the increased cylinder pressures, the engine structure, sealing (e.g., head gasket), and lubrication (e.g., connecting rod bearings must support higher cylinder pressures with the same bearing areas) must be improved. Cylinder pressures also increase piston/ring friction, and an additional challenge is to keep the increase to a minimum. These changes require careful engineering but increase engine cost only slightly.

Improving Thermodynamic-Cycle Efficiency by Optimizing Combustion and Emissions for Maximum Efficiency

The combustion process and its phasing relative to piston motion are important determinants of thermodynamic-cycle

[1] Truly equal performance involves nearly equal values for a large number of measures such as acceleration (e.g., 0-60 mph, 30-45 mph, 40-70 mph, etc.), launch (e.g., 0-30 mph), gradability (steepness of slopes that can be climbed without transmission downshifting), maximum towing capability, and others. In the usage herein, equal performance means 0-60 mph times within 5 percent. This measure was chosen because it is generally available for all vehicles. The equal-performance constraint is important because vehicle FC can always be reduced by lowering vehicle performance. Thus objective comparisons of the cost-effectiveness of different technologies for reducing FC can be made only when vehicle performance remains equivalent.

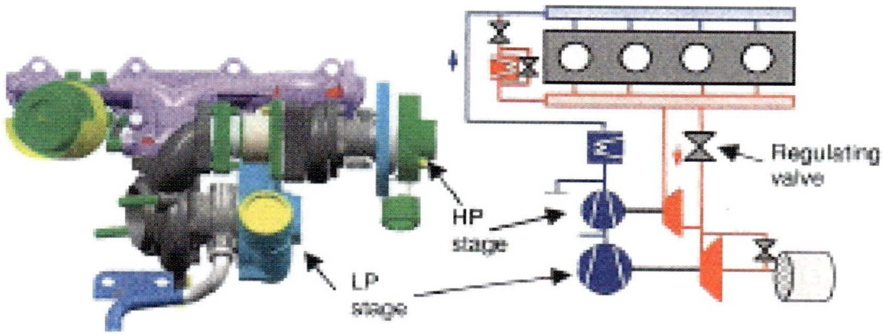

FIGURE 5.1 Schematic of two-stage turbocharger system. HP, high pressure; LP, low pressure. SOURCE: Joergl et al. (2008). Reprinted with permission from SAE Paper 2008-01-0071, Copyright 2008 SAE International.

efficiency. However, the combustion process also plays the key role in the engine-out emissions. As a result, optimizing combustion to minimize FC and emissions simultaneously requires careful analysis of the interactions between fuel spray dynamics, in-cylinder fluid motions resulting from the interactions of the intake flow with the piston bowl shape (i.e., combustion chamber), gas temperature history, and chemical reactions of the fuel. As fuel composition evolves from entirely petroleum based to a mixture of petroleum and bio-sourced components in the next decade to reduce petroleum dependence and increase sustainability, it is critical that understanding of combustion be increased. It is believed that advanced combustion research with tools such as three-dimensional computational fluid dynamic computer codes, including spray and combustion as well as coordinated experiments in highly instrumented engines with optical access for advanced laser-based tools, will improve understanding of combustion in the longer term. This improved understanding is critical to reducing exhaust emissions without compromising engine efficiency and along with new technologies discussed later should enable reductions in FC.

Reducing Engine Friction

Friction sources in engines are journal bearing friction, valve-train friction, and piston assembly friction. In the past 10 to 15 years, all significant sliding interfaces in valve trains have been replaced by rolling interfaces, which minimize friction. Connecting rod, camshaft, and main bearing friction is hydrodynamic, thus coming primarily from lubricating oil shear processes. This friction has been reduced by the use of lower viscosity lubricants. Therefore, the largest remaining friction sources in both CI and SI engines is that due to the piston assembly. Friction from this assembly comes from both piston skirt-to-wall interactions as well as piston ring-to-wall interactions. Both skirt and ring friction can be decreased by improved cylinder-bore roundness, which depends on both cylinder block design and associated thermal distortions as well as bore distortion due to mechanical loading by the preloaded cylinder head attachment bolts. Rounder bores under hot and loaded conditions allow lower ring tension, which in turn decrease ring-to-wall friction. Coatings to reduce ring friction are also being developed, although it is not yet clear whether such coatings can be both friction reducing and sufficiently durable. Piston skirt friction can be reduced by improved skirt surface coatings. Most current pistons have proprietary skirt coatings, but new materials are continuously being studied to further reduce skirt-to-wall friction.

Reducing Accessory Loads

Engine loads to drive accessories include those for coolant pump, oil pump, alternator, air-conditioning compressor, power-steering pump, etc. Electric-motor-driven coolant pumps are being considered because they can be turned off or run slowly during engine warm-up and at other conditions when coolant flow can be reduced without engine damage, thus reducing fuel use to drive the electrical alternator. Two-mode mechanical water pumps are also being developed that require less power to drive at part-load engine conditions but still provide more coolant flow at high-load conditions. Oil pumps, like coolant pumps, are sized for maximum engine power conditions and are hence oversized for part-load, low-speed conditions. Two-mode oil pumps are being developed and becoming available.

Exhaust Emissions Control of CI Diesel Engines

The most critical aspect of increasing the use of CI diesel engines in the United States to take advantage of their excellent efficiency is the development and production of technologies that can enable these engines to meet the 2010 and post-2010 exhaust emissions standards. As noted above, CI diesel engines without emission controls have very low

FC characteristics. So the challenge for CI engines is to reduce emissions into compliance without losing the excellent fundamental CI low FC. This challenge is in contrast to the case of the SI gasoline engines, for which reducing FC is the major issue. As noted earlier, in the 2009 model year 13 new CI diesel vehicles were announced for introduction to the U.S. market (Diesel Forum, 2008). These vehicles have been developed to meet the 2010 emissions standards, and so whatever efficiency deterioration has occurred as a result of applying the combustion and exhaust aftertreatment technologies necessary to meet the standards is reflected by the fuel economy of these vehicles. Data from the 2009 VW Jetta indicated that the fuel consumption reduction between the diesel and gasoline versions of the Jetta expected from earlier (e.g., 2006) models has been retained, in spite of the significantly reduced emissions, although this result may not hold true for all the new diesel models. As a result, the overall choice between investing in SI gasoline engine technologies to reduce the SI gasoline fleet FC on the one hand and replacing some SI gasoline engines with CI diesel engines on the other hand will rest on the total cost for emissions-compliant CI diesel engines and their remaining FC advantage after emissions control measures are implemented. In addition to the specific FC tradeoffs between SI and CI FC, business decisions on whether to tool up CI engines also depend heavily on the availability of investment capital in an industry undergoing drastic financial problems as well as expectations of the willingness of buyers to invest in CI engines, with which they are largely unfamiliar or have out-of-date perceptions.

Combustion System Technologies

The direction for CI diesel combustion system technology development has been toward more premixed combustion and away from traditional CI diesel engine diffusion-type combustion. Diffusion-type combustion tends to generate both high NO_x and high particulate matter (PM) engine-out emissions because diffusion flames tend to stabilize at a nearly stoichiometric local mixture ratio that is characterized by high temperatures and resultant high NO_x formation. Surrounding this local stoichiometric diffusion flame are rich local fuel mixtures whose thermal and mixture environment also cause high PM formation. Higher levels of dilution by means of large amounts of EGR as well as earlier injection and longer ignition delays reduce both average and local temperatures as well as allowing more mixing time, thus making the local fuel-air ratios much leaner. This combination of lower temperatures and locally leaner mixtures minimizes the extent of diffusion flame occurrence and thereby reduces both NO_x and PM emissions. The combustion strategies that utilize this approach have been given many different names in the literature, including PCI (premixed compression ignition) (Iwabuchi et al., 1999), PCCI (premixed-charge compression ignition) (Kanda et al., 2005), LTC (low-temperature combustion) (Pickett and Siebers, 2004), and others. All these partially homogeneous charge strategies drive the combustion process in the direction of HCCI (homogeneous-charge compression ignition) (Ryan and Callahan, 1996). The term HCCI in its purest form refers to virtually homogeneous rather than partially homogeneous charge.

To utilize these premixed forms of combustion, a number of measures are used to reduce temperatures and improve mixing of the charge. The simplest and most effective measure is increased EGR, as noted above. In addition to increased EGR, lowering compression ratio also reduces mixture temperatures and, as a bonus, allows increasing engine power without exceeding cylinder-pressure design limits. Lower compression ratios make developing acceptable cold-start performance more challenging in spite of improved glow plugs and glow plug controls.

Technologies being developed to support this move in combustion technology toward premixed low-temperature combustion are cylinder-pressure-based closed-loop control; piezo-actuated higher-pressure fuel injectors; two-stage turbocharger systems; and combinations of high- and low-pressure EGR systems.

Cylinder-Pressure-Based Closed-Loop Combustion Control Technologies

Cylinder-pressure-based closed-loop combustion control technologies enable operating the engine closer to the low-temperature limit without encountering misfire or excessive hydrocarbon and carbon monoxide (HC/CO) emissions. This technology is especially important in the North American market, where the variation of North American diesel fuel ignition quality (i.e., cetane number) is greater than in Europe. This large cetane number variability makes combustion control more difficult especially for more dilute, lower-temperature combustion strategies. The FC impact of cylinder-pressure-based closed-loop combustion control is 0 to 5 percent. However, since certification fuels are well controlled, the efficiency impact would not be observed on the drive cycle for vehicle emissions certification, but only in customer use when poor ignition quality fuels are encountered in the marketplace.

Piezo-Triggered Common-Rail Fuel Injectors

Piezo-actuated common-rail fuel injectors are being developed aggressively by the global diesel fuel-injection system suppliers (e.g., Bosch, Continental, Delphi, and Denso). These injectors open faster and more repeatably than do solenoid-actuated injectors, thereby enabling more injections per combustion event. The latest generations of these injectors designed on direct-acting principles entered low-volume production for the 2009 model year in European passenger cars. Multiple injections per combustion cycle allow lower combustion noise (i.e., diesel knock) and more precise control of mixing and local temperatures than is possible with a single injection per cycle. This additional level of control is useful to maximize the benefits of premixed low-temperature combustion. In addition to combustion control, multiple-injection capability is used to enable post-combustion injections, which have been used as part of the engine control strategy used to trigger and sustain regeneration of particulate filters.

EGR Issues

Using increased EGR levels to reduce mixture temperatures to suppress formation of NO_x and PM creates two major difficulties in addition to the points mentioned above. First, the levels of EGR at idle and part-load conditions typical of urban and extra-urban driving can reach 60 to 70 percent. This means that with normal high-pressure EGR, only 30 to 40 percent of the engine air flow is going through the turbocharger with the remainder recirculated back through the engine. As a result, the turbine generates less torque and the ability of the turbocharger to boost intake pressure is severely hampered. Low inlet pressures lead to lower cylinder charge masses, causing richer mixtures and thus increasing PM formation as well as making it more difficult for post-combustion oxidation of both PM and HC/CO due to lower oxygen availability.

The second difficulty associated with very high EGR levels is that EGR cooling requirements increase. EGR cooling is extremely important because EGR enters the EGR cooler at exhaust temperatures. Mixing this hot EGR with intake air, which is already heated through compression in the turbocharger compressor, leads to hot inlet mixtures. Hot inlet mixtures negate some of the potential of lowering NO_x and PM formation through lower mixture temperatures. Therefore, high EGR levels require larger and more effective EGR coolers. Not only do these larger coolers present packaging difficulties in already crowded engine compartments, but they also are subject to fouling through condensation of heavy hydrocarbons and water vapor present in the EGR stream, which form deposits inside the EGR cooler decreasing their cooling efficiency (Styles et al., 2008).

High- and Low-Pressure EGR Systems

In most CI diesel engines, EGR is supplied to the intake manifold directly from the exhaust manifold before the turbo. This approach provides high-pressure, high-temperature exhaust gas to the intake manifold. Thus this type of system is called an HP (for high-pressure) system. The HP approach is simple in principle because the exhaust manifold pressure is normally slightly higher than the intake manifold pressure. Thus EGR can be passed directly from the exhaust manifold into the intake manifold at a rate controlled by both the EGR flow control valve and the pressure difference between the exhaust and intake manifolds. This approach was inexpensive and effective in the early days of CI engine emissions control. However, as emission standards tightened, more EGR was needed, resulting in the hot intake mixture problem noted above. Partly to avoid the hot-EGR and EGR cooler fouling problems, low-pressure (LP) EGR systems have been developed (Keller et al., 2008).

In low-pressure systems, exhaust gas is taken from the exhaust system downstream of the particulate filter. As a result, particulates and heavy hydrocarbons have been removed. In addition, these exhaust gases are much cooler since energy has been removed by expanding the gases down to atmospheric pressure through the turbocharger turbine and by heat transfer in the exhaust piping leading to the particulate filter. As a result, these cooler, cleaner low-pressure exhaust gases now have to be pumped back up the intake boost pressure by passing them through the turbocharger compressor and subsequently through the charge cooler. EGR systems combining both high-pressure and low-pressure circuits have been developed and put into production on light-duty vehicles (e.g., the 2009 VW Jetta) (Hadler et al., 2008).

Variable Valve Timing

Some suggestions have been put forth that variable valve timing (VVT) mechanisms may provide opportunities for improved usage of EGR as well as other emissions control functionality (Bression et al., 2008) for CI engines. However, the current consensus from advanced development groups at OEMs and consulting firms is that VVT for CI diesels provides little or no benefit and therefore is not cost effective.

Exhaust Aftertreatment Technologies

HC/CO Control

The control of HC/CO has traditionally been relatively easy for CI engines due to the relatively low levels of these constituents emitted from conventional CI diesel combustion, in spite of relatively low exhaust temperatures. However, that situation has changed as the CI diesel combustion process has been modified to reduce combustion-gas temperatures, which reduces exhaust temperatures even further. As the combustion temperatures have been reduced, HC/CO emissions have risen. The diesel oxidation catalyst (DOC) was introduced around 1996 to reduce hydrocarbon emissions and in turn to reduce the soluble organic fraction of the dilute particulate matter. As a result of the reduced exhaust temperatures noted above, the DOC is being moved closer and closer to the turbocharger outlet to increase the temperature of the catalyst to increase its conversion efficiency. This packaging trend need not significantly increase costs but such minimal cost increases are only possible when other vehicle changes provide the opportunity to modify the engine compartment packaging to allow space for close-coupling the DOC. In addition, oxidation catalyst coatings are being added to diesel particulate filters (DPFs) and NO_x storage catalysts for additional HC/CO control.

Particulate Control

Particulate filter control of emissions from CI diesel engines is presently in use by vehicle manufacturers in Europe and the United States. These particulate filters are quite effective, filtering out 90 to 99 percent of the particulates from the exhaust stream, making CI diesel engines more attractive from an environmental impact point of view. Obviously, particulates accumulate in the filters and impose additional back pressure on the engine's exhaust system, thus increasing pumping work done by the engine. This increase in pumping work increases fuel consumption. In addition, there is a second fuel economy decrement caused by the additional fuel required to regenerate the filter by oxidizing retained particulates. The low exhaust temperatures encountered in light-duty automotive applications of these filters are insufficient to passively oxidize the accumulated particulates. As a result, temperatures must be increased by injecting fuel (most frequently in the engine cylinder after combustion is over) to be oxidized, raising the temperature of the cylinder gases. These hot gases then pass from the cylinder out into the exhaust system and then downstream to the particulate filter to oxidize the particulates retained in the filter. To achieve sufficiently rapid regeneration for practical use in light-duty vehicles (e.g., in around 10 to 15 minutes), exhaust gases must be raised to 625 to 675°C.

Engine control algorithms for filter regeneration not only must sense when the filters need to be regenerated and bring about the regeneration without overheating the filter, but also these algorithms must contend with other events like the driver turning off the vehicle while regeneration is underway, thus leaving an incompletely regenerated filter. When the vehicle is then restarted, the control algorithms must appropriately manage either completion of the regeneration or start of a new filling and regeneration cycle. These algorithms have become quite sophisticated, with the result that particulate filter systems are quite reliable and durable.

NO_x Control

There are two approaches to aftertreatment of NO_x emissions: NO_x storage and reduction catalysts (NSC), which are also called lean NO_x traps (LNT) (Myoshi et al., 1995), and selective catalytic reduction devices.

NO_x Storage Catalysts

NO_x storage catalysts utilize a typical monolith substrate that has both barium and/or potassium as well as precious metal (e.g., platinum) coatings. These coatings adsorb NO_x from the exhaust gas stream to form nitrates, thus storing the NO_x in the catalyst. As NO_x is adsorbed from the exhaust, adsorption sites on the surface of the coating fill up. Once all the coating sites have adsorbed NO_x, the NSC is no longer effective at adsorbing additional NO_x, which then passes right through the NSC. Therefore, at some point before the catalyst is filled, the NSC must be regenerated to purge the adsorbed NO_x and free the sites to adsorb the next wave of NO_x. By supplying the NSC with a rich exhaust stream containing CO and hydrogen, the CO and H_2 molecules desorb the NO_x from the catalyst surface and reduce the NO_x to N_2, H_2O, and CO_2. Therefore, like the particulate filter, the NSC operates in a cyclic fashion, first filling with NO_x from the lean diesel exhaust (i.e., an oxidizing atmosphere) and then being purged of NO_x in a rich exhaust (i.e., a reducing atmosphere) that, with the help of precious metals also part of the catalyst surface coating, reduces the NO_x back to N_2.

Accordingly, application of an NSC to any engine that has a lean exhaust stream like diesel engines requires that periodically (every 30 to 60 seconds depending on the size of the catalyst and the operating condition of the engine) the engine system must create a rich exhaust stream for 10 to 15 seconds to clear the catalyst surface of NO_x, thus preparing it to adsorb the next wave of NO_x. One approach to creating the required rich exhaust stream in the engine cylinder is by throttling the engine to reduce airflow, thus enriching the mixture in the cylinder. Although gasoline engines operate quite happily with rich mixtures, operating a CI diesel engine with a rich mixture without forming excessive particulate and hydrocarbon emissions is quite challenging. If the combustion process is carried out at sufficiently low temperatures, particulate formation is minimized, but both hydrocarbon emissions and FC increase significantly during this brief rich operation.

An additional difficulty with NSCs is that the catalyst coatings preferentially adsorb sulfur compounds from the exhaust. These sulfur compounds originate mostly from the sulfur in the fuel. This sulfur takes up the adsorbing surface sites on the catalyst, leaving no sites to adsorb NO_x. This sulfur adsorption, termed sulfur poisoning, is problematic even with today's low-sulfur (<15 ppm) diesel fuel. Some of the sulfur in the exhaust gases may also come from the engine lubricating oil. Thus the NSC must also be periodically regenerated to clear out the adsorbed sulfur. Sulfur forms a much stronger bond with the catalyst surface than does NO_x and as a result, sulfur regeneration requires not only a rich exhaust stream but also higher temperatures like ~650°C rather than the typical 200 to 300°C temperatures adequate for NO_x regeneration. While the sulfur regeneration does not need to be done nearly as frequently as NO_x regeneration, sulfur regeneration also causes a FC penalty.

The current NO_x aged conversion capability of NSCs is around 70 percent. Early attempts to develop NSCs had difficulty achieving even 50 percent aged conversion efficiency in spite of ~80 percent for a fresh NSC. Extensive development on catalyst test benches indicated that excessive temperatures, particularly during sulfur regeneration, caused the observed deterioration in conversion efficiency. Recently, two factors have enabled improvements. First, newer catalyst formulations have been developed to allow sulfur regeneration at somewhat lower temperatures. Second, empirical models of catalyst behavior have been developed and incorporated into the engine controller. The combined effect of these two developments has enabled increasing aged conversion efficiency to ~70 percent. In the summer of 2008, VW released the 2009 Jetta TDI for the U.S. market which utilizes an NSC and meets Tier 2, Bin 5, as well as LEV II emissions standards, enabling VW to sell the vehicle in all 50 states and Canada. A schematic of the aftertreatment system used on this vehicle is shown in Figure 5.2.

Selective Catalytic Reduction

Selective catalytic reduction (SCR) was originally developed for stationary power plants but is now being applied to heavy-duty truck CI engines in Europe (Müller et al., 2003) and in the United States in 2010. SCR was also introduced in the United States in 2009 on some Mercedes, BMW, and VW vehicles. This system, called BlueTec, was jointly developed by all three manufacturers. SCR works by having ammonia in the exhaust stream in front of a copper-zeolite or iron-zeolite SCR catalyst. The ammonia gets stored on the catalyst surface where it is available to react with the NO_x over the catalyst converting the NO_x into N_2 and water. To provide ammonia to the exhaust stream, a liquid urea-water mixture is injected into the exhaust sufficiently upstream of the SCR catalyst unit and before a mixer, to allow time for vaporization and mixing of the urea and creation of ammonia from the urea, which is an industrial chemical used primarily as a fertilizer. In the fertilizer application, urea is relatively inexpensive, but for use with an SCR system, it must be considerably more pure and as a result is more expensive. SCR systems tend to have NO_x conversion efficiencies of 85 to 93 percent or more without the increased engine-out hydrocarbon emissions and FC resulting from NSC regenerations. As a result, vehicles using SCR have better FC characteristics at equivalent emission levels than those using NSC systems.

When urea is used to provide the ammonia, the urea-water mixture that is injected into the exhaust stream must be carried on board the vehicle. The amount of urea that needs to be supplied to the SCR catalyst depends on the level of NO_x in the exhaust and therefore depends on driving conditions, but for light-duty vehicles it is a small fraction of the fuel flow. Initial discussions regarding the possibility of using an SCR-urea approach to NO_x aftertreatment for the U.S. market were met with concern on the part of the EPA that there was considerable risk that drivers would not keep their urea tanks filled thus rendering the system ineffective. However, together with EPA oversight, vehicle manufacturers have developed systems to monitor the supply of urea in the urea tank, which will not allow the engine to restart more than a small number of times (e.g., 20) when the urea supply starts running out, following appropriate warnings to the driver. As a result of such safeguards, the EPA has approved the certification of the 2009 vehicles using the SCR-urea approach to NO_x aftertreatment. One example of an SCR-urea-based exhaust aftertreatment system is illustrated in Figure 5.3.

Combined NSC and SCR Systems

Another strategy that has been proposed is to use a system in which the NSC is followed by SCR without external urea addition. It is well known that under some operating conditions with the appropriate washcoat formulation, NSCs can convert NO_x to ammonia, which is undesirable for an NSC-only system and hence must be cleaned up before exiting the exhaust system. However, by following the NSC with SCR without urea injection, which is generally called passive SCR, SCR will capture and store the ammonia generated by the NSC and use it to reduce NO_x. Since the amount of am-

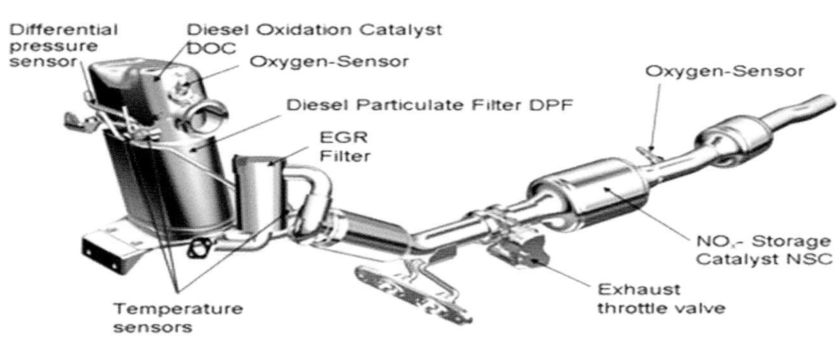

FIGURE 5.2 Exhaust aftertreatment system on the 2009 VW Jetta using NO_x storage and reduction catalyst technology for control of NO_x.
SOURCE: Courtesy of Volkswagen AG.

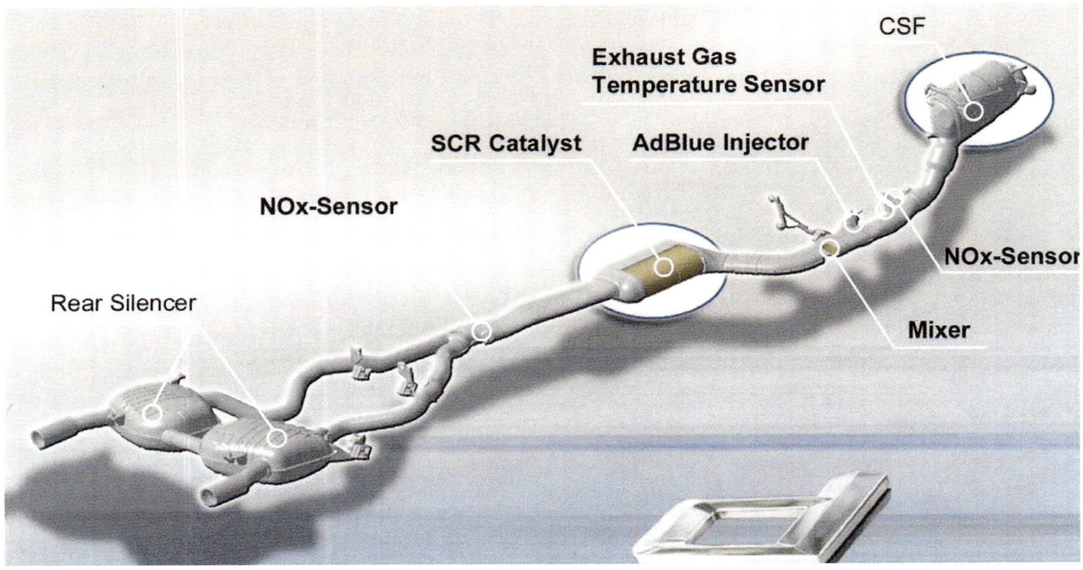

FIGURE 5.3 Schematic of a BMW exhaust aftertreatment system with selective catalytic reduction (SCR) for NO_x control using urea (called AdBlue) addition. The catalyzed soot filter (CSF) is close-coupled to the engine. SOURCE: Mattes et al. (2008). Reprinted with permission.

monia generated by the NSC is not large, the passive SCR unit will have low conversion efficiencies but can be a useful supplement to the NSC system. This approach has been used by Mercedes in its Blue-Tec I system used in Europe.

Choosing Between NSC and SCR Systems

There are both cost and functionality differences between NSC and SCR systems which would influence which choice an OEM might make for NO_x aftertreatment with CI engines. NSC systems use much more PGM (platinum group metals) than do SCR systems. (The SCR unit itself uses no PGM.) As a result, NSC system costs increase faster with increasing engine displacement than do SCR systems. Thus, from a cost point of view, NSC systems would be chosen for smaller displacement engines for which the current 70 percent NO_x conversion efficiency of the NSC is sufficient to reduce engine-out NO_x levels to below the Bin 5 emissions standards. As engine displacement is increased and engine-out NO_x emissions increase, there is an engine displacement above which the 70 percent conversion efficiency of NSCs is insufficient and the higher (approximately 85 to 93 percent) conversion efficiency of SCR is required. If PGM commodity prices are sufficiently low, NSC systems costs for larger displacement I4 engines (e.g., 2.5 to 2.8 L) might be lower than those for SCR systems for those same engines, but NO_x conversion efficiencies might not be high enough to meet the standards. Thus, the engine displacement above which an OEM would choose SCR rather than the NSC is not simply a cost-based decision.

FUEL CONSUMPTION REDUCTION POTENTIAL

CI Fuel Consumption Reduction Advantage

In a study for the EPA (EPA, 2008), Ricardo, Inc., carried out full system simulation (FSS) to assess the FC and CO_2 impact of many of the technologies expected to enable reduced FC by 2020. FSS calculations were made for the 2007 model-year light-duty vehicle fleet for a set of vehicles representing five vehicle classes. Combinations of technologies deemed to be complementary were applied to baseline vehicles considered to be representative of each class. For the selected combinations of power train and vehicle technologies, final drive ratios were varied to find the ratios that enabled performance equivalent to the baseline vehicles based on a comprehensive set of performance measures while minimizing FC. CI diesel power trains were evaluated among the combinations of technologies considered. Results for the CI diesel power train CO_2 emissions and FC versus the baseline vehicles for three of the five vehicle classes are summarized in Table 5.1. CI power trains were not applied to the other two vehicle classes, but the results for the three classes for which CI engines were evaluated are considered representative of all classes.

As indicated in Table 5.1, for the three vehicle classes considered, the average reduction in CO_2 emissions was about 23 percent and the corresponding average reduction in FC was 33 percent when the baseline 2007 model year SI power trains were replaced with CI power trains utilizing DCT6, EACC, HEA, and EPS. The 2009 VW Jetta was introduced with a 6-speed DSG (VW's name for DCT6) transmission.

TABLE 5.1 Estimated CO_2 and Fuel Consumption Reductions for Three EPA Vehicle Classes, as Determined from Full System Simulation (FSS)

Vehicle	Technology Package	Major Features	SI to CI Downsize Ratio	Combined CO_2 Emissions g/mi.	Combined Fuel Consumption gal/100 mi.	Combined CO_2 Reduction	Combined Fuel Consumption Reduction
Full-size car	Baseline	3.5-L V6 gasoline SI, AT5		356	4.051	Baseline	Baseline
	5	2.8-L I4 diesel, DCT6, EACC, HEA, EPS	80%	273	2.707	23.3%	33.2%
Small MPV	Baseline	2.4-L I4 gasoline SI, DCP, EPS, AT4		316	3.596	Baseline	Baseline
	5	1.9-L I4 diesel, DCT6, EACC, HEA, EPS	79%	247	2.449	21.8%	31.9%
Truck	Baseline	5.4-L V8, gasoline SI, CCP, AT4		517	5.883	Baseline	Baseline
	5	4.8-L V8 diesel, DCT6, EACC, HEA, EPS	89%	391	3.877	24.4%	34.1%
		Average CI diesel versus gasoline				23.2%	33.0%

NOTE: See Chapters 2 and 8 for more information on FSS. To determine the FC reductions, the CO_2 emissions results taken from EPA (2008) were converted to volumetric FC using conversion factors from EPA (2005). AT5, lockup 5-speed automatic transmission; AT4, lockup 4-speed automatic transmission; CCP, coordinated cam phasing; DCP, dual (independent) cam phasing; DCT6, dual-clutch 6-speed automated manual transmission; EACC, electric accessories (water pump, oil pump, fans); EPS, electric power steering; HEA, high-efficiency alternator.
SOURCE: Based on EPA (2008).

Note also that CI engines were downsized in displacement by an average of about 83 percent from the SI engines they replaced. Tables 7.13, 7.15, and 7.18 from EPA (2008) for small MPVs, full-size cars, and trucks, respectively, indicate that these CI engine-powered vehicles with DCT6 transmissions provided equivalent performance to the vehicles with larger-displacement original SI engines and transmissions.

The 2007 model-year baseline vehicles were equipped with 4- and 5-speed automatic transmissions. As noted above, the 33 percent FC reduction indicated in Table 5.1 reflected DCT6 transmissions and more efficient engine accessories as well as the engine change. To estimate the separate effect of replacing SI engines and transmissions by CI engines with equivalent transmission technology and without advanced accessories, a European database of 2009 vehicles was analyzed. Using vehicles that are offered with 5- and 6-speed transmissions for both SI and CI engines, an estimate was derived of the reduction in FC from replacing SI engines with CI engines at equivalent vehicle performance without the effect of simultaneously converting from 4- and 5-speed automatics to DCT6 transmissions. The data used for this estimate are plotted in Figure 5.4 and shown in tabular form in Table 5.A.1 in the annex at the end of this chapter.

Figure 5.4 indicates that the average FC reduction for this vehicle subset was about 25 percent. Therefore, the FC reductions achievable from engine replacement alone without a simultaneous transmission change to DCT6 (and EACC with HEA) would be about 25 percent.

Fuel Volumetric Energy Effect

It should be noted that part of the volumetric FC benefit of CI diesel engines stems from the differences in volumetric energy content between gasoline and diesel fuels. The energy content of a gallon of diesel fuel is about 11 percent higher than that of gasoline. While this factor can be an advantage for drivers if diesel fuel is selling at gasoline prices or lower, the carbon dioxide emissions advantage for the diesel would be less than would be indicated by the volumetric FC advantage of the CI diesel engine. As indicated in Table 5.1, the CO_2 reduction advantage for CI engines is about 10 percent less than their FC reduction advantage.

Fuels for CI Engines

The performance and emissions of diesel engines are also influenced by the fuel characteristics and fuel quality. Although fuel is not a focus of this report, several relevant characteristics for performance and emissions are important in connection with their influence on engine performance,

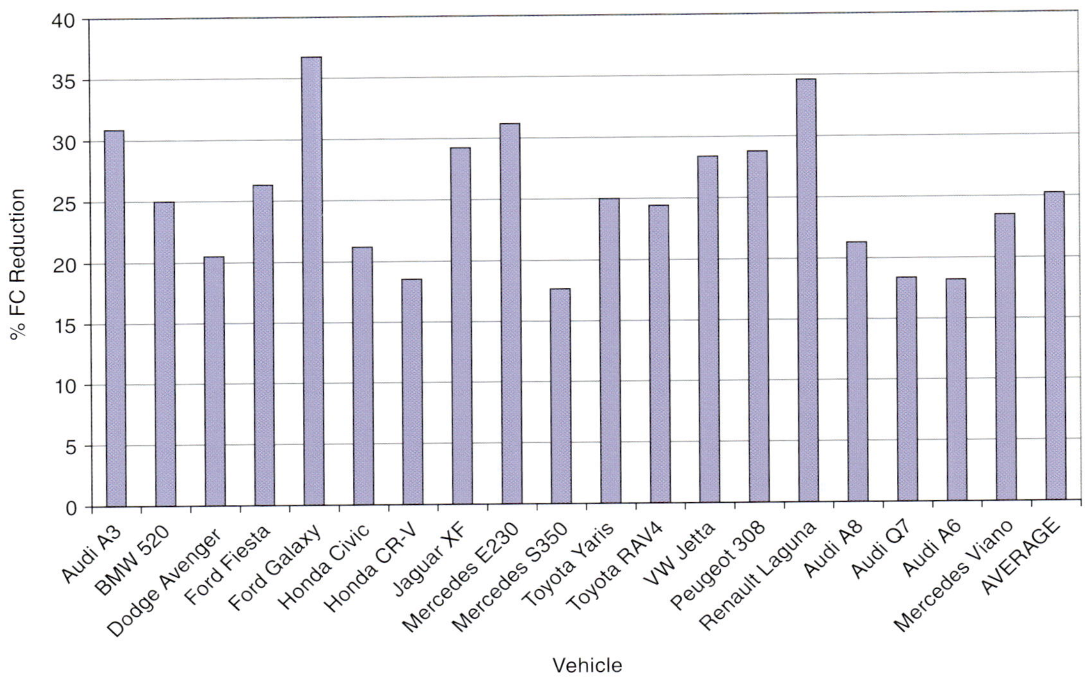

FIGURE 5.4 Percent reduction of fuel consumption (FC) on the NEDC driving cycle for a subset of 2009 European vehicle platforms offered with both SI and CI engines. The subset was selected from a larger set of 2009 vehicle platforms offered with both SI and CI engines by including only those platforms for which 0-62 mph (0-100 km/mile) times were within 5 percent, which was considered to be equivalent performance. The data used to construct this figure are shown in Table 5.A.1 in the annex at the end of this chapter.

efficiency, and emissions. These characteristics are *cetane number* (a measure of fuel self-ignition in the CI cycle—important in cycle efficiency, but also in low-temperature operation), *density/heating value* (a measure of volumetric energy content), *lubricity* (important for fuel system wear and durability), and *sulfur level* (important for proper operation of the engine exhaust aftertreatment system).

In the U.S. market, there is only one diesel fuel suited for on-road transportation; its characteristics are specified by the ASTM Standard D975. Most state regulations require the enforcement of these specifications. In the EU, where light-duty CI diesel passenger cars are widespread and about half the new cars are powered by diesel engines, the diesel fuel is specified by the EN590 standard. There are significant differences between the EU and the ASTM standards. The EU fuel has much higher cetane (e.g., 52 versus 40-48), the fuel density is limited to a minimum to assure adequate energy density (no limit exists in the ASTM standard), and the lubricity is better. In terms of fuel sulfur, European fuel has similar levels to U.S. fuels, for which sulfur level is regulated by the 2006 EPA standards to 15 ppm or less.

In the near future, most diesel passenger cars in the United States will be imports from Europe. Their engines have been adapted for use of U.S. diesel fuel, and the manufacturers do not expect to encounter performance and emission issues connected with the fuel, as long as fuel specifications are enforced and quality is adequate. Cylinder-pressure-based closed-loop control, as discussed earlier and utilized in one of the new 2009 CI diesel vehicles, can adjust for market variability in the cetane number of the fuel and provide compensation over the entire operating engine map. The lower lubricity of the U.S. diesel fuel requires protective coatings for the high-pressure pump in the fuel injection system. As noted earlier, the ultralow level of sulfur in the fuel regulated to less than 15 ppm is a necessary enabler for the efficient and durable operation of the exhaust aftertreatment system. Nonetheless, all OEMs marketing CI diesel vehicles in the North American (NA) market have concerns over the seasonal and regional variability of diesel fuel as well as the enforcement of fuel quality.

At present, the ASTM D975 fuel standard allows up to 5 percent biodiesel blend stock in the fuel provided the blend stock meets the characteristics of the ASTM standard. The European OEMs exporting diesel vehicles to the United States have stated that their engines are robust to this fuel blend and that performance and emissions are not affected as long as the blend is at or under 5 percent. For the European market, the manufacturers may allow up to 7 percent FAME (fatty acid methyl ester), plus up to an additional 3 percent hydrogenated biofuel. The difference in the proportion allowed by the European OEMs for the U.S. market versus for the European market is due to their concern over the quality and stability of American blend stock and the variety of feedstocks, including soy, recycled used oils, fats, etc.

Efficiency Improvements from Transmissions

The transmission technology utilized in the FSS results shown in Table 5.1 was a dual-clutch 6-speed (automated manual) transmission (DCT), which is a very efficient design concept. Transmissions used for CI diesels must be designed to handle their larger torque, which may reduce their efficiencies slightly due to larger gears, bearings, and seals. DCTs are already in production for smaller displacement CI engines (e.g., 2009 VW Jetta). The most challenging aspect of designing DCTs with the higher torque capacities needed for larger displacement CI engines is providing adequate cooling for their wet clutches (i.e., oil-cooled clutches). Dual-mass flywheels, which reduce drive train vibration, thus reducing heat-generating clutch slippage, will be used. Nonetheless, it is not presently known when such DCT units will be available with 500-650 N-m torque capacities for larger CI engines.

Expected transmission-based CI vehicle efficiency improvements beyond those already comprehended by the use of the DCT6 transmissions are estimated at 1 to 2 percent for downspeeding the engine by increasing the number of discrete speed ratios beyond six. The increased number of ratios allows keeping the average engine speed lower while still maintaining equal performance, which is why this approach is called "downspeeding." Another 2 to 3 percent is expected from reduced transmission internal losses.

Overall Fuel Consumption Reduction Potential

The FC reduction potential via replacement of SI gasoline power trains by base-level CI power trains is illustrated by Table 5.1 (i.e., ~33 percent) for CI engines with advanced transmissions (plus EACC, HEA, and EPS) and by Figure 5.4 for engine replacement alone (i.e., ~25 percent). Additional technical improvements, as noted earlier, from downsizing, thermodynamic improvements, friction reduction, and engine accessory improvements, are being developed and will be implemented. CI engines with these technologies implemented are termed advanced-level CI engines. Transmission improvements are also possible.

Based on interactions with OEMs, consulting companies, review of the technical literature, and the judgment of the committee, estimates of the overall FC reduction potential from these advanced-level technology areas are presented in Table 5.2. For the ranges shown, the 10 percent for engine technologies alone and 13 percent for vehicles applies to larger vehicles with automatic transmissions. For smaller vehicles with manual transmissions and engine displacements less than 1.5 L, cost constraints are likely to reduce the extent of downsizing and the potential would be about 6 percent for engine alone and 7 percent for vehicle due to elimination of not only the gain from automatic transmission efficiency

TABLE 5.2 Estimated Fuel Consumption Reduction Potential for Advanced-Level CI Power Trains Compared to Base-Level CI Power Trains

Item	Average Reduction (%)	Min	Max
Large Vehicles			
Downsizing	4	3	5
Downspeeding	1.5	1	2
Friction reduction	1.5	1	2
Combustion improvement	3	2	4
Total engine improvement	10		
Accessory improvement	1	0.5	1.5
Transmission loss reduction	2	1.5	2.5
Combined engine and transmission potential	13		
Item	**(%) Reduction**	**Min**	**Max**
Small Vehicles (<1.5 L)			
Downsizing	1	0	2
Downspeeding	0.5	0	1
Friction reduction	1.5	1	2
Combustion improvement	3	2	4
Total engine improvement	6		
Accessory improvement	1	0.5	1.5
Thermal management	0	0	0
Transmission loss reduction	0	0	0
Combined potential	7		

NOTE: The values shown for the combined potential do not show a range. It is tempting to use the sum of the minimum values for the lower limit of the range and the sum of the maximum values for the upper end of the range. However, this would be inappropriate because no original equipment manufacturer is likely to simultaneously achieve either the minimum or the maximum for all items. Therefore, a realistic range for the combined potentials is about ±1 percent.

improvement (−2 percent) but also some of the gains from downsizing (−3 percent) and downspeeding (−1 percent).

TECHNOLOGY READINESS/SEQUENCING

In 2003, J.D. Power estimated the CI light-duty market share would reach 16 percent by 2015 (Peckham, 2003). However, the fuel price run-up of 2007-2008 caused a significant negative price differential between diesel and gasoline fuel (i.e., diesel fuel more expensive than gasoline) due to a global shortage of distillate/diesel fuel. This negative price differential has probably interfered with the growth of CI diesel vehicle sales. Even with the large fuel price reduction resulting from the economic slowdown of 2008 to 2009, the negative price differential has gone away slowly. Table 5.3 provides a brief summary of the average U.S. gasoline-to-diesel price differential evolution between May 2008 and June 2009. From Table 5.3 it can be seen that the negative price differential decreased substantially (from 54 cents/gal, or 15 percent, to 11 cents/gal, or 5.2 percent) between May 2008 and May 2009. Between May 2009 and June 2009, gasoline prices increased more than diesel (~45 cents/gal versus 17 cents/gal) causing a shift to a positive price differential. Whether this positive price differential remains when global economic activity returns to normal levels can only be guessed. The current positive price differential in combination with the new national fuel economy standards announced May 19, 2009, may strengthen interest in CI diesel vehicles, but it remains to be seen if the predicted U.S. CI diesel market share of 16 percent will be reached by 2015.

Application of CI technology into the NA market to reduce fuel consumption involves two steps. The first step is the introduction of vehicles with optional base-level CI power trains. The second step is the improvement of these CI power trains to advanced-level ones by implementation of the advanced technologies whose potential gains are indicated in Table 5.2.

The first step is underway now, as noted earlier in this chapter, as demonstrated by the introduction of a large number of vehicles for the 2009 model year. However, these vehicles primarily use versions of CI already in production for the European market. The decisions that put these introductions into product plans occurred several years earlier when it became clear first that there was encouraging development of technology enabling compliance with the 2010 Tier 2, Bin 5 and LEV II emissions standards for modified versions of these existing engines, and second that market conditions were supportive of such introductions due to increasing concern with the rise in both the price of fuel and in greenhouse gas (GHG) emissions. Had these conditions continued, it seems likely that additional vehicles beyond those announced for 2009 would have been introduced in model years 2010 and 2011. However, as noted earlier, as the petroleum price rose and fell during 2008, the unfavorable differential between gasoline and diesel fuel grew and then decreased, leaving potential CI vehicle buyers uncertain about future fuel prices. As a result, the pace of introduction of vehicle platforms with CI power trains for the NA market based on engines already in production is likely to decrease due to reduced market demand because of the fuel-price differential history as well as lower fuel prices in general. In addition, the global economic slowdown and the associated reduced tooling capital availability caused by the global auto industry's economic problems will also have a major impact on decisions about tooling new CI power trains for those OEMs that do not already have appropriately sized CI engines in production. Appropriately sized engines would be those with displacements suitable for the classes of vehicles whose fuel consumption reduction would have the largest impact on OEMs' specific fleet CAFE values.

Therefore, the second step, introduction into the market of CI technologies that could reduce light-duty fuel consumption beyond that shown in Table 5.1, will likely follow two paths. The first path is the introduction of the advanced-level technologies listed in Table 5.2 into post-2009 vehicles that were newly introduced in the 2009 model year. It is expected that this will occur in vehicles for model-years 2011-2014. This estimate is based on several factors. First, it is known that these technology areas are currently under development based on meetings with several OEMs. Second, European OEMs that are introducing CI-powered vehicles in the North American market in 2009 will also be preparing for Euro 6 emissions regulations that will take effect in 2014. Since Euro 6 NO_x requirements are less stringent than Tier 2, Bin 5 and LEV II emissions technologies to be used for Euro 6 will have already been developed to meet the U.S. requirements. As a result, it is expected that European OEM engineering resources in the 2009-2011 time frame will be partly applied

TABLE 5.3 Comparison of U.S. Average Gasoline and Diesel Fuel Prices Between May 2008 and June 2009

Date	Gasoline Cost ($/gal)	Diesel Cost ($/gal)	Gasoline to Diesel Cost Difference (cents)	Diesel to Gasoline Cost Difference (percent)
May 9, 2008	3.613	4.149	−54	−14.8
May 9, 2009	2.078	2.185	−11	−5.15
June 1, 2009	2.524	2.352	+17	+7.3

SOURCE: EIA (2009b).

to realizing some of the efficiency gains summarized in Table 5.2. For the OEMs active in the European market, this timeline is compatible with tax incentives expected in 2011 for early introduction of vehicles meeting Euro 6 as well as with the next European fleet CO_2 reduction target in 2012.

The second path for introduction of the advanced-level technologies summarized in Table 5.2 is their introduction simultaneously with new CI power trains in the period 2014-2020. These advanced-level versions will be required for market competitiveness for these new vehicles since the OEMs introducing CI vehicles between 2009 and 2011 will probably have already implemented advanced-level technology features. For example, BMW has already introduced an engine with two-stage turbocharging, one of the key features of the advanced-technology level. However, the pace of introduction of these vehicles with newly tooled CI engines will follow the new market conditions based on the economic recovery of global economies and the related automobile markets.

In addition, California Air Resources Board (CARB) LEV III standards are expected for 2013. The LEV III emissions levels currently under discussion would be very challenging. So OEMs will be developing technologies to enable their diesel products to meet LEV III and associated regulations. Studies at European OEMs with development vehicles using emissions control technologies developed to meet Tier 2, Bin 5 standards indicate that these technologies need additional development to achieve proposed LEV III requirements. As a result, it is expected that there will be some fuel consumption increase in order to meet the new standards.

In summary, the following technology sequencing is envisioned:

- For OEMs with existing CI engines, vehicles introduced in 2009 will be joined by additional models from 2011 to 2014, with base-level or advanced-level technology features depending on each OEM's particular marketing strategy.
- During the period 2015-2020, it is expected that development efforts for these OEMs will be focused on further reduction of power train cost and fuel consumption to achieve the upper limits of the ranges shown in Table 5.2.

For OEMs without existing CI engines with displacements in the range that would have the biggest impact on improving their CAFE values (e.g., V6 engines with displacements around 3.5 L for SUV and pickup trucks), new engines may be developed and put into production if three conditions are met. First, overall light-duty markets in the 2010-2012 period must improve sufficiently from those of 2009 to generate improved corporate financial health and required tooling capital. Second, a favorable customer perception of CI power trains must evolve based on the 2009-2012 CI vehicles already in the market. These new engines would probably be introduced in both base-level and advanced-level technology versions in order to both be technologically competitive with advanced-level technology products already in the market and to achieve market volumes necessary to justify the tooling investment. Third, fuel prices must increase from late 2009 levels but without significant negative price differential between gasoline and diesel in order to provide potential customers with sufficient incentive to offset the additional prices that must be charged for CI engines.

TECHNOLOGY COST ESTIMATES

There are a number of complexities in making cost estimations for CI engines to replace SI engines. The first of these involves selecting the appropriate displacement for the CI engine. This is important because CI engine costs depend significantly on their displacement for two primary reasons. First, the configuration and cost of their exhaust aftertreatment systems depend on engine displacement since component substrate (e.g., oxidation catalyst, particulate filter) volume is proportional to engine displacement and precious metal washcoat weights applied to the substrates are proportional to substrate volume. In addition to washcoat factors, NSC (NO_x storage catalyst) and urea-SCR-based NO_x reduction systems have different relationship multipliers to engine displacement. This is because urea-SCR-based systems use much less PGM compared to NSC-based systems, thus decreasing the rate at which costs increase with displacement.

Second, the degree of downsizing employed for the CI engine determines the cost and complexity of the air system for the engine. Maximum downsizing corresponding to advanced-level CI engines requires two-stage turbo systems, which cost about twice those of base-level single-stage turbo systems.

The cost of the engine structure and mechanical parts of CI engines depends less on displacement since smaller engines have all the same parts as larger displacement ones. These parts all require the same casting, fabrication, and machining processes and differ primarily in the amount of raw materials used, which has a relatively small influence on total cost. In the present work, no displacement-based adjustment was made to the cost estimates for the basic engine structure and parts.

Engine Sizing Methodology

The engine sizing methodology developed for this work is based on current and future product development directions. Two CI engine configurations have been considered, namely, base-level engines and advanced-level engines, as discussed above in the subsection titled "Overall Fuel Consumption Reduction Potential." Performance of a given vehicle depends primarily on the combined effect of the torque curve of the engine, the transmission characteristics (e.g., speed ratio range and internal efficiency), and final drive ratio. For base-level CI engines, a maximum specific torque density of 160 N-m/L is assumed. This level is achievable with single-stage turbo systems and, for example, is the level achieved by the Tier 2, Bin 5-compliant 2009 VW Jetta. The CI engines considered in the Ricardo, Inc., FSS analysis (EPA, 2008) from which the fuel consumption reduction values in Table 5.1 were determined had base-level technology features with single-stage turbo systems.

For advanced-level CI engines, a specific maximum torque density of 200 N-m/L is assumed. This level allows downsizing from base-level CI engines, thereby enabling additional fuel consumption reductions. The Tier 2 Bin 5 compliant 2009 BMW 335d with two-stage turbocharging achieves over 192 N-m/L and the Mercedes OM651 recently introduced in Europe achieves 233 N-m/L, and so the 200 N-m/L assumed for the advanced-level technology CI engine is considered realistic.

Based on the results from the full system simulation vehicle simulations carried out by Ricardo, Inc., for the EPA (EPA, 2008) (see Table 5.1) for 2007 model-year midsize MPV, full-size car, and truck-class vehicles, base-level CI engines displacing about 83 percent of the SI engines they replaced achieved equivalent vehicle performance when combined with advanced DCT6s (6-speed dual-clutch transmissions). It is therefore assumed that base-level CI engine displacement is about 83 percent of that of the 2007 model-year SI engine being replaced. Similarly, advanced-level CI engines having displacements about 80 percent of those of base-level CI engines can maintain equivalent vehicle performance. This is because the maximum torque of a base-level CI engine of displacement δ would be about $160 \times \delta$ N-m. Since the base-level maximum specific torque of 160 N-m/L is 80 percent of the 200 N-m/L for the advanced-level CI engine, the appropriately sized advanced-level CI engine would have 80 percent of the displacement of the base-level engine (i.e., 80 percent $\times\ \delta$). Then peak specific torque of the advanced-level CI sized at 80 percent would be equal to that of the base-level (i.e., $200 \times (80 \text{ percent} \times \delta) \approx 160 \times \delta$). With equal maximum torque, the advanced-level CI engine would enable equivalent vehicle performance.

Cost Estimation Methodology

The cost estimations from the sources considered in the present work (Martec Group, Inc., 2008; EPA, 2008, 2009; Duleep, 2008/2009) are then compared with those used by the NHTSA in its final rulemaking for 2011 (DOT/NHTSA, 2009). The Martec study used a BOM (bill of materials) approach based on technology packages consisting of combinations of components that fit together technically and made sense from a marketing point of view. BOM is also discussed in Chapter 3. This assessment was made by OEMs and suppliers with which Martec met. Martec then developed component-by-component costs and described the resultant BOM and cost sets in extensive detail. The resultant BOMs included not just the CI engine hardware added or SI hardware subtracted but also additional components that, in the judgments of the OEMs and suppliers, were necessary to make fully functional vehicles meeting both emissions standards and customer expectations. Martec reviewed the resultant cost tables with both the OEMs and the suppliers to reach consensus. It is often said by OEMs that cost numbers provided by suppliers are lower than what OEMs actually have to pay, while suppliers counter that the costs that OEMs say they have to pay include more content than that quoted by the supplier. It is hoped, therefore, that the approach used by Martec to reach consensus avoided this potential confusion and provided more correct estimates. Finally, the Martec study was carried out in 2007-2008—more recently than the years (2002-2006) on which the EPA (2009) estimates were based or the period covered (2005-2008) in Duleep (2008/2009) estimates.

To avoid the rather subjective issue of cost reductions over the production life of components, Martec developed cost estimates assuming very large production volumes so that all volume-related learning could be considered already reflected by its cost estimates. For some existing components, like common rail injection systems, global production volumes are already high enough to exceed the Martec volume threshold, and cost estimates for these items would automatically include cost reductions from high-volume learning. On the other hand, it is not expected that the CI diesel engines used for the NA market alone will exceed that volume threshold before 2020. However, since many of these engines will also be produced for the European Union (EU) market, whether by EU OEMs or by U.S. domestic OEMs that produce such engines for their EU products, the combined EU, U.S., and Canadian volumes may reach the 500,000-unit threshold. Thus the volume thresholds required to realize high-volume earnings will consist of combined EU and NA volumes for a number of the engines in the CI diesel fleet. It is expected that volumes will reach the 500,000-unit threshold primarily for the engines sold in the highest volumes in the EU (e.g., ~1.6 L). Thus for some of the smaller engine displacements likely to have low volumes in the U.S. market (e.g., <1.5 L) as well as for larger engines (e.g., 4.0-4.5 L) used in vehicles not marketed at high volume in the EU (e.g., large SUVs and pickups), the 500,000-unit volume target may not be reached by 2020 and costs will remain somewhat higher. To that extent, some of the Martec CI cost increment estimates could be too low.

The cost estimates developed in the present work were derived primarily from the Martec study (Martec Group, Inc., 2008). This choice was made for the reasons stated above. In addition, the Martec report included detailed specification of the exhaust aftertreatment system configuration, sizing, and PGM washcoat loadings. This type of information was not included in EPA (2008, 2009) studies or in Duleep (2008/2009). In addition, the Martec report described the commodity cost basis used, thus allowing modification of those costs in the present work to reflect recent decreases in commodity pricing for PGMs.

Base-Level Engine Technology Cost Estimates

Incremental CI diesel engine cost estimates developed in the present study for replacing 2007 model-year SI gasoline engines with equivalent performance CI diesels are summarized in Tables 5.4, 5.5, and 5.6. Appendix G contains the same information for full-size body-on-frame pickup trucks.

Emissions Systems Cost Estimates

Since the exhaust emissions systems are a significant fraction of the cost for CI diesel power trains, the brief entries in Tables 5.4 and 5.5 are described in more detail in Table 5.6. Note that the entries in Tables 5.4 and 5.5 reflect choices made for NO_x aftertreatment technologies. For the midsize sedan, it was assumed that the 70 percent aged conversion efficiency currently achievable with NSC-based systems would be sufficient for emissions compliance through the year 2020. Using the spreadsheet from which the cost estimates shown in Table 5.6 were obtained, it was also determined that for a 2.0-L CI engine for a midsize sedan, the NSC system is a lower cost approach ($688) than is a urea-SCR-based system ($837). As a result, Table 5.6 contains no cost estimates for the SCR-urea system for the midsize sedan. This choice could be changed depending on success in meeting LEV III requirements with NSC-based systems and changes in PGM commodity prices. However, for the heavier SUV, SCR-urea with its capability for 85 to 93 percent conversion efficiency will be required for emissions compliance. As a result, there are no entries in Table 5.6 for NCS NO_x aftertreatment for the SUV since it is assumed that SCR technology will be used.

Commodity prices were quite volatile between 2004 and 2008 (Martec Group, Inc., 2008), making product planning for CI diesel vehicles quite challenging. To illustrate the impact of PGM (platinum group metals consisting of platinum, palladium, and rhodium) commodity price volatility, Table 5.6 includes estimates for the precious metal wash coats used in the catalysts in separate rows labeled PGM loading. In addition, two columns are shown for each of the two reference vehicles. Columns two and four correspond to the PGM prices in November 2007 used in the Martec study (Martec Group, Inc., 2008). The estimates in columns three and five illustrate emissions systems costs based on PGM prices from April 2009 computed in the present study. These latter costs were used for the aftertreatment system cost estimates in Tables 5.4 and 5.5 because they are considered more representative of the post 2009 period. Obviously, this price situation must be monitored, since it is unlikely to remain at April 2009 levels until 2020. For the sedan with an advanced-level downsized 1.6-L engine, emissions system cost between November 2007 and April 2009 dropped 30 percent. Note that the catalyst volumes for the cost computation for the downsized 1.6-L engine were not reduced from the 2.0-L sizes since the 1.6-L engine must produce the same power

TABLE 5.4 Committee's Estimates of Incremental Cost of CI Diesel Engine over a Baseline SI Gasoline Engine for Replacing SI 2.4-L MPFI DOHC Four-Valve Engines in Midsize Sedans (e.g., Malibu, Accord) with Base-Level 2.0-L I4 CI Engines

50-State-Saleable ULEV II 2.0-L DOHC CI Diesel Engine Baseline: SI Gasoline 2.4- L MPFI DOHC 4V I4	Estimated Cost vs. Baseline ($)
Common rail 1,800 bar piezo-actuated fuel system with four injectors (@$75), high-pressure pump ($250), fuel rail, regulator, and fuel storage upgrades plus high-energy driver upgrades to the engine control module. Credit for SI content deleted ($32)	675
Variable-geometry turbocharger (VGT) ($250) with electronic controls, aluminum air-air charge air cooler, and plumbing ($125)	375
Upgrades to electrical system: starter motor, alternator, battery, and the 1-kW supplemental electrical cabin heater standard in Europe ($59)	125
Cam, crank, connecting rod, bearing, and piston upgrades, oil lines ($50) plus NVH countermeasures to engine ($40) and vehicle ($71)	161
HP/LP EGR system to suppress NO_x at light and heavy loads; includes hot side and cold side electronic rotary diesel EGR valves plus EGR cooler and all plumbing	215
Emissions control system including the following functionality: diesel oxidation catalyst (DOC), catalyzed diesel particulate filter (CDPF), NO_x storage catalyst (NSC), EGR catalyst, passive SCR. Stoichiometric MPFI emissions and evaporative systems credit ($245). See Table 5.6 for a detailed breakdown of the emissions control system components leading to the total shown here.	688
On-board diagnostics (OBD) and sensing including an electronic throttle control ($25), four temperature sensors (@$13), wide-range air-fuel ratio sensor ($30), two pressure-sensing glow plugs (@17), two conventional glow plugs (@$3), and Delta-P sensor for DPF ($25). Credit for two switching O_2 sensors (@$9).	154
Total variable cost with credits for SI parts removed. Excludes any necessary transmission, chassis, or driveline upgrades.	2,393

NOTE: The credit for downsizing from V6 to I4 included in the Martec Group, Inc. (2008) study was not used in the committee's estimates since baseline 2007 midsize sedan SI gasoline engines were not V6 but 2.4-L I4 engines. Cost estimates for aftertreatment systems reflect April 2009 prices for platinum group metals.

TABLE 5.5 Committee's Estimates of Incremental Cost of CI Diesel Engine over a Baseline SI Gasoline Engine for Cost Estimations to Replace SI MPFI DOHC Four-Valve 4.0- to 4.2-L Six-Cylinder Engine in a Midsize Body-on-Frame SUV (e.g., Explorer, Durango) with a 3.5-L V6 DOHC CI Engine

50-State-Saleable ULEV II 3.5-L V6 DOHC CI Diesel Engine Baseline: SI Gasoline DOHC 4V 4.0-4.2-L Six Cylinder	Estimated Cost vs. Baseline ($)
Common rail 1,800 bar piezo-actuated fuel system with six injectors (@$75), high-pressure pump ($270), fuel rail, regulator and fuel storage upgrades plus high-energy driver upgrades to the engine control module. Credit for MPFI content deleted ($48).	911
Variable-geometry turbocharger (VGT) ($350) with electronic controls, water-air charge air cooler, circulation pump, thermostat/valve and plumbing ($135)	485
Upgrades to electrical system: starter motor, alternator, battery, and the 1.5-kW supplemental electrical cabin heater standard in Europe ($99)	167
Cam, crank, connecting rod, bearing, and piston upgrades, oil lines ($62) plus NVH countermeasures to engine ($47) and vehicle ($85)	194
HP/LP EGR system to suppress NO_x at light and heavy loads; includes hot side and cold side electronic rotary diesel EGR valves plus EGR cooler and all plumbing	226
Emissions control system including the following functionality: DOC, CDPF, selective catalytic reduction (SCR), urea dosing system ($363). Stoichiometric MPFI emissions and evaporative systems credit ($343). See Table 5.6 for a detailed breakdown of the emissions control system components leading to the total shown here.	964
On-board diagnostics (OBD) and sensing including four temperature sensors (@$13), wide-range air-fuel ratio sensor ($30), NO_x sensor ($85), two pressure-sensing glow plugs (@17), four glow plugs (@$3), and Delta-P sensor for DPF ($25). Credit for four switching O_2 sensors (@$9)	227
Total variable cost with credits for SI parts removed. Excludes any necessary transmission, chassis, or driveline upgrades.	3,174

NOTE: The credit for downsizing from V8 to V6 included in Martec Group, Inc. (2008) was not used here because the baseline 2007 SI engine was a V6, not the V8 assumed in Martec Group, Inc. (2008). Aftertreatment system cost estimates reflect April 2009 prices for platinum group metals.

TABLE 5.6 Cost Estimates for Exhaust Emissions Aftertreatment Technologies Capable of Enabling Tier 2, Bin 5 Compliance

Item	Midsize Car (e.g., Malibu) Catalytic Device Sizing Based on 2 L (Nov. 2007 PGM prices)	Midsize Car (e.g., Malibu) Catalytic Device Sizing Based on 2 L (Apr. 2009 PGM prices)	Midsize SUV (e.g., Explorer), Catalytic Device Sizing Based on 3.5 L (Nov. 2007 PGM prices)	Midsize SUV (e.g., Explorer), Catalytic Device Sizing Based on 3.5 L (Apr. 2009 PGM prices)
DOC 1				
Monolith and can	$52	$52	$52	$52
PGM loading	$174	$139	$210	$200
DOC 2				
Monolith and can	Not used	$0	$52	$52
PGM loading	Not used	$0	$73	$70
EGR catalyst				
Monolith and can	$7	$7	Not used	Not used
PGM loading	$22	$13	Not used	Not used
Coated DPF				
Advanced cordierite brick and can	$124	$124	$270	$270
PGM loading	$160	$131	$29	$26
NSC system				
Catalyst brick and can	$114	$114	Not used	Not used
PGM loading	$533	$314	Not used	Not used
SCR-urea system				
SCR brick and can	$39	$39	$274	$274
Urea dosing system	Passive SCR	Passive SCR	$363	$363
Stoichiometric gasoline emissions and evaporative system credit	–$245	–$245	–$343	–$343
Emissions System Total	$980	$688	$980	$964

NOTE: The significant impact of platinum group metals (PGM) commodity prices is illustrated by the difference between the costs in columns 2 and 4 (based on November 2007 prices) and the costs in columns 3 and 5 (based on April 2009 prices).

output as the 2.0-L engine, requiring that exhaust gas flow rates remain virtually unchanged. For the SUV, a smaller 10 percent emissions system cost drop was observed due to the lower PGM usage with SCR-urea aftertreatment for out-of-engine NO_x control for the SUV. With SCR-urea systems, only the SCR device contains no PGM. As can be observed from examination of the entries in Table 5.6, DOC1, DOC2, and the coated DPF (called CDPF) all utilize PGM washcoats. As noted earlier, the spreadsheet used to generate the aftertreatment cost estimates shown in Table 5.6 is available for recomputing the aftertreatment system cost estimates should PGM commodity prices change significantly.

Finally, there is a technology choice involved in DPF systems. The four substrate options currently available for particulate filters are silicon carbide (Si-C), conventional cordierite, advanced cordierite, and acicular mullite. Conventional cordierite is used for most nonparticulate filter substrates (e.g., DOC and NSC catalysts), whereas Si-C has been the predominant choice for light-duty DPF usage in Europe. Conventional cordierite is less expensive and lower in mass than Si-C. On the other hand, Si-C has much higher thermal conductivity and strength, which are very favorable properties for withstanding regeneration without local hot spots causing thermal stress cracking and ultimate failure of the filter. As a result of these property differences, Si-C filters are typically filled (i.e., loaded) with about twice the amount of particulate (e.g., 8-9 g/L) during vehicle operation before regeneration is carried out, whereas conventional cordierite filters must be regenerated after about half that loading (e.g., 4-5 g/L) of particulate.

There are two results from this difference. First, conventional cordierite-based filter systems tend to require more frequent regenerations with associated FC increases. Second, since during regeneration fuel is injected into the engine cylinder during the expansion stroke with the piston part way down the cylinder to raise the temperature of the gases by partial oxidation of this regeneration fuel in the cylinder and completion of oxidation of that fuel in the oxidation catalyst, some fuel from the high-pressure spray reaches the cylinder wall and some of that fuel escapes past the piston rings down into the crankcase, where it dilutes the lubricating oil with fuel. This dilution requires more frequent oil changes to protect engine durability. Since frequency of oil changes is a marketing attribute, the choice of substrate has multiple implications, namely cost, durability, mass, and oil-change interval.

Advanced cordierite is emerging as a compromise between the properties of Si-C and conventional cordierite (Tilgner et al., 2008). Therefore, for the purpose of this report, it has been assumed that new DPF applications will utilize advanced cordierite (as was assumed for the estimates in the Martec [2008] report) and that existing Si-C applications will be converted to advanced cordierite for the next design and development cycle. Thus the cost estimates shown in Table 5.7 are based on the use of advanced cordierite for DPF monoliths.

Finally, acicular mullite has recently been introduced to the market. This new material has a number of properties that are potentially advantageous for exhaust filtration. First, this material appears to have lower pressure drop than the other materials due to higher porosity. According to material property specifications (Dow, 2009), this higher porosity and lower pressure drop remain when catalytic coatings are applied. As a result, it may be possible to integrate additional exhaust aftertreatment system components (e.g., combining SCR and DPF units into one component), thus reducing system cost, packaging volume, and complexity. The first production application of this material is expected in 2011, after which its technical potential and cost tradeoff relative to other materials will become clearer.

TABLE 5.7 Comparison of CI Engine Cost Estimates from Different Sources and the Committee's Estimates

Source	I4 CI Engine ($)	V6 CI Engine	Engine Sizing Methodology Specified	Aftertreatment System Configurations and PGM Loadings	PGM Cost Basis	Dollar Basis
Martec Group Inc. (2008)	2,361	3,465	Partially	Yes	Nov. 2007	2007
EPA (2009)	2,052	2,746	Yes	Configuration, yes; sizing-loading, no	Not specified	2007[a]
Duleep (2008/2009)	1,975	2,590	No	Configuration, yes; sizing-loading, no	Not specified	2008
DOT/NHTSA (2009)[b]	2,667	3,733	Partially	Assumed to be based on those of Martec Group, Inc. (2008)	Nov. 2007	2007
NRC (2010)[c]	2,393	3,174	Yes	Yes[d]	Apr. 2009	2007

[a]EPA 2009 estimates provided were for dollar-year-basis 2002 for engine and 2006 for aftertreatment. The numbers shown have been corrected by applying the ratios of the yearly producer's price index (1.0169 for 2002 to 2007 and 1.0084 for 2006 to 2007). However, significant technology development has taken place since 2002, and so it is likely that technology-based component specifications and associated costs have changed.

[b]Costs from Tables IV-21, IV-22, and IV-23 of DOT/NHTSA (2009) were divided by 1.5 to convert from RPE (retail price equivalent) to cost.

[c]NRC (2010) refers to the present report. The CI engine costs are for base-level specifications. Detailed breakdowns of the committee's cost estimates are given in Tables 5.4 and 5.5.

[d]The spreadsheet used to compute aftertreatment system costs for the present work utilizes the configuration, sizing, and washcoat loadings included in the December 2008 version of the Martec Group, Inc. (2008) study.

Comparison of Cost Estimates with Those of Other Sources

The cost estimates from Martec Group, Inc. (2008), EPA (2009), and Duleep (2008/2009) are summarized in Table 5.7. From the left, the columns show:

- The cost estimate source;
- The cost estimates for replacing the baseline I4 SI engines in 2007 model-year midsize sedans (e.g., Malibu, Camry) with CI engines;
- The cost estimates for replacing the baseline six-cylinder SI engines in 2007 model-year midsize SUVs (e.g., Explorer, Trailblazer) with V6 CI engines;
- Whether the sources include details on how the displacements for the replacement CI engines were chosen;
- Whether the sources include details on exhaust aftertreatment system configurations, component sizing, and catalyst washcoat loading;
- What is the timing basis for PGM commodity costs;
- What is the dollar basis year.

Present Cost Estimates Compared to Martec Estimates

Although the cost estimates developed in the present study were based on the estimates from Martec Group, Inc. (2008), a number of revisions were made to the Martec estimates. First, the Martec estimates assumed that the 2-L four-cylinder CI engine replaced a V6 SI engine in the midsize sedan vehicle. As a result, Martec included a downsizing credit resulting from the savings from the elimination of two cylinders and their associated parts. Whether or not it is appropriate to include such a credit depends on what baseline vehicle is assumed. Because of the timing of the EISA that motivated the present study, the baseline vehicles for the present study are 2007 model-year vehicles. The vehicle class that would utilize the 2.0-L CI engine, namely the 2007 midsize sedan (e.g., Malibu, Camry), typically used a four-cylinder 2.4-L SI engine with 4/5-speed automatic transmission. Therefore, for the present study, the downsizing credit for reducing the number of cylinders was excluded from the cost estimate since a four-cylinder CI engine would replace a four-cylinder SI engine. This increased the estimate from the Martec value of \$2,361 by \$310 to \$2,671. Second, the Martec cost estimates were based on November 2007 commodity prices for the precious metals used in the exhaust aftertreatment system washcoats. Based on the detailed exhaust aftertreatment system specifications provided in the Martec (2008) report, the committee constructed a spreadsheet to compute the exhaust aftertreatment system costs, and April 2009 rather than November 2007 PGM prices were used. This change was made to reflect the significant commodity price deflation since November 2007. The difference amounted to \$292, which lowered the cost estimate from \$2,671 to \$2,379. Finally, an additional pressure-sensing glow plug was added to provide OBD backup for the single pressure-sensing glow plug assumed in the Martec BOM (replace 1 ceramic glow plug @\$3 with pressure-sensing glow plug @\$17 for net increase of \$14). That brought the present estimate to the \$2,393 shown in Tables 5.4 and 5.7.

For the SUV case, the Martec analysis assumed that a 3.0-L V6 CI engine would replace a V8 SI engine. As is discussed above for the I4 case, for the case of a baseline 2007 midsize SUV (e.g., Explorer, Trailblazer), the baseline SI engine was a 4.0- to 4.2-L six-cylinder engine rather than the V8 assumed in the Martec analysis. Therefore, the downsizing credit from V8 to V6 used in the Martec analysis (\$270) was not included for the present analysis, increasing the cost estimate from \$3,465 to \$3,735. The Martec analysis assumed a two-stage turbo system for the 3.0-L V6 engine system. For the comparisons in Table 5.7, only the 3.5-L base-level technology engine was included to be compatible with the packages assumed in EPA (2009) and Duleep (2008/2009). Therefore, the air system cost from the Martec analysis was reduced for the present analysis by replacing the two-stage turbo system cost estimate (\$1,030) with that for a single-stage system (\$485). That reduced the estimate from \$3,735 to \$3,190. Finally, the increase in displacement from the Martec 3.0-L displacement to the 3.5 L of the present analysis along with the use of the April 2009 PGM prices rather than the November 2007 PGM prices used by Martec reduced the aftertreatment system cost from \$980 to \$964, which in turn reduced the total V6 SUV replacement cost from \$3,190 to the \$3,174 shown in Tables 5.5 and 5.7.

Present Cost Estimates Compared to EPA Estimates

The EPA cost estimate shown in Table 5.7 for the I4 CI replacement for the 2.4-L SI engine is \$2,052, which is \$341 less than the committee's estimate of \$2,393. Using detailed breakdowns of the EPA estimates (EPA, 2009), one major difference is the cost credits used in the EPA breakdown for parts removed from the SI engine. The EPA estimate for the gasoline fuel system removed was \$240 (\$165 for injectors and rail and \$75 for fuel pump and vapor recovery (Evap) system, whereas that used for the present work from Martec Group, Inc. (2008) was \$32 for the injection system and \$37 for the Evap canister and purge valve (included within the \$245 emissions system credit). The fuel pump for the gasoline system is actually replaced by the low-pressure supply pump for the CI fuel system, which is very similar to the gasoline pump, and so there should be no credit for that item. The injectors and rail are extremely high-volume commodity items sold by suppliers at close to cost because of the strong global competition for such parts. Therefore, the \$32 credit used for those items is considered representative. The difference between the EPA estimate and the committee's estimate for the fuel system and vapor recovery is thus \$240 versus \$69. The EPA assumed a \$75 credit for ignition system parts removed from the SI engine. The pencil coils used in 2007 ignition systems are again extremely high-volume commodity items. The ignition control drivers used in such systems are up-integrated into the ECM, and so there is effectively no savings from their removal. For the CI engine, a glow plug and wire is required for each cylinder, so the SI to CI ignition cost difference was considered $0. There were other differences in the individual item estimates between the EPA estimate and that from the present estimate as well. The EPA estimate for the turbocharger system was less than that of the present study ($181 versus $375). The EPA estimate for emissions controls appeared to reflect a somewhat different approach to emission control, with more emphasis on aftertreatment and less emphasis on in-cylinder combustion-based control of emissions. This approach is illustrated by the EPA choice of a urea-SCR strategy for NO_x aftertreatment while that for the present approach was an NSC-based approach. The present approach also included an HP/LP EGR system, whereas the EPA system did not. The HP/LP EGR system will lower engine-out emissions, whereas the NSC NO_x conversion efficiency is lower than that of the urea-SCR approach, as noted earlier in the discussion of NO_x aftertreatment system technologies. As a result, the EPA emissions system cost estimate was significantly higher than that from the present work ($1,220 versus $903 ($688 for aftertreatment plus $215 for HP/LP EGR)). The urea-SCR subsystem cost in the EPA estimate versus that for the NSC in the present study was $670 versus $428, and the EPA CDPF cost was estimated at $480 versus $255 for the present study. No information was available concerning CDPF substrate volume or PGM loading to understand the source of these differences in more detail. The present study assumed that the aftertreatment system would also require an EGR catalyst ($20) to control EGR cooler fouling, and a passive SCR catalyst ($39), which would provide a small amount of NO_x reduction on the US06 test using the small amount of ammonia produced by the NSC at the higher load conditions of the US06 test rather than urea from a separate system like that in the urea-SCR system. OEMs will make the choice of emissions control strategy based on many factors, including cost, durability, customer convenience, and packaging. In addition to cost differences, the urea-SCR approach requires finding space to package a urea supply tank, which is more problematic in a smaller vehicle like the midsize sedan than for a larger vehicle like an SUV. As noted earlier, the 2009 VW Jetta utilizes a system very much like the system assumed in the present study. The other area in which different components were assumed by the EPA was for OBD and sensing. The present study assumed four temperature sensors ($52) and two pressure-sensing glow plugs ($34), which were not included in the EPA system. As noted earlier in discussions about combustion technologies, the closed-loop cylinder-pressure sensing system is beneficial for minimizing engine fuel consumption and emissions when different fuels of widely different cetane ratings are encountered in the market place, although the benefits of this technology will not show up on the EPA certification tests because those are conducted using standardized certification fuels for which the engines are calibrated during development.

As shown in Table 5.7 for the V6 midsize SUV case, the EPA estimate for replacing the SI engine with a CI engine was $2,746, which was $694 greater than that for the I4 CI engine substitution. The corresponding increment as determined in the present study was $781. The differences between the detailed items in the two cost estimates remain similar to those already discussed for the I4 case, and since the total cost differences were similar, the details are not discussed here. However, for the V6, both estimates assumed the urea-SCR approach for NO_x aftertreatment.

Present Cost Estimates Compared to EEA (Duleep) Estimates

The EEA (Duleep, 2008/2009) variable cost estimate for replacing the 2.4-L SI engine with a 2.0-L CI engine (Table 5.7) was $1,975. This total consisted of $1,145 for the engine and $830 for emissions control. The present study's engine cost estimate was $1,336. One of the larger differences between these two estimates was for the turbo system—EEA estimated a total of $280 and the Martec-based present study's estimate was $250 for the VGT turbo with electronic controls and $125 for the intercooler and plumbing, for a total turbo system cost of $375, or $95 above the EEA estimate. Also, the EEA estimate did not include a cabin heater, which is standard with CI diesel vehicles and which Martec estimated at $59. For exhaust emissions control, the differences between the EEA estimates and the Martec-based estimates used in the present study were also significant. EEA assumed an integrated DPF and NSC unit (called DPNR), which is proprietary to Toyota. All other OEMs are using separate DPF and NSC units. The EEA estimate assumed $730 for the DPNR unit, but no cost basis was specified for the PGM prices or loadings. The present study assumed $688 (see Table 5.6) based on April 2009 PGM prices for the separate DPF and NSC units. EEA assumed $60 for the EGR system and cooler, whereas the present study estimated $215 for an HP/LP EGR system (for details see Table 5.4). As noted in earlier discussion of emissions control technology, a combined HP/LP EGR system has many advantages for reducing engine-out NO_x, thus reducing the NO_x conversion requirements for the aftertreatment system. The LP EGR system requires several control valves and cooler in addition to those for the HP EGR system. The 2009 VW Jetta has such an HP/LP EGR system. For oxidative cleanup of the exhaust (e.g., unburned HC, CO, and soluble particulates), a DOC (diesel oxidation catalyst) is used. EEA assumed $50 for the DOC. Again, no information was provided about volume, PGM loading, or PGM cost basis for the EEA estimate. The present study assumed $52 for the monolith and housing and $139 for the PGM washcoat cost based on April 2009 PGM prices. The emissions control system cost estimate differences then totaled $227.

For the V6 SUV case, the EEA estimate was $2,590, whereas that of the present study was $3,174. The EEA estimate for the engine was $1,715 versus $1,983 for the present study. Of the $268 difference, the majority is explained by the lack of a cabin heater in the EEA estimate and inclusion of the cabin heater for the present study at $99 (more costly than that of the midsize sedan I4 vehicle because of the larger cabin volume for the midsize SUV with the V6) and the air system (turbocharger and intercooler) for which EEA estimated $365 versus $485 for the present study. The remainder of the difference was due to emissions control. Again, one of the main differences was the use of an HP/LP EGR system for the present study as included in the Martec BOM but not in the EEA estimate ($86 difference). In addition, the present study included the use of a second DOC ($122) included in the Martec BOM that was worked out in collaboration with OEMs and suppliers.

Present Cost Estimates Compared to NHTSA Estimates

According to the NHTSA final ruling for 2011 (DOT/NHTSA, 2009), costs for CI engines and DCT6 transmissions were also derived from the Martec estimates. For the 2.0-L I4, the NHTSA number from Table 5.7 is $2,667, whereas the corresponding number from the present study is $2,393. Most of the difference between these estimates is due to the $292 reduction in aftertreatment system costs used in the present study and derived from using April 2009 PGM prices rather than the November 2007 prices reflected in the Martec numbers presumably used by the NHTSA. It is not known whether the NHTSA estimate includes the downsizing credit or not.

The NHTSA cost estimate of $5,600 retail price equivalent ($3,733 cost) from Tables IV-21, IV-22, and IV-23 (DOT/NHTSA, 2009) for the larger vehicle classes (e.g., large car versus subcompact, compact, and midsize car) is assumed to derive from the Martec cost estimate of $3,465 for V6 diesel (Martec Group, Inc., 2008, p. 37). The corresponding value for the V6 CI engine from the present study was $3,174. A significant portion of the $559 difference between the NHTSA estimates and those of the present work is due to the inclusion in the Martec, and presumably also in the NHTSA, estimates of two-stage turbocharger systems that for the present study correspond to advanced-level engine technology, as described in the section "Engine Sizing Methodology." As noted above, the costs from the present work that were used in Table 5.7 were those for the base-level technology configuration. The base level was assumed to use single-stage VGT turbo systems and the advanced level to use two-stage turbo systems. The cost estimate from the present work, which is included in Table 5.7, is for the base-level CI engine. Including the two-stage turbo system in the cost estimate from the present study would increase the estimate from $3,174 to $3,719, leaving a difference between the NHTSA estimate and the present estimate of about $14.

There are also other differences between the assumptions made in the present study and those of the Martec study. For the engine sizing methodology used herein, the baseline six-cylinder engine for the midsize vehicle class of about 4.2 L downsized by the assumed 83 percent is 3.5 L, whereas the Martec study assumes 3.0 L. According to the costing methodology used in the present study, the increase of displacement from 3.0 L to 3.5 L increases cost (entirely as a result of aftertreatment systems cost) from $921 to $964. Subtracting this difference from the engine cost estimate of $3,174 increases the cost differential between the NHTSA estimate and the present study from $14 to $57. As for the remaining difference, there is insufficient information in the NHTSA report to understand the sources of this difference, although it is less than 10 percent, which is well within the uncertainty of these cost estimates in general.

Advanced-Level CI Engine Cost Estimates

Cost estimates for the technologies necessary to raise base-level CI engines to advanced-level engines inherent in the gains described in Table 5.2 are listed in Table 5.8.

Advanced-Level Transmission Cost Estimates

There seems to be an emerging consensus that dual-clutch automatically shifted manual transmissions (DCTs) offer a very attractive combination of efficiency and driver satisfaction with acceptable cost. In the Ricardo, Inc., FSS studies for the EPA (EPA, 2008), CI engines were combined with DCT6 units for the simulations, as noted in earlier discussions of Table 5.1. For that reason, it was assumed for the present analysis that the CI replacements for SI engines would use DCTs. Transmission technologies are discussed in Chapter 7, which considers non-engine vehicle technologies. Cost estimates for advanced transmissions used for this committee's work are also shown there and are summarized in Table 7.10.

Summary of Total SI to CI Power Train Replacement Cost Estimates

The total estimated costs to replace 2007 model-year SI power trains with base-level and advanced-level CI power trains for the example midsize sedan and midsize SUV vehicles indicated in Tables 5.4 and 5.5 are summarized in Table 5.9.

FINDINGS

Based on a combination of analysis and engineering judgment applied to information collected from many sources, the committee's key findings are as follows regarding tech-

TABLE 5.8 Committee's Estimates of Incremental Costs to Implement Advanced-Level Diesel Developments (downsizing, thermodynamic improvements, friction reduction, and engine accessory improvements) Whose Estimated Potential for Reducing Fuel Consumption Is Summarized in Table 5.2

Item	Midsize Car (e.g., Malibu) 1.6-Liter I4	Midsize SUV (e.g., Explorer) 2.8-Liter V6	Comment
Downsize engines from 2-L I4 to 1.6-L I4 and from 3.5-L V6 to 2.8-L V6	\$50	\$75	Higher load capacity rod bearings and head gasket for higher cylinder pressures (~\$12.50/cylinder)
Two-stage turbocharger system	\$375	\$545	Additional air flow control valves, piping, cost of additional turbo, water-to-air intercooler with separate pump, control valve
Dual-pressure oil pump	\$5	\$6	Switchable pressure relief valve for high or low oil pressure
Non-recirculating low-pressure (LP) fuel pump	\$10	\$12	Variable output LP pump controlled by high-pressure (HP) pump output
Cylinder pressure sensors	—	—	Two pressure-sensing glow plugs, one to sense fuel property differences, second to provide on-board diagnostics durability backup for first, already included for both I4 and V6 in Tables 5.3 and 5.4
Low-pressure exhaust gas recirculation (EGR)	—	\$95	Additional piping (~\$20) and valves (e.g., integrated back pressure and LP EGR rate ~\$75), much more difficult to package for V6 engine with underfloor diesel particulate filter, cost for I4 already included in Table 5.4
Direct-acting HP (maximum injection pressures >2,000 bar) piezo injectors	\$80	\$120	\$20/injector, benefits derived from combination of higher rail pressure and more injector controllability
Total	\$520	\$853	

TABLE 5.9 Estimated Total Costs to Replace 2007 Model-Year SI Power Trains with Base- and Advanced-Level CI Power Trains for Example Midsize Sedan and Midsize SUV-Type Vehicles

	Base-Level CI Engine	Advanced-Level CI Engine
Midsize Sedan		
I4 engine	\$2,393 (Table 5.4) or \$2,400 (when rounded to nearest \$50)	\$2,913 (Tables 5.4 and 5.8) or \$2,900 (when rounded to nearest \$50)
DCT6/7[a] transmission	\$140-\$400 (Table 7.10)	\$140-\$400 (Table 7.10)
Total	\$2,550-\$2,800 (when rounded to nearest \$50)	\$3,050-\$3,300 (when rounded to nearest \$50)
Midsize SUV		
V6 engine	\$3,174 (Table 5.5) or \$3,150 (when rounded to nearest \$50)	\$4,027 (Tables 5.5 and 5.8) or \$4,050 (when rounded to nearest \$50)
DCT6/7 transmission	\$140-\$400 (Table 7.10)	\$140-\$400 (Table 7.10)
Total	\$3,300-\$3,550 (when rounded to nearest \$50)	\$4,150-\$4,450 (when rounded to nearest \$50)

[a]Note that the higher of the two estimates shown in Table 7.10 is for a 6/7-speed dual-clutch transmission (DCT). In accordance with the potential fuel consumption reduction gains discussed in Table 5.2 due to transmission improvements, it was assumed that 7-speed versions would be used. Due to the wide range of cost estimates for DCTs as discussed in Chapter 7, no adjustment was made for the higher torque requirements of the V6 CI.

nology combinations for reducing the fuel consumption of 2007 model-year SI gasoline engine vehicles by equipping them with advanced CI diesel power trains.

Finding 5.1: By a joint effort between OEMs and suppliers, new emissions control technology has been developed to enable a wide range of light-duty CI engine vehicles to meet the 2010 Tier 2, Bin 5, LEV II emissions standards.

Finding 5.2: Replacing 2007 model year MPFI SI gasoline power trains with base-level CI diesel engines with advanced dual-clutch (automated manual) transmissions (DCTs) (6-speed) and more efficient accessories packages can reduce fuel consumption by an average of about 33 percent (or reduce CO_2 emissions by about 23 percent) on an equivalent vehicle performance basis. Advanced-level CI diesel engines with advanced DCTs could reduce fuel consumption by about an additional 13 percent for larger vehicles and by about 7 percent for small vehicles with engine displacements less than 1.5 L.

Finding 5.3: The characteristics of CI diesel engines that enable their low fuel consumption apply over the entire vehicle operating range from city driving to highway driving, hill climbing, and towing. This attribute of CI diesel engines is an advantage when compared with other technology options that are advantageous for only part of the vehicle operating range (e.g., hybrid power trains reduce fuel consumption primarily in city cycle/city driving).

Finding 5.4: The identified advanced-level technology improvements to CI diesel engines are expected to reach market in the 2011-2014 time frame, when advanced technology additions to SI gasoline engines will also enter the market. Thus, there will continue to be a fuel consumption and cost competition between these two power train systems. For the period 2014-2020, further potential fuel consumption reductions for CI diesel engines may be offset by fuel consumption increases due to engine and emissions system changes required to meet stricter emissions standards (e.g., LEV III).

Finding 5.5: CI diesel engine market penetration will be strongly influenced both by the incremental cost of CI diesel power trains above the cost of SI gasoline power trains and by the price differential of diesel fuel relative to gasoline. The estimated incremental cost differential for base-level and advanced-level I4 CI diesel engines to replace 2007 model-year midsize sedan SI gasoline engines ranges from \$2,400 (base level) to \$2,900 (advanced level). For base-level I4 engines combined with DCTs, power train replacement cost is estimated at \$2,550 to \$2,800 and for advanced-level I4 power trains is estimated at \$3,050 to \$3,300 (both rounded to the nearest \$50). For midsize 2007 model-year SUVs, the estimated cost for replacement of SI gasoline engines with base-level and advanced-level V6 CI diesel engines ranges from \$3,150 (base level) to \$4,050 (advanced level) (both rounded to the nearest \$50). For V6 CI engines combined with DCTs, the estimated V6 CI power train replacement cost increment over 2007 model-year SI power trains is \$3,300 to \$3,550 (base level), and the advanced-level power train incremental cost is \$4,200 to \$4,500 (both rounded to nearest \$50). These costs do not include the retail price equivalent factor.

REFERENCES

Bression, G., D. Soleri, S. Savy, S. Dehoux, D. Azoulay, H.B-H. Hamouda, L. Doradoux, N. Guerrassi, and N. Lawrence. 2008. A study of methods to lower HC and CO emissions in diesel HCCI. SAE Paper 2008-01-0034. SAE International, Warrendale, Pa.

Diesel Forum. 2008. Available at http://www.dieselforum.org/DTF/news-center/pdfs/Diesel%20Fuel%20Update%20-%20Oct%202008.pdf.

DieselNet. 2008. February 22. Available at http://www.dieselnet.com/news/2008/02acea.php.

DOT/NHTSA (Department of Transportation/National Highway Traffic Safety Administration). 2009. Average Fuel Economy Standards Passenger Cars and Light Trucks—Model Year 2011. Docket No. NHTSA-2009-0062, RIN 2127-AK29, March 23. Washington, D.C.

Dow. 2009. Available at http://www.dow.com/PublishedLiterature/dh_02df/0901b803802df0d2.pdf? filepath=automotive/pdfs/noreg/299-51508.pdf&fromPage=GetDoc.

Duleep, K.G. 2008/2009. Diesel and hybrid cost analysis: EEA versus Martec, presentation made to NRC Committee, February 25, 2008, updated June 3, 2009.

EIA (Energy Information Administration). 2009a. Light-Duty Diesel Vehicles: Efficiency and Emissions Attributes and Market Issues. February. Available at http://www.eia.doe.gov/oiaf/servicerpt/lightduty/execsummary.html.

EIA. 2009b. Diesel Fuel Prices. Available at http://tonto.eia.doe.gov/oog/info/gdu/gasdiesel.asp. Accessed May 9, 2009, and June 5, 2009.

EPA (U.S. Environmental Protection Agency). 2005. Document 420-F-05-001. Available at http://www.epa.gov/otaq/climate/420f05001.htm.

EPA. 2008. A Study of Potential Effectiveness of Carbon Dioxide Reducing Vehicle Technologies. Report 420r80040a. Revised June.

EPA. 2009. Updated cost estimates from those in U.S. EPA, 2008. Committee e-mail communications with EPA, May 27 and May 28.

Hadler, J., F. Rudolph, R. Dorenkamp, H. Stehr, T. Düsterdiek, J. Hilzendeger, D. Mannigel, S. Kranzusch, B. Veldten, M. Kösters, and A. Specht. 2008. Volkswagen's new 2.0 L TDI engine fulfills the most stringent emission standards, 29th Vienna Motor Symposium.

Iwabuchi, Y., K. Kawai, T. Shoji, and Y. Takeda. 1999. Trial of new concept diesel combustion system—Premixed compression-ignition combustion. SAE Paper 1999-01-0185. SAE International, Warrendale, Pa.

Joergl, Volker, P. Keller, O. Weber, K. Mueller-Haas, and R. Konieczny. 2008. Influence of pre-turbo catalyst design on diesel engine performance, emissions, and fuel economy. SAE Paper 2008-01-0071. SAE International, Warrendale, Pa.

Kanda, T., T. Hakozaki, T. Uchimoto, J. Hatano, N. Kitayama, and H. Sono. 2005. PCCI operation with early injection of conventional diesel fuel. SAE Paper 2005-01-0378. SAE International, Warrendale, Pa.

Keller, P.S., V. Joergl, O. Weber, and R. Czarnowski. 2008. Enabling components for future clean diesel engines. SAE Paper 2008-01-1530. SAE International, Warrendale, Pa.

Martec Group, Inc. 2008. Variable Costs of Fuel Economy Technologies. Prepared for Alliance of Automobile Manufacturers, June 1; amended September 26 and December 10.

Mattes, Wolfgang, Peter Raschl, and Nikolai Schubert. 2008. Tailored $DeNO_x$ concepts for high-performance diesel engines. Second $MinNO_x$ Conference, June 19-20, Berlin.

Müller, W., et al. 2003. Selective catalytic reduction—Europe's NO_x reduction technology. SAE 2003-01-2304. SAE International, Warrendale, Pa.

Myoshi, N., et al. 1995. Development of new concept three-way catalyst for automotive lean-burn engines. SAE Paper 95809. SAE International, Warrendale, Pa.

NRC (National Research Council). 2002. Effectiveness and Impact of Corporate Average Fuel Economy (CAFE) Standards. National Academy Press, Washington, D.C.

Peckham, John. 2003. How JD Power/LMC calculate 16% light-duty diesel sales share for North America. Diesel Fuel News, October 13.

Pickett, L.M., and D.L. Siebers. 2004. Non-sooting, low flame temperature mixing-controlled DI diesel combustion. SAE Paper 2004-01-1399. SAE International, Warrendale, Pa.

Ryan, T.W., and T.J. Callahan. 1996. Homogeneous charge compression ignition of diesel fuel. SAE Paper 961160. SAE International, Warrendale, Pa.

Styles, D., J. Giuliano, J. Hoard, S. Sluder, J. Storey, S. Lewis, and M. Lance. 2008. Identification and control of factors that affect EGR cooler fouling. 14th Diesel Engine-Efficiency and Emissions Research Conference, Dearborn, Mich.

Tilgner, Ingo-C., T. Boger, C. Jaskula, Z.G. Pamio, H. Lörch, and S. Gomm. 2008. A new diesel particulate filter material for passenger cars: Cordierite diesel particulate filters for the new Audi A4 V6 TDI, 17. Aachener Kolloquium Fahrzeug- und Motorentechnik, p. 325.

ANNEX

Table 5.A.1 shows the data used in Figure 5.4 for the percentage reduction of fuel consumption in 2009 European vehicle platforms offered with both SI gasoline engines and CI diesel engines in configurations that provide virtually equal performance (i.e., 0 to 100 km/h acceleration times within 5 percent between SI and CI).

TABLE 5.A.1 Data Used in Figure 5.4

Vehicle	% FC Reduction
Audi A3	30.88
BMW 520	25.00
Dodge Avenger	20.51
Ford Fiesta	26.32
Ford Galaxy	36.73
Honda Civic	21.21
Honda CR-V	18.52
Jaguar XF	29.25
Mercedes E230	31.18
Mercedes S350	17.65
Toyota Yaris	25.00
Toyota RAV4	24.42
VW Jetta	28.38
Peugeot 308	28.79
Renault Laguna	34.62
Audi A8	21.30
Audi Q7	18.38
Audi A6	18.18
Mercedes Viano	23.53
AVERAGE	25.25

6

Hybrid Power Trains

INTRODUCTION

Hybrid vehicles achieve reduced fuel consumption by incorporating in the drive train, in addition to an internal combustion (IC) engine, both an energy storage device and a means of converting the stored energy into mechanical motion. Some hybrids are also able to convert mechanical motion into stored energy. In its most general sense, the storage device can be a battery, flywheel, compressible fluid, elastomer, or ultra capacitor. The means of converting energy between storage and mechanical motion is through the use of one or more motors/generators (e.g., electric, pneumatic, hydraulic). In motor mode, these devices convert stored energy into mechanical motion to propel the vehicle, and in generator mode, these devices convert vehicle motion into stored energy by providing part of the vehicle braking function (regeneration). Similarly, a fuel cell vehicle is also a hybrid in which the internal combustion engine is replaced by the fuel cell, but this system will likely need supplemental energy storage to meet peak power demands and to allow the fuel cell to be sized for the average power requirement.

In this chapter, hybrid vehicle designs employing an internal combustion engine and battery-energy storage are considered. Battery electric and fuel cell vehicles (BEVs and FCVs) are also briefly discussed as other alternative power trains.

Hybrid electric vehicles incorporate a battery, an electric motor, and an internal combustion engine in the drive train. In its most effective implementation this configuration permits the IC engine to shut down when the vehicle is decelerating and is stopped, permits braking energy to be recovered, and permits the IC engine to be downsized and operated at more efficient operating points. It should be emphasized that the benefits of hybrids are highly dependent on the drive cycle used to measure fuel consumption. For example, a design featuring only idle-stop operation, which shuts off the internal combustion engine when the vehicle is stopped, will demonstrate a large improvement on the city cycle portion of the Federal Test Procedure (FTP), where stop-start behaviors are simulated, but virtually no improvement on the highway cycle.

In addition to the introduction of an electric motor, hybrid designs may include the functions of idle-stop and regenerative braking, and the IC engine is frequently downsized from that in its equivalent conventional vehicle. As shown in Table 6.A.1 in the annex at the end of this chapter, for a hybrid vehicle, these operational and physical changes alone or in combination can result in an increase in fuel economy (mpg) of between 11 and 100 percent or a decrease in fuel consumption (gallons per 100 miles driven) of between 10 and 50 percent, depending on the vehicle class, as is discussed below in this chapter. Hybrid vehicles are the fastest-growing segment of the light-duty vehicle market, although they still make up less than 3 percent of the new car market in the United States.

HYBRID POWER TRAIN SYSTEMS

As stated above, hybrid vehicles are defined as having an internal combustion engine and one or more electric machines that in some combination can provide tractive force to propel the vehicle. An exception to this definition is the simple idle-stop design, which provides no electrically derived tractive force. Depending on the architectural configuration of the motors, generators, and engine, hybrid designs fall into three classes—parallel, series, and mixed series/parallel. The third design is commonly known as power split architecture. Schematics of these architectures are shown in Figures 6.1, 6.2, and 6.3. Within each class there are variations of implementation. Broadly defined, the series hybrid uses the internal combustion engine for the sole purpose of driving a generator to charge the battery and/or powering an electric drive motor. The electric motor provides all the tractive force. Energy flows from the IC engine through the generator and battery to the motor. In the parallel and mixed series/parallel designs, the IC engine not only charges the battery but also is mechanically connected to the wheels and, along with the electric motor, provides tractive power.

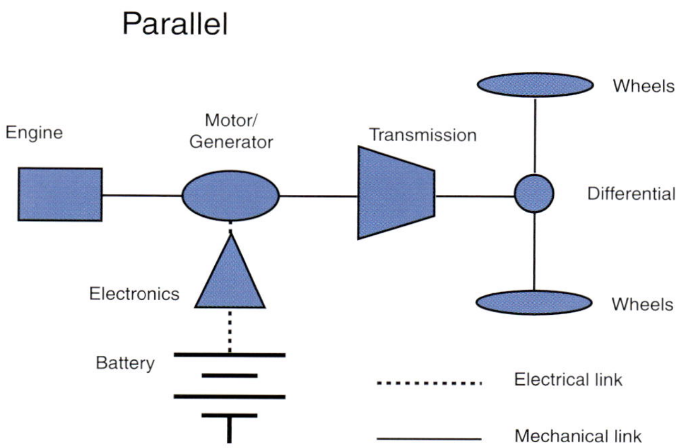

FIGURE 6.1 Schematic of parallel hybrid power train configuration.

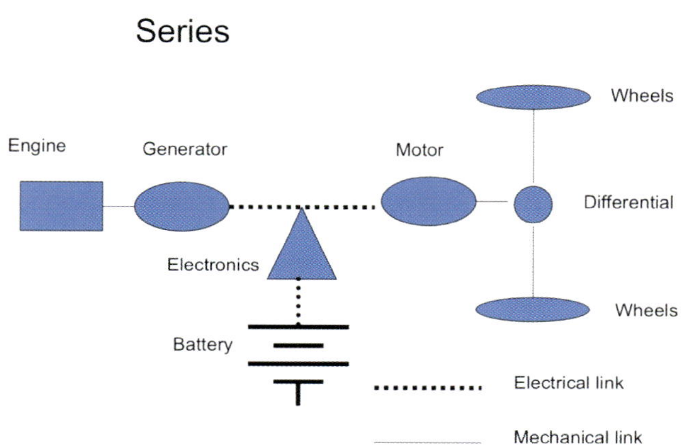

FIGURE 6.2 Schematic of series hybrid power train configuration.

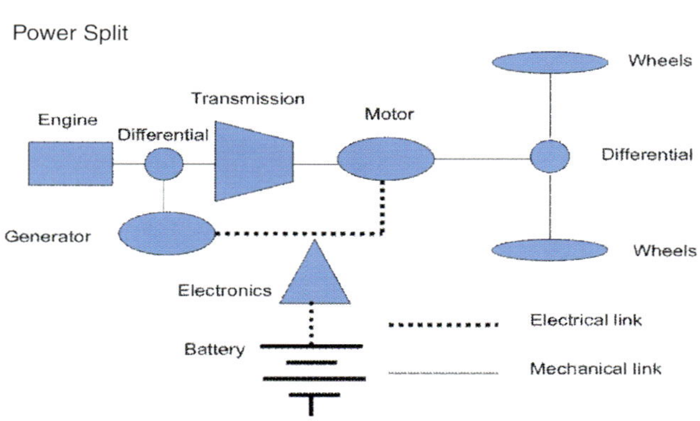

FIGURE 6.3 Schematic of power-split hybrid power train configuration.

Hybrid vehicles are further differentiated by the relative sizes of the IC engine, battery, and motor. Some of the more common variants of these broad classes are described in the following paragraphs. In all cases an economically and functionally significant component of the system is the power electronic subsystem necessary to control the electrical part of the drive train.

The hybridization of diesel (compression ignition; CI) vehicles is expected to have somewhat lower efficiency benefits than hybridization of gasoline vehicles, in part because conventional CI vehicles already exhibit lower fuel consumption than comparable gasoline vehicles. Further, CI vehicles also have very low fuel consumption at idle, making the benefits of idle-stop less attractive. Conventional CI power trains are more expensive than their gasoline counterparts (see Tables 5.4, 5.5, and 5.6), which, when added to the cost of hybridization, makes a CI hybrid power train very expensive for the additional fuel consumption reductions provided over and above just moving to a hybrid or CI power train alone. As a result, it is unlikely that original equipment manufacturers (OEMs) will offer a wide array of CI hybrids. The most likely levels of CI hybridization will be idle-stop and, perhaps, some mild hybrids. Idle-stop will not provide much fuel consumption reduction on the city driving portion of the FTP test cycle, upon which the judgments in this report are based. However, OEMs may still offer such technologies since they provide in-use fuel consumption reductions. In Europe, a number of new diesel hybrid vehicles have been announced for production in 2010 or 2011, especially for larger and heavier vehicles (e.g., Land Rover).

There are numerous hybrid vehicles now in production, and the committee believes it is more representative to quote actual data rather than analyze the effectiveness of each design to estimate fuel consumption benefits. This is preferable to having the committee and its consultants estimate fuel consumption benefits through simulations. It is assumed that the production vehicles are designed to meet customer expectations, including acceleration, passenger space, and adequate trunk space. The average fuel consumption of production hybrid HEVs was determined from fuel economy data supplied by Oak Ridge National Laboratory and included as Table 6.A.1 in the annex at the end of this chapter.

Belt-Driven Alternator/Starter

In the belt-driven alternator/starter (BAS) design, sometimes known as a micro or mild hybrid, the starter and generator of a conventional vehicle are replaced by a single belt- or chain-driven larger machine, capable of both starting the engine and generating electric power. In some BAS designs, in addition to the new belt-driven starter generator, the original geared-to-flywheel starter is retained for cold starts. Fuel consumption is reduced by turning off and decoupling the engine at idle and during deceleration. In some designs, particularly those that have replaced the belt with a chain for increased torque transmission, both electric vehicle launch and some degree of braking energy regeneration are possible.

This mode of operation is known as idle-stop, and while not technically qualifying as a hybrid since the motor/generator provides no or little tractive power, it is included in this chapter for completeness. Idle-stop designs reduce fuel consumption by up to 6 percent in urban driving with SI engines (Ricardo, Inc., 2008). For SI engines having variable valve timing to reduce inlet throttling loss the benefit may be less than 6 percent. For CI engines, the benefit of idle-stop drops to about 1 percent because CI engines are more efficient at idle due to their lack of inlet throttling.

The BAS design is not quite as simple as it first appears. Maintaining hydraulic pressure in the automatic transmission is necessary for smooth and rapid restart, and safety issues related to unexpected restart must be considered. The company ZF has designed a transmission that provides a means of maintaining hydraulic pressure using a "hydraulic impulse storage device" that appears to address the transmission problem (Transmission Technology International, 2008), which is also addressed in existing designs by an electrically driven hydraulic pump.

Full Hybrid

The full hybrid (HEV) has sufficient electrical energy storage and a powerful enough electric motor to provide significant electrical assist to the IC engine during acceleration and regeneration during braking. There are several architectural approaches to achieving a full hybrid, the three in current production being the integrated starter/generator (ISG) or integrated motor assist (IMA), the power split, and the two-mode. These are all parallel or power split designs. The HEV may also provide a limited electric-only range if the battery capacity and motor size are sufficient.

The ratio of electric to mechanical power provided for propulsion of an HEV varies with driving conditions and the state of charge of the battery. This operational feature is accomplished with sophisticated computer controls. Commercially available HEVs such as the Toyota Prius, Honda Civic, Nissan Altima, or Ford Escape can support a limited all-electric range at limited speeds. In these vehicles the battery is operated in a charge-sustaining (CS) mode; that is, the state of charge (SOC) of the battery is allowed to vary over a very narrow range, typically 15 to 20 percent, to ensure long battery life. The IC engine operates over a narrow speed/load range to improve efficiency, and regeneration is employed to recover braking energy. According to Toyota, as shown in Figure 6.4, the contributions of stop-start, regenerative braking, and engine modifications to fuel consumption improvements are approximately 5, 10, and 30 percent, respectively.

ISG/IMA Hybrid

In the ISG/IMA design, the starter and generator are replaced by a larger electrical machine connecting the engine and transmission. These vehicles generally use a larger

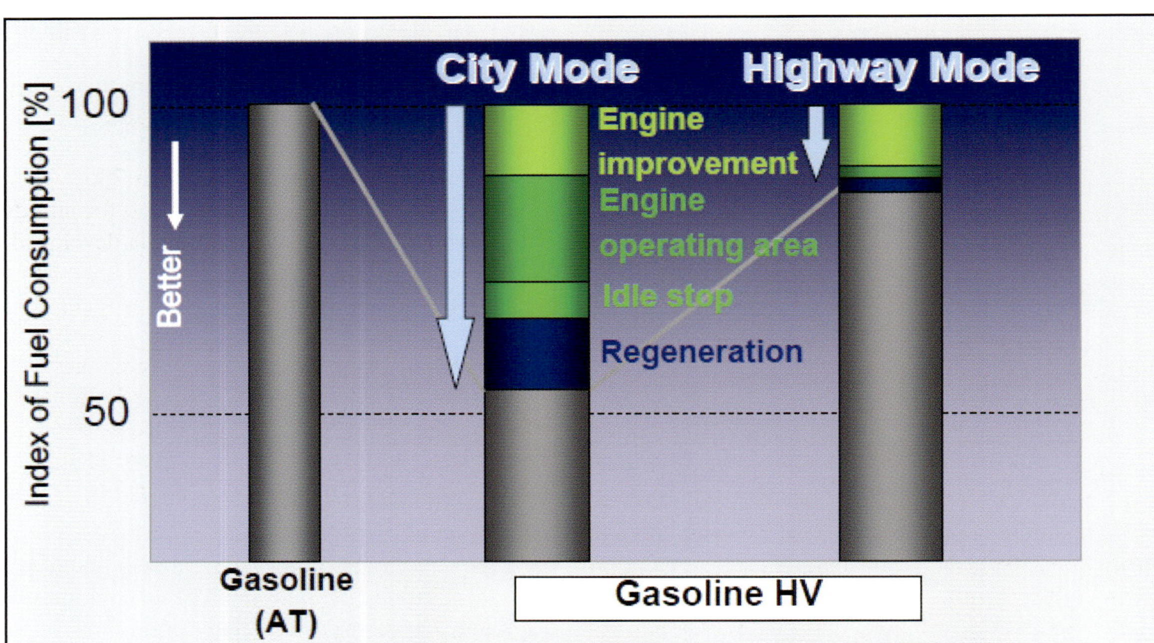

FIGURE 6.4 Individual technology contributions to fuel consumption in hybrid electric vehicles. SOURCE: Fushiki and Wimmer (2007). Reprinted with permission.

battery and a higher voltage (e.g., 140 V) than the BAS. Additionally, the motor/generator and battery are powerful enough to provide electrical launch from a stop and the ability to support some degree of electric-only travel. In its simplest form the ISG is mechanically fixed to the IC engine crankshaft, but in some designs a second clutch isolates the engine and the electrical machine to enable larger regeneration of braking energy (Dan Hancock, General Motors, personal communication, November 30, 2007). When incorporating an effective regenerative braking system, the ISG hybrid achieves a fuel consumption reduction of 34 percent in the combined driving cycle, as demonstrated by the Honda Civic. A part of the improved fuel consumption comes from vehicle modifications, including the use of a smaller, more efficient SI engine.

Power-Split Hybrid

The power-split hybrid design, typified by the Toyota Prius, the Ford Escape, and the Nissan Altima, incorporates a differential gear set that connects together the IC engine, an electrical generator, and the drive shaft. The drive shaft is also connected to an electric motor. This mechanical configuration incorporating the addition of a generator provides the flexibility of several operational modes. In particular the wheels can be driven by both the IC engine and the electric motor, with the motor's power coming from the generator, not the battery. The car is thus driven in both series and parallel modes simultaneously, which is not a possible mode for the ISG design. This operational mode allows the IC engine operation to be optimized for maximum reduction in fuel consumption. The vehicles that use this power split design show a range of fuel consumption reduction from 10 to 50 percent. The low end of this range is the Toyota Lexus, the design of which is optimized for performance, not low fuel consumption. In Chapter 9, where the committee estimates fuel consumption benefits for vehicle classes, the Lexus is not used in the range of benefits for the power split design. This gives the fuel consumption benefits from the power split design a range of 24 to 50 percent.

General Motors (GM) is working with BMW and Chrysler on a different split hybrid architecture that uses the so-called two-mode system (Grewe et al., 2007). This also splits the power flow from the engine but uses more clutches and gears to match the load to the drive and minimize electrical losses. The claim is that by using multiple gears the drive is more efficient in real-world driving situations and reduces fuel consumption when towing a trailer or driving at high speed. Toyota is using a similar approach with one or two gears in its latest hybrid systems. The fuel consumption reduction for the two-mode power split design, characterized by the Chevrolet Tahoe and Saturn Vue, ranges from 25 to 29 percent. However, the committee thinks that other implementations of the two-mode system could provide a maximum fuel consumption benefit of about 45 percent.

Series Hybrid

The series HEV is configured with the engine driving a generator providing electric power to charge the battery. The wheels are driven by an electric motor powered from the battery. The only function of the IC engine is to charge the battery while driving. Because there is no mechanical connection between the IC engine and the wheels, the motor and the battery must be sized for the vehicle's full torque and power requirements. The advantages of this configuration are that a smaller engine can be used since it is not required to provide the power needed for acceleration, and the engine can be optimized with respect to fuel consumption. At present the only OEM planning a series hybrid is GM, which is proposing it as a plug-in hybrid electric vehicle (PHEV).

Plug-In Hybrid

The principal difference between the previously described HEV variants and the PHEV is that the latter is fitted with a larger battery that can be charged from the electric utility grid ("plugged in") and that operates in a charge-depleting mode; that is, the state of charge of the battery is allowed to vary over a much larger range, 50 percent being typically proposed. The significant fuel consumption benefit is obtained during urban driving when the vehicle can be driven on electric power only. Once the all-electric range has been achieved and the battery discharged to its lowest allowable state of charge, the vehicle is operated in the charge-sustaining mode and differs little from the HEV. A small industry has developed around the conversion of the Prius power-split HEVs to PHEVs by supplementing the battery and modifying the control electronics.

PHEVs require a much larger battery than other hybrids (4 to 24 kWh)[1] depending on the desired electric-only range. There has been much activity related to PHEVs since the committee inaugurated its work in 2007. The General Motors Volt mentioned above is planned for introduction in 2010 provided that a suitable battery is developed (Tate et al., 2009). The Volt currently is expected to be launched late in 2010 as a 2011 model. Toyota has also announced plans for a plug-in hybrid for 2011, although it will be built on a Prius platform using its power split architecture (Fushiki and Wimmer, 2007). In addition to the Volt and the Prius, the Volkswagen Golf PHEV is expected in 2010 and Ford's Escape SUV PHEV is due out to the general public in 2012. A PHEV in China went on sale to the public in China early in 2010.

While the micro and ISG hybrids offer some improvement in fuel consumption for a relatively modest cost, it is

[1]The Energy Independence and Security Act of 2007 defines a plug-in hybrid as a light-, medium-, or heavy-duty vehicle that draws motive power from a battery with a capacity of at least 4 kilowatt-hours and can be recharged from an external source of electricity.

the power-split HEV and PHEV architectures that promise a significant improvement. The PHEV also offers the long-term potential for displacing fossil fuels with other primary energy sources such as nuclear or renewable sources of electricity, depending on the fuel source of the electric grid from which the PHEV draws electricity.

Battery Electric Vehicles

The prospect for widespread introduction of full-performance all-electric vehicles depends on significant advancements of the battery technologies discussed above, and the commercial viability of these vehicles depends on a battery cost breakthrough. Advances in electric motors, power electronics, and batteries for automotive applications, which have resulted from the development and production of hybrid vehicles, have renewed interest in the development of battery electric vehicles. However, the cost, low energy density, and required charging time of batteries will continue to constrain the introduction of BEVs. The high low-speed torque performance of electric motors gives the BEV a potential acceleration advantage over conventional internal combustion engine-powered vehicles, and this can be an attractive feature for some customers

A review of zero-emission vehicle technology commissioned by the California Air Resources Board (CARB) concluded that commercialization (tens of thousands of vehicles) of full-performance battery electric vehicles would not occur before 2015 and that mass production (hundreds of thousands of vehicles) would not occur before 2030 (Kalhammer et al., 2007). These projections were based on the continued development of lithium-ion (Li-ion) battery technology leading to reduced cost, higher energy densities, and reduced charging times, all of which allow greater range. They pointed to a possible role for a limited range, city electric vehicle (CEV), which could meet the requirements of a majority of household trips. However, recent BEV introductions suggest that progress in the technology and acceptance of Li-ion batteries may be more rapid than the CARB study concluded.

Early commercial application of Li-ion battery technology to vehicles includes the Tesla Roadster, a high-performance sports car. This vehicle, of which about 1,000 have been sold, has a fuel consumption of 0.74 gal/100 miles (energy equivalent basis, EPA combined city/highway).[2] The manufacturer claims a range of 244 miles (also EPA combined city/highway) and a useful battery life of more than 100,000 miles.[3] The base price of $128,000 indicates the continuing problem of battery cost when used in near full-performance vehicles. Tesla has announced that it will produce and sell, at about half the price of the Roadster, a five-passenger BEV, the Tesla S, with a range of 160, 230, or 300 miles, depending on optional battery size.[4] Nissan has also announced production of its Leaf EV, a five-passenger car with a range of 100 miles.[5] This vehicle has a Li-ion battery with a total storage capacity of 24 kWh.

Within the horizon of this study, the most likely future for large numbers of battery electric vehicles in the United States is in the limited-range, small-vehicle market. Range extended electric vehicles (hybrids and PHEVs) are more likely to satisfy the electricity-fueled full-performance—market, from both cost and technological considerations, over the next 15 years.

BATTERY TECHNOLOGY

In spite of the significant progress that battery technology has experienced in the last 20 years, the battery is still the most challenging technology in the design of hybrid vehicles. Figure 6.5 illustrates the dramatic difference between the energy densities of today's commercial batteries and gasoline, diesel fuel, ethanol, compressed natural gas, and hydrogen. At the time of this report, all production hybrid vehicles used batteries employing nickel-metal-hydride (NiMH) chemistry. It is anticipated that the NiMH battery will be replaced by Li-ion batteries in the near future. The acceptability of today's hybrid vehicles has been shown to be strongly dependent on the price of gasoline, as evidenced by the rapid growth of hybrid sales in 2008, when gasoline prices were high, and the fact that hybrid sales dropped dramatically in early 2009 when prices returned to lower values. The key to improving the competitive position of hybrid vehicles of the HEV and PHEV types is the commercial development of batteries with parameters that are substantially better than those of today's batteries, leading to reduced cost and size. The required parametric improvements are as follows:

- Higher cycle life at increased SOC variation,
- Higher energy density,
- Higher power density, and
- Lower cost.

Figure 6.6 shows the desirable characteristics of batteries suitable for the HEV, the PHEV, and the all-electric (EV or BEV) vehicles. The HEV uses electric propulsion primarily as an assist to the IC engine, thus requiring a battery with a high power capability but relatively little energy capacity, i.e., a high power to energy (P/E) ratio. To preserve battery life and maintain the capacity to recover charge through regenerative braking, the battery is cycled over a relatively small state of charge. This mode of operation is known as charge sustaining (CS). The PHEV is expected to provide

[2] California Air Resources Board (2009), available at http://www.driveclean.ca.gov.

[3] Tesla Motors (2009), available at http://www.teslamotors.com/display_data/teslaroadster_specsheet.pdf; IEEE Vehicular Technology, March 2010.

[4] See http://news.cnet.com/tesla-motors-ceo-model-s-is-cheaper-than-it-looks/.

[5] See http://www.nissanusa.com/leaf-electric-car/tour.jsp#/details.

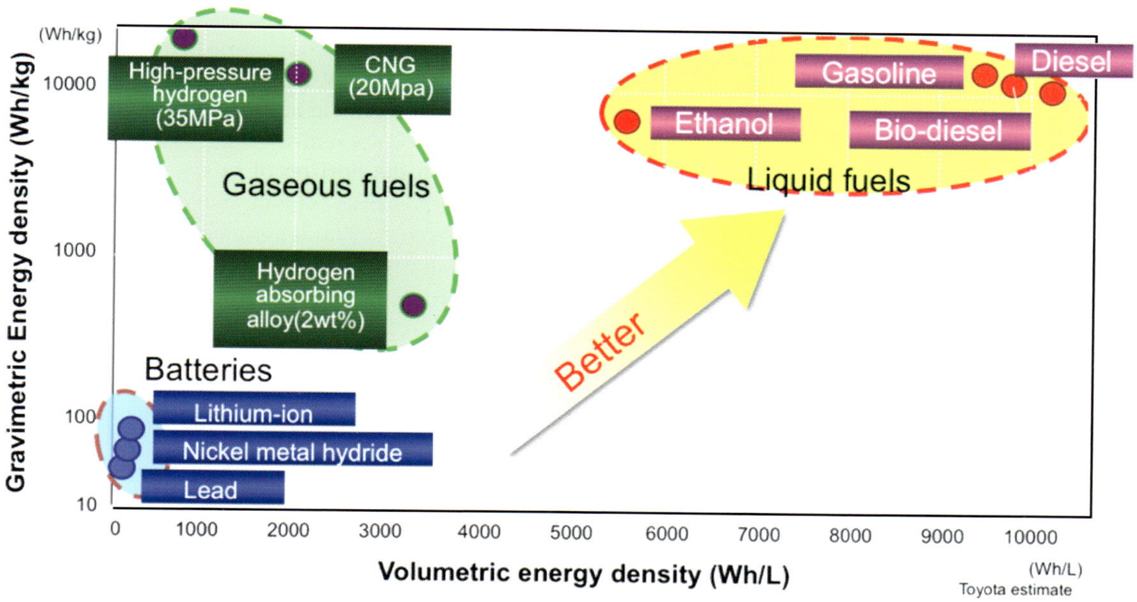

FIGURE 6.5 Volumetric and gravimetric energy densities of different energy storage mechanisms. SOURCE: Fushiki and Wimmer (2007). Reprinted with permission.

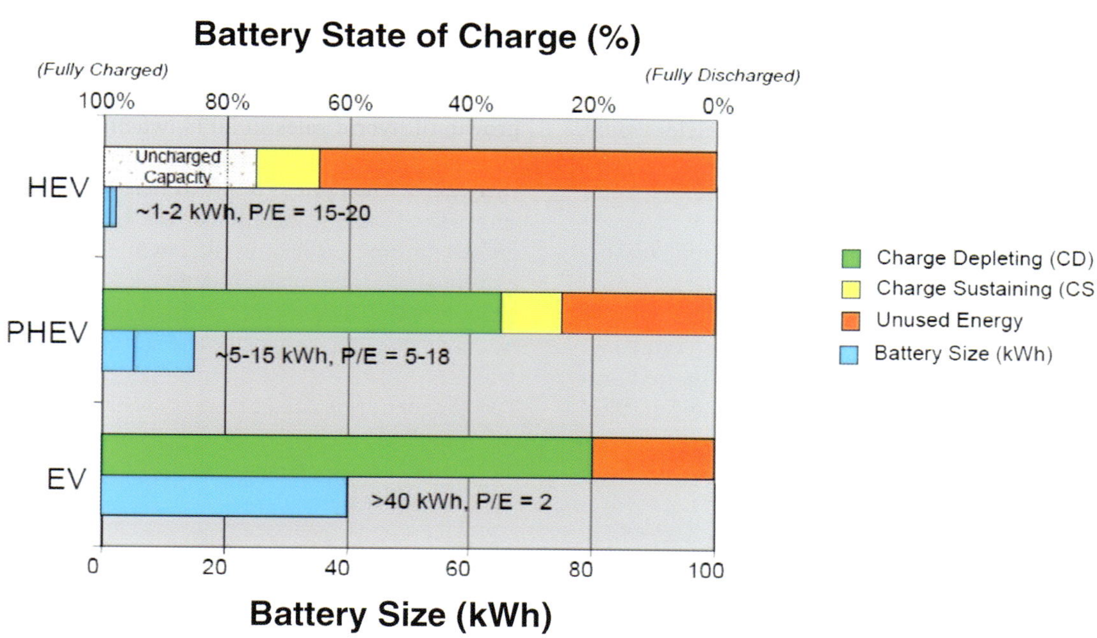

FIGURE 6.6 Energy capacity, state-of-charge variation, and relative power density to energy density ratios for batteries applicable to full-hybrid (HEV), plug-in hybrid (PHEV), and all-electric (EV) vehicles. The units of P/E are kW/kWh. SOURCE: Amine (2007).

some degree of electric-only range. Its battery must therefore contain sufficient energy to provide this range. The battery may be allowed to expend all of its stored energy to achieve this range goal, in which case the battery is said to be operated in the charge-depleting (CD) mode. The power requirement of this battery is not much different from that of the HEV battery, but because of the higher energy requirement, the P/E ratio is smaller. The BEV requires an even higher energy capacity battery than the PHEV, the value depending on the desired driving range. Since the BEV has no IC engine, its battery cannot be charged during driving, and therefore it cannot operate in a CS mode. In all cases the SOC variation is limited to a specified range by the vehicle manufacturer to preserve battery cycle life. Figure 6.6 shows typical ranges for the HEV, PHEV, and BEV. Thus the usable energy is less than the battery rated (or "nameplate") capacity.

Despite substantial improvements in the packaging and performance of lead-acid batteries, their energy and power densities are still considerably inferior to those of NiMH. And while other chemistries, like Li-air, have theoretically better performance than Li-ion, their development is not at a stage where one could envision them in practical automotive applications within the timeline of this study. Therefore the committee considers only NiMH and Li-ion as chemistries of interest here.

NiMH Batteries

The highest-performance battery currently available in commercially significant quantities for HEVs and PHEVs uses NiMH chemistry. Despite significant improvements in lifetime and packaging, these batteries are still expensive, heavy, and in application are restricted to a SOC range of about 20 percent to preserve battery cycle life. Because of their relatively poor charge/discharge efficiency, special consideration must be given to their thermal management. The NiMH chemistry also exhibits a high rate of self-discharge.

The most technically advanced NiMH battery used in the Toyota Prius has a weight of 45 kg and an energy capacity of 1.31 kWh. This results in a usable energy of approximately 0.262 kWh when applied with a SOC variation of 20 percent.

Li-Ion Batteries

The most promising battery technologies are those employing various Li-ion chemistries. Characteristics of the more common lithium-based cell compositions are shown in Table 6.1. The column heads denote the common abbreviation for the different chemistries: NCA (nickel-cobalt-aluminum), LFP (lithium-iron-phosphate), MS (manganese-spinel), MNS (manganese-nickel-spinel), and MN (manganese-nickel). The first entry gives the detailed composition of the anode and cathode materials, with the positive (cathode) material shown first. The second entry gives the gravimetric energy density of the chemistry in milliampere-hours/gram (mAh/g), the third entry shows the open-circuit terminal voltage when the cell is 50 percent depleted (50 percent state of charge), and the fourth entry gives the area specific impedance (ASI) as measured during a 10-second pulse at the 5C rate, which is indicative of the battery's ability to provide power necessary for acceleration. The relative safety of the different chemistries is given in the fifth entry. The safety of using Li-ion batteries has received considerable attention since the 2006 recall of Li-ion batteries used in laptops. In some of the chemistries, particularly those using a cobalt (Co)-based cathode, failure can occur due to overheating or separator failure. This problem is well known, and safety is a characterizing parameter common to all the Li systems. Some manufacturers believe they can solve the safety problem through careful monitoring and charge control. Relative cost among the different Li chemistries is shown in the seventh entry, although at this time the absolute cost of all is considerably higher than the cost for NiMH. The last entry in Table 6.1 indicates the state of the technology. Pilot scale indicates that cells are currently being manufactured in sufficient quantities for testing in vehicle fleets of limited size. Development means that the chemistry is well controlled, but the production of practical cells is anticipated and under development. Research indicates just that—the chemistry is still a subject of research, and the production of cells using the chemistry has not been demonstrated to an extent sufficient to anticipate their use.

TABLE 6.1 Comparative Characteristics and Maturity of Lithium-Ion Battery Chemistries

	Battery System				
	NCA-Graphite	LFP-Graphite	MS-TiO	MNS-TiO	MN-Graphite
Electrodes					
Positive	$LiNi_{0.8}Co_{0.15}Al_{0.05}O_2$	$LiFePO_4$	$LiMn_2O_4$	$LiMn_{1.5}Ni_{0.5}O_4$	$Li_{1.2}Mn_{0.6}Ni_{0.2}O_2$
Negative	Graphite	Graphite	$Li_4Ti_5O_{12}$	$Li_4Ti_5O_{12}$	Graphite
Capacity, mAh/g					
Positive	155	162	100	130	275
Negative	290	290	170	170	290
Voltage, 50% state of charge	3.6	3.35	2.52	3.14	3.9
ASI for 10-s,	25	25	9.2	100	25
Safety	Fair	Good	Excellent	Excellent	Excellent
Life potential	Good	Good	Excellent	Unknown	Unknown
Cost	Moderate	Moderate	Low	Moderate	Moderate
Status	Pilot scale	Pilot scale	Develop.	Research	Research

NOTE: NCA, Ni-Co-Al; LFP, Li-Fe-PO_4; MS-TiO, Mn(Spinel)-Ti-O; MNS-TiO, Mn-Ni(Spinel)-Ti-O; MN-Graphite, Mn-Ni-Graphite.

The relative gravimetric energy densities of Li-ion, NiMH, and Pb-acid are approximately 4, 2, and 1, respectively. An additional advantage of the Li systems is their high cell potential, approximately 3 times that of NiMH. This means that 66 percent fewer Li-ion cells are required to achieve a given battery voltage. The ecologically benign materials in the Li-ion systems are also an advantage. A disadvantage of Li-ion cells is that the requirement for cleanliness in the manufacturing environment is considerably more severe than for NiMH cells (Zempachi Ogumi, Kyoto University, personal communication, December 8, 2008). This increases manufacturing costs. Another critical issue is how the performance of Li-ion batteries is impacted by low and high temperatures (Amine, 2007; Reilly, 2007; Andermann, 2007).

The first three columns in Table 6.1—NCA-Graphite, LFP-Graphite, and MS-TiO—represent the most promising Li-ion systems currently under development. The NCA-graphite chemistry is used by JCS/SAFT in its VL41M module that has undergone dynamometer testing in a Toyota Prius at Argonne National Laboratories (ANL) (Rousseau et al., 2007). The lithium-iron phosphate (LFP) system is currently receiving a great deal of attention because of its stability, potentially lower material costs, and its application in power tools. Its development is being aggressively pursued by A123 and Enerdel. The manganese-spinel-lithium-titanate system (MS-TiO) is the safest of any being studied because of the mechanical stability of the spinel structure, but its cell voltage is considerably lower than those of the NCA and LFP systems. However, it has the highest charge/discharge efficiency, and it is predicted to be the lowest-cost system.

To put in perspective the merits of the Li-ion battery relative to NiMH, consider the requirements for a 20-mile all-electric range PHEV. According to an ANL study (Nelson et al., 2007), which assumed a 100 to 10 percent SOC range, the required battery capacity for its assumed vehicle is 6.7 kWh. For an MS-TiO battery the calculated weight is 100 kg. If an NiMH battery were used, with a SOC range of 20 to 80 percent and a gravimetric energy density one-half that of the MS-TiO system, the committee estimates that it would require a capacity of 10.35 kWh and weigh 300 kg.

The needs of HEVs and PHEVs are quite distinct, as shown in Figure 6.6. HEVs need high power density and long cycle life over a very small excursion of the SOC. For example the Prius battery has a nominal rating of 1.3 kWh but it uses only 260 Wh in +/−10 percent excursions around 50 percent SOC. On the other hand, the larger energy requirement of the PHEV argues for a battery with a higher energy rating and the capability of deeper cycling. The Volt, the PHEV being developed by GM, uses a 16-kWh battery to meet its advertised all-electric range of 40 miles. This is a substantial challenge to achieve at acceptable weight, volume, and cost. The Li-ion chemistry comes closest to meeting it, given the present state of battery development. It should be noted that the Volt is designed to use only 8 kWh by operating from 80 percent to 30 percent SOC.

POWER ELECTRONICS

The term *power electronics* refers to the semiconductor switches and their associated circuitry that are used to control the power supplied to the electrical machines or to charge the battery in an HEV or PHEV. For purposes of driving electric motors these circuits function as an inverter, changing the battery direct voltage into an alternating voltage of controlled amplitude and frequency. For charging the propulsion battery they function as a controlled rectifier, changing the ac voltage of the machine to the dc value required by the battery. The direction of power flow is either into or out of the battery, depending on vehicle mode of operation. Plug-in hybrids also require power electronic circuits to convert the ac main voltage to a precise dc voltage to charge the propulsion battery.

Power electronic circuits known as dc/dc converters change the propulsion battery dc voltage to the dc voltage appropriate to charging the accessory battery (i.e., the standard 12 V battery retained to power vehicle accessories). A dc/dc converter may also be used to increase system efficiency by stepping up the propulsion battery voltage before it is supplied to the inverter. The latest Toyota Prius uses such a design.

Both inverter and dc/dc converter technologies are well developed for industrial and other applications. The special problems for hybrid vehicles are cost, cooling, and packaging. Although the ambient environment for automotive electronics is much harsher than that in industrial or commercial applications, the cost in the automotive application is required to be lower. Figure 6.7 illustrates the improvement over a 10-year period in the volumetric power density of the motor drive inverter for Toyota's hybrid product line. The significant improvement after 2005 is due in large measure to the increased switching frequency made possible by the higher-speed motor and higher voltage introduced in 2005. These changes reduce the physical size of magnetic components and improve the utilization of silicon devices. Both these consequences result in improved packaging density.

ROTATING ELECTRICAL MACHINES AND CONTROLLERS

With the possible exception of microhybrids, all vehicles use permanent magnet alternating current motors. Since the battery capacity is the key limitation for hybrid vehicles, electrical machine efficiency is of paramount importance. Most systems employ "buried magnet" rotating machine configurations with expensive rare-earth high-strength magnets. GM and Honda are using flat wire for the armature winding to increase efficiency. Although rectangular conductors are common for large machines, their use in relatively small machines shows the extent to which manufacturers are going to get better efficiency. Rotating machine technologies and designs are well developed, and the automotive applica-

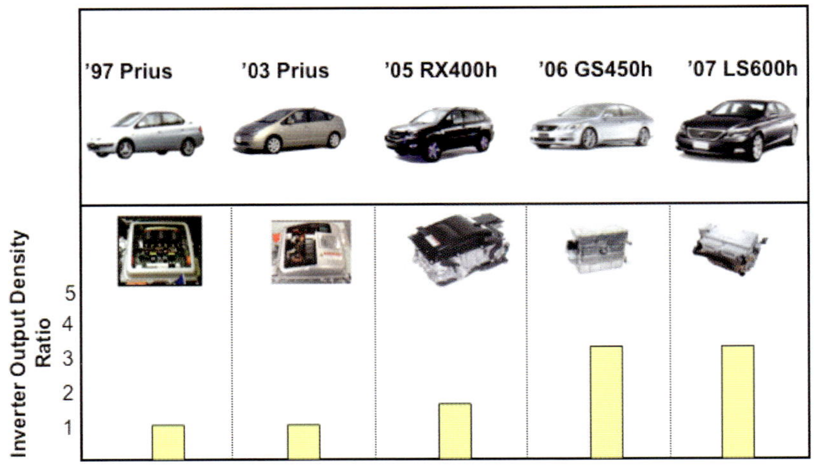

FIGURE 6.7 Evolution of hybrid drive inverter volumetric power density. SOURCE: Fushiki and Wimmer (2007). Figure used with permission of Toyota.

tion challenge is to lower their manufacturing cost. Because rotating machines are such a mature component, the cost of their manufacture in high volumes is driven principally by the cost of materials. Thus their cost is relatively unresponsive to technology developments. Major improvements in volumetric power density can be achieved by increasing the speed of the motor. This volumetric improvement results in materials reduction but generally also in increased losses. High-speed motors also require a gear set to match the mechanical speed required of the drive train. While the design of the motor/inverter system is an optimization problem, no technology breakthroughs that would radically improve the state of the art are foreseen. Figure 6.8 illustrates the improvement in volumetric power density that Toyota has achieved by increasing the speed of the electric motor in its hybrid vehicles.

Computers have been used to control emissions and optimize efficiency of conventional power trains. In addition to engine control, controllers in hybrid vehicles monitor the state of charge of the battery and determine power flows to and from the battery and engine. The control task is more complex for the PHEV where there is a greater opportunity to optimize the tradeoff between electric and IC engine use with respect to fuel consumption. One suggested approach is to have the controller predetermine the propulsion profile from expected route data provided by the driver or an off-board wirelessly connected server. Vehicle computers are powerful enough to handle these tasks, and no technical problems are expected.

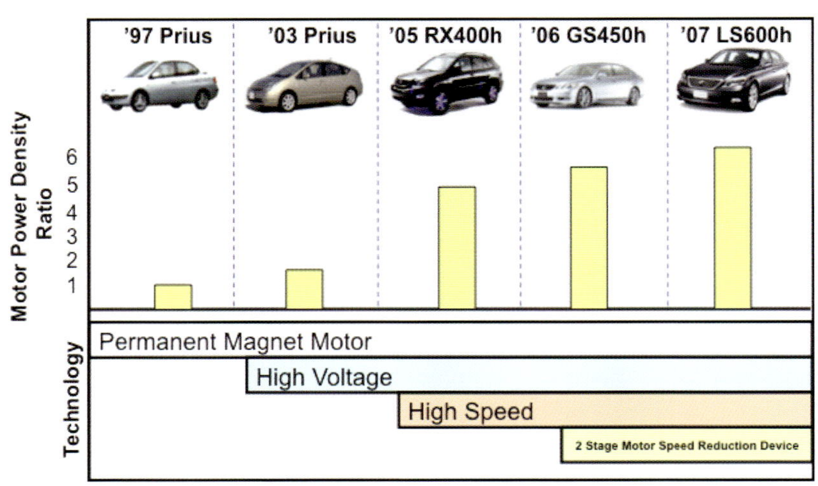

FIGURE 6.8 Evolution of the volumetric power density of electric motors used in Toyota's hybrid vehicles. SOURCE: Fushiki and Wimmer (2007). Figure used with permission of Toyota.

COST ESTIMATES

The objective in determining costs of new technologies is understanding their factory cost. The factory cost is the direct cost to the OEM of replacing existing production technology A by technology B. It is determined as follows:

1. Take the price (B) that a supplier charges the OEM for technology B;
2. Add the engineering cost (C) to the OEM of integrating technology B into a vehicle;
3. Add the cost (D) of any parts that the OEM makes in-house to implement the technology (labor cost plus factory overhead, plus amortization of required new investment); and
4. Subtract the cost (A) of technology A similarly calculated.

The *factory cost* is then $B + C + D - A$.

The cost estimates have been validated by soliciting feedback from a number of U.S. and Japanese OEMs and suppliers. The costs presented here are a consensus that the numbers are "about right." The costs of hybrid technologies vary depending on the degree of hybridization, from a low cost in the case of the BAS design, to a very high cost for a series PHEV. It should be noted that the factory cost definition used here includes engineering costs and other part costs, including labor and overhead, for integrating the technology. Using the studies described in Chapter 3, the committee developed a different markup factor for hybrids that relates the definition of factory cost to RPE. Although different studies use different definitions and allocations for items such as profit, vehicle warranty, corporate overhead, transportation, marketing, and dealer costs, the committee concluded that the factory markup for hybrids should be on the order of 1.33 rather than 1.5 for factory cost to RPE. The committee's justification for using an RPE of 1.33 for hybrids is that the factory cost estimates it developed already include engineering costs and other part costs, including labor and overhead, for integrating the technology. Using a cost multiplier of 1.5 would double count these costs.

As an example of the process, Table 6.2 shows an estimated breakdown of the factory cost of a "mature" Prius—a Prius-type drive that has benefited from the learning curve and has an annual production volume in excess of 100,000 units. The additional components and their estimated OEM costs from the supplier are listed. The committee also lists the cost decrement of items, such as the automatic transmission, that will be removed from the baseline vehicle, a Toyota Corolla in this case. The net cost increase for the mature Prius is then calculated as $3,385.

Next the committee projects costs for 5-year increments to 2025, as shown in Table 6.3. Percentage cost reductions

TABLE 6.2 Factory Cost Estimation Process Applied to a Mature Prius-type Hybrid Vehicle in U.S. Dollars

20 kW	Factory Cost ($B + C + D - A$)
Motor/generator/gears	1,100
Control electronics + dc/dc (1.2 kW)	1,100
Battery (NiMH 21 kW)	1,000
Electrical accessories	100
Electric PS and water pump	200
Automatic transmission	–850
Regenerative brakes	250
Electric A/C	300
Engine downsize	–120
Starter and alternator	–95
High-voltage cables (Martec 500 V)	200
Body/chassis/special components	200
Total	3,385

TABLE 6.3 Projections of the Future Factory Cost of a Mature Prius-type Hybrid in U.S. Dollars

	Factory Cost ($B + C + D - A$)				
20 kW	Cost Reductions (%)	2008	2015	2020	2025
Motor/generator/gears	5	1,100	1,050	990	940
Control electronics + dc/dc (1.2 kW)	15	1,100	940	800	680
Battery (NiMH 21 kW, Li-ion Martec)	15	1,000	850	720	720
Electrical accessories	5	100	90	90	85
Electric PS and water pump	5	200	190	180	170
Automatic transmission	0	–850	–850	–850	–850
Regenerative brakes	5	250	240	230	210
Electric A/C	10	300	270	240	220
Engine downsize	0	–120	–120	–120	–120
Starter and alternator	0	–95	–95	–95	–95
High-voltage cables (Martec 500 V)	10	200	180	160	150
Body/chassis/special components	10	200	180	160	150
Total		3,385	2,925	2,505	2,260

appropriate for each component are used. For example, expected reductions are on the order of 15 percent for each 5-year period for the battery and control electronics, 5 percent for the electrical machines, and no change in cost for the mature components such as engine downsizing, and the alternator.

A similar analysis has been done for the other hybrid classes, and the summary results are shown in Table 6.4. It should be noted that future costs for PHEVs and EVs are highly uncertain due to the uncertainties in future battery chemistries and tradeoffs between power and energy. Li-ion batteries for consumer electronics are a commercial technology, and costs have gone down along the learning curve. However, many OEMs and battery suppliers are expecting large cost reductions for Li-ion batteries with increasing applications in vehicles. Among its provisions related to fuel economy, the Energy Independence and Security Act of 2007 requires periodic assessments by the National Research Council of automobile vehicle fuel economy technologies. Thus, follow-on committees will be responsible for responding to this legislative mandate, including the periodic evaluation of PHEVs, EVs, and other technologies and how these technologies can help meet new fuel economy standards.

TABLE 6.4 Retail Price Estimates for Various Types of Hybrids Projected to 2025 (using an RPE of 1.33)

Vehicle	2009 ($)	2015 ($)	2020 ($)	2025 ($)
Prius-type power split	4,500	3,900	3,300	3,000
BAS/12 V	670	570	490	440
BAS/42 V	1,500	1,200	1,100	1,000
ISG 12 kW/144 V	2,900	2,500	2,100	2,000
Prius-type PHEV 10 (Li-ion battery)	8,800	7,600	6,500	5,900
Series PHEV 40 (Li-ion battery)	13,000	11,000	9,800	8,900
HEV crossover (V6)	6,900	6,000	5,200	4,700
Large SUV/pickup (V8)	8,700	7,500	6,400	5,700

FUEL CONSUMPTION BENEFITS OF HYBRID ARCHITECTURES

As noted earlier, the average fuel consumption of production hybrid HEVs was determined from fuel economy data supplied by Oak Ridge National Laboratory and included as Table 6.A.1 in the annex at the end of this chapter. For several specific models, these data were compared to data from conventional (nonhybrid) vehicles of approximately similar performance and physical specifications, and the results are shown in Table 6.5. As mentioned earlier, a significant contribution to the fuel consumption benefit of hybrid vehicles is due to modifications to the engine, body, and tires. For example, the fuel economy of the Prius is significantly influenced by engine improvements and optimized operating area. The 2007 model-year version of the Saturn Vue hybrid, which used a BAS design, exhibits a 25 percent improvement in fuel economy on the FTP cycle, but approximately half of that improvement is due to vehicle modifications, including a more aggressive torque converter lockup and fuel cutoff during vehicle deceleration (D. Hancock, General Motors, personal communications, November 30, 2007).

The Oak Ridge data did not include information on the Honda Accord, which was discontinued in 2007. The Accord has a motor/generator of 15 kW in motoring mode and a slightly higher 15.5 kW in regenerative mode (J. German, Honda, personal communication, February 28, 2008). The motor generator has high-energy-density magnets in an interior configuration. It also has flat wire windings that provide better packing density compared to round wire. The NiMH battery has 132 cells with a nominal voltage and energy of 144 V and 0.87 kWh, respectively (Iijima, 2006). Honda calls the system an integrated motor assist.

Plug-In Hybrids

The rules for assigning fuel economy ratings to plug-in hybrids are currently being developed by SAE (revision of J 1711). Thus the committee cannot predict at this time what

TABLE 6.5 Comparison of Fuel Economy, Fuel Consumption, Performance, and Physical Specifications of Hybrid and Comparable SI Engine-Powered Vehicles

Architecture	Volume Trunk	EPA Test (mpg, combined)	Fuel Consumption (gal/100 mi)	EPA Test Car Weight	Acceleration (Consumer Reports, mph/sec) 0 to 30	0 to 60	45 to 65	Edmund's MSRP Price
Prius								
Prius/Corolla	1.33	1.64	0.61	1.13	1.06	1.07	1.05	1.36
Prius/Camry	1.07	2.00	0.50	0.87	1.03	1.10	1.03	1.09
Honda Civic								
Civic hybrid/Civic SI	0.83	1.51	0.66	1.00	1.22	1.16	1.22	1.45
Chevy Tahoe 4WD								
Tahoe 4WD Hybrid/Tahoe 4WD SI	N/A	1.53	0.65	1.00	1.15	1.07	0.96	1.30

the official fuel economy rating of a specific PHEV design will be. At the time of this writing only two PHEVs have been announced for production—the GM Volt, which is expected to have a 40-mile range on battery alone, and the Toyota plug-in Prius, which will have a 12-mile all-electric range and the ability to cruise at highway speeds under all electric power.[6] GM has announced that LG Chem of Korea will be supplying the Volt's Li-ion battery.

FUEL CELL VEHICLES

Fuel cell vehicles have the potential to significantly reduce greenhouse gas emissions (depending on how hydrogen is produced) as well as U.S. dependence on imported oil over the long term. However, fuel cell vehicle technologies have technical challenges that are severe enough to convince the committee that it is unlikely such vehicles will be deployed in significant numbers within the time horizon of this study.

A recent report (NRC, 2008) states that under the following set of very optimistic assumptions, 2 million fuel cell vehicles could be part of the U.S. fleet in 2020:

- The technical goals are met and consumers readily accept such vehicles.
- Policy instruments are in place to drive their introduction.
- The necessary hydrogen production, supply, distribution, and fueling infrastructure is present.
- Oil prices are at least $100/barrel by 2020.
- Fuel cell vehicles are competitive on the basis of life-cycle cost.

Although the committee agrees with that study's conclusions under these optimistic assumptions, it believes that achieving them is unlikely. Almost every major OEM has a fuel cell vehicle program, and several have deployed limited fleets of experimental vehicles. These fleets invariably represent limited mission, localized experiments, city buses, or postal vehicles, for example. Through interviews and presentations, the committee can find little evidence that a commercially viable fuel cell light-duty vehicle will be available in significant numbers by 2020. The Japanese auto industry will not decide to pursue a commercial development program until 2015, thus making a 2020 introduction date very difficult. The committee confirmed this target decision date with Japan's NEDO, Japanese academics, and the OEMs themselves. All current fuel cell vehicle research assumes stored hydrogen as the fuel. The monumental difficulty of providing the necessary hydrogen distribution infrastructure is another factor mitigating against the presence of fuel cell vehicles in significant numbers by 2020.

For fuel cells, in spite of hundreds of millions of dollars having been devoted to their development by vehicle builders, equipment suppliers, and government organizations, there remain significant problems requiring technical and economic resolution, including the following:

- Higher cost of fuel cells compared to other energy converters,
- Lack of a hydrogen distribution infrastructure,
- Need for a low carbon source of hydrogen (biomass or water electrolysis using electricity produced with low emissions),
- Need to demonstrate acceptable durability and reliability, and
- Weight and volume of an on-board hydrogen storage tank sized for a range of 300 to 400 miles.

Because of these factors, the committee does not expect wide use of fuel cell vehicles before 2025.

FINDINGS

Finding 6.1: The degree of hybridization can vary from minor stop-start systems with low incremental costs and modest reductions in fuel consumption (i.e., the most basic stop-start systems may have a fuel consumption benefit of up to about 4 percent at an estimated incremental retail price equivalent (RPE) cost of $670 to $1,100) to complete vehicle redesign (e.g., Prius) and downsizing of the SI gasoline engine at a high incremental RPE cost ($3,000 to $9,000) and with significant reductions in fuel consumption. A significant part of the improved fuel consumption of production hybrid vehicles comes from vehicle modifications such as low-rolling-resistance tires, improved aerodynamics, and the use of smaller, more efficient SI engines.

Finding 6.2: In the next 10 to 15 years, improvements in hybrid vehicles will occur primarily as a result of reduced costs for hybrid power train components and improvements in battery performance such as higher power per mass and volume, increased number of lifetime charges, and wider allowable state-of-charge ranges.

Finding 6.3: During the past decade, significant advances have been made in lithium-ion battery technology. When the cost and safety issues associated with Li-ion batteries are resolved, they will replace NiMH batteries in HEVs and PHEVs. A number of different Li-ion chemistries are being studied, and it is not yet clear which ones will prove most beneficial.

Finding 6.4: Given the high level of activity in lithium-ion battery development, plug-in hybrid electric vehicles will be commercially viable and will soon enter at least limited production. However, improving the cost-effectiveness of PHEVs depends on the cost of fuel and whether significant reductions in battery cost are achieved.

[6]See http://www.reuters.com/article/pressRelease/idUS238743+09-Sep-2009+PRN20090909.

Finding 6.5: The practicality of full-performance battery electric vehicles (i.e., with driving range, trunk space, volume, and acceleration comparable to those of internal combustion-powered vehicles) depends on a battery cost breakthrough that the committee does not anticipate within the time horizon considered in this study. However, it is clear that small, limited-range, but otherwise full-performance battery electric vehicles will be marketed within that time frame.

Finding 6.6: Although there has been significant progress in fuel cell technology, it is the committee's opinion that fuel cell vehicles will not represent a significant fraction of on-road light-duty vehicles within the next 15 years.

REFERENCES

Amine, K. 2007. Advanced high power chemistries for HEV applications. Presentation to the National Research Council Committee on the Assessment of Technologies for Improving Light-Duty Vehicle Fuel Economy, November 27, Washington, D.C.

Andermann 2007. Lithium-ion batteries for hybrid electric vehicles: Opportunities and Challenges. Presentation to the National Research Council Committee on the Assessment of Technologies for Improving Light-Duty Vehicle Fuel Economy, October 25, Washington, D.C.

California Air Resources Board. 2009. DriveClean. Available at http://www.driveclean.ca.gov. Accessed June 29, 2009.

Fushiki, S., and B. Wimmer. 2007. Perspectives from Toyota. Presentation to the National Research Council Committee on the Assessment of Technologies for Improving Light-Duty Vehicle Fuel Economy, November 27, Washington, D.C.

Grewe, T.H., B.M. Conlon, and A.G. Holmes. 2007. Defining the General Motors 2-mode hybrid transmission. SAE Paper 2007-01-0273. SAE International, Warrendale, Pa.

Iijima, T. 2006. Development of hybrid system for 2006 compact sedan. SAE Paper 2006-01-1503. SAE International, Warrendale, Pa.

Kalhammer, F.R., B.M. Kopf, D.H. Swan, V.P. Roan, and M.P. Walsh. 2007. Status and Prospects for Zero Emissions Vehicle Technology. Report presented to ARB Independent Expert Panel, April 13, State of California Air Resources Board, Sacramento.

Nelson, P., K. Amine, and H. Yomoto, 2007. Advanced lithium-ion batteries for plug-in hybrid-electric vehicles. Paper presented at 23rd International Electric Vehicle Symposium, December, Anaheim, Calif.

NRC (National Research Council). 2008. Transitions to Alternative Transportation Technologies: A Focus on Hydrogen. The National Academies Press, Washington, D.C.

Reilly, B. 2007. Battery Technologies. Presentation to the National Research Council Committee on the Assessment of Technologies for Improving Light-Duty Vehicle Fuel Economy, October 25, Washington, D.C.

Ricardo, Inc. 2008. A Study of Potential Effectiveness of Carbon Dioxide Reducing Vehicle Technologies. Prepared for the U.S. Environmental Protection Agency. EPA420-R-08-004. Contract No. EP-C-06-003. Work Assignment No. 1-14. Ann Arbor, Mich. Available at http://www.epa.gov/oms/technology/420r08004a.pdf. Accessed June 29, 2009.

Rousseau, A., N. Shidore, R. Carlson, and P. Nelson. 2007. Research on PHEV battery requirements and evaluation of early prototypes. Paper presented at the Advanced Automotive Battery Conference, May 17, Long Beach, Calif.

Tate, E.D., M.O. Harpster, and P.J. Savagian. 2009. The electrification of the automobile: From conventional hybrid, to plug-in hybrids, to extended-range electric vehicles. SAE International 1(April):156-166.

Tesla Motors, Inc. 2009. Tesla Roadster Spec Sheet. Available at http://www.teslamotors.com/display_data/teslaroadster_specsheet.pdf. Accessed June 29, 2009.

Transmission Technology International. 2008. ZF eight-speed hybrid. September, p. 10.

ANNEX

TABLE 6.A.1 Performance of Production Hybrid Vehicles from 2009 CAFE Certification Data

				Volume	EPA Fuel Economy (unadjusted mpg)			EPA Test	Acceleration (Consumer Reports)			Edmund's MSRP
Make	Type	Model	Drive	Trunk	City	Comb.	Hwy	Car Weight	0 to 30 mph, sec.	0 to 60 mph, sec.	45 to 65 mph, sec.	Price
Toyota	Split	Highlander Hybrid	4WD	NA	35	35	35	5000	3.4	8.2	5	$34,700
Toyota		Highlander	4WD	NA	21	25	31	4750	3	8	5.1	$29,050
Toyota	Split	Prius	FWD	16	67	66	65	3250	3.8	10.6	6.2	$22,000
Toyota		Corolla	FWD	12	35	40	49	2875	3.6	9.9	5.9	$16,150
Toyota		Camry	FWD	15	27	33	44	3750	3.7	9.6	6	$20,195
Toyota		Yaris	FWD	13	37	42	49	2625	4.1	11.4	6.9	$13,765
Toyota	Split	Camry Hybrid	FWD	11	44	46	48	4000	3.5	8.5	5.1	$26,150
Toyota		Camry	FWD	15	27	33	44	3750	3.7	9.6	6	$20,195
Toyota		Camry	FWD	15	25	30	40	3875	3.3	7.1	4.4	$24,215
Ford	Split	Escape Hybrid	FWD	NA	45	44	43	4000	NA	NA	NA	$29,645
Ford		Escape	FWD	NA	26	30	39	3625	NA	NA	NA	$21,645
Ford		Escape	FWD	NA	23	27	36	3625	NA	NA	NA	$24,465
Ford	Split	Escape Hybrid	4WD	NA	37	37	37	4250	4.1	10.7	5.8	$31,395
Ford		Escape	4WD	NA	24	28	35	3875	3.3	10	6.4	$23,395
Ford		Escape	4WD	NA	22	26	33	3875	3	7.9	5.2	$26,215
Saturn	Parallel	Aura Hybrid	FWD	16	33	39	48	NA	NA	NA	NA	$26,325
Saturn		Aura	FWD	16	28	34	47	4000	3.4	9.4	6.9	$22,655
Saturn		Aura	FWD	16	21	26	36	4000	2.8	6.6	4.3	$27,250
Saturn		Vue Hybrid	FWD	NA	32	37	45	4000	4.2	10.9	7.3	$28,160
Saturn		Vue	FWD	NA	24	28	37	4000	NA	NA	NA	$23,280
Saturn		Vue	FWD	NA	21	25	33	4250	NA	NA	NA	$26,435
Honda	Parallel	Civic Hybrid	FWD	10	55	59	65	3125	4.4	11.7	7.3	$23,650
Honda		Civic	FWD	12	33	39	51	3125	3.6	10.1	6	$16,305
Nissan	Parallel	Altima Hybrid	FWD	10	47	47	47	3750	3.1	7.6	4.4	$26,650
Nissan		Altima	FWD	15	29	34	43	3500	3.2	8.1	5	$19,900
Mazda	Split	Tribute Hybrid	FWD	NA	45	44	43	NA	NA	NA	NA	$28,175
Mazda		Tribute	FWD	NA	26	30	39	NA	NA	NA	NA	$21,790
Mazda		Tribute	FWD	NA	23	27	36	NA	NA	NA	NA	$23,055
Mazda	Split	Tribute Hybrid	4WD	NA	37	37	37	NA	4.1	10.7	5.8	$29,925
Mazda		Tribute	4WD	NA	24	28	35	NA	3.3	10	6.4	$23,545
Mazda		Tribute	4WD	NA	22	26	33	NA	3	7.9	5.2	$24,805
Mercury	Split	Mariner Hybrid	FWD	NA	45	44	43	NA	NA	NA	NA	$30,090
Mercury		Mariner	FWD	NA	26	30	39	NA	NA	NA	NA	$22,650
Mercury		Mariner	FWD	NA	23	27	36	NA	NA	NA	NA	$23,660
Mercury	Split	Mariner Hybrid	4WD	NA	37	37	37	NA	4.1	10.7	5.8	$31,840
Mercury		Mariner	4WD	NA	24	28	35	NA	3.3	10	6.4	$24,400
Mercury		Mariner	4WD	NA	22	26	33	NA	3	7.9	5.2	$25,410

continued

TABLE 6.A.1 Continued

				Volume	Official EPA (unadjusted mpg)			EPA Test	Acceleration (Consumer Reports)			Edmund's MSRP
Make	Type	Model	Drive	Trunk	City	Comb.	Hwy	Car Weight	0 to 30 mph, sec.	0 to 60 mph, sec.	45 to 65 mph, sec.	Price
Chevrolet	Parallel	Malibu Hybrid	FWD	15	33	39	48	3875	4.1	10.3	6.9	$25,555
Chevrolet		Malibu	FWD	15	27	33	43	3750	3.4	9.4	7	$21,605
Chevrolet		Malibu	FWD	15	23	28	40	NA	3	8.1	5.1	NA
Lexus	Split	RX 400h Hybrid	2WD	NA	NA	NA	NA	NA	NA	NA	NA	NA
Lexus		RX 350	2WD	NA	20	22	25	4250	NA	NA	NA	$37,700
Lexus		RX 350	2WD	NA	20	22	25	NA	NA	NA	NA	NA
Lexus	Split	RX 400h Hybrid	4WD	NA	NA	NA	NA	NA	2.9	7.4	4.6	NA
Lexus		RX 350	4WD	NA	22	26	32	4500	NA	NA	NA	$39,100
Lexus		RX 350	4WD	NA	22	26	32	NA	2.7	7.3	4.8	NA
Lexus	Split	GS 450h Hybrid	RWD	9	28	31	35	4500	2.5	5.9	3.9	$56,550
Lexus		GS 350	RWD	13	24	28	37	4000	NA	NA	NA	$45,000
Lexus	Split	LS 600h L	AWD	12	25	27	30	5500	NA	NA	NA	$106,035
Chevrolet	Split	Tahoe Hybrid	RWD	NA	27	28	30	6000	NA	NA	NA	$50,455
Chevrolet		Tahoe	RWD	NA	15	19	27	6000	NA	NA	NA	Premium
Chevrolet		Tahoe	RWD	NA	17	20	27	5500	NA	NA	NA	$39,315
Chevrolet	Split	Tahoe Hybrid	4WD	NA	27	28	30	6000	3.9	9.6	5.5	$53,260
Chevrolet		Tahoe	4WD	NA	15	18	26	6000	3.4	9	5.7	$41,025
Chevrolet	Split	Silverado Hybrid	RWD	NA	27	28	30	NA	NA	NA	NA	$38,020
Chevrolet		Silverado	RWD	NA	17	21	27	5500	NA	NA	NA	NA
Chevrolet		Silverado	RWD	NA	18	21	27	5000	NA	NA	NA	$26,915
Chevrolet	Split	Silverado Hybrid	4WD	NA	27	28	30	NA				$41,170
Chevrolet		Silverado	4WD	NA	17	20	27	5500				NA
Chevrolet		Silverado	4WD	NA	18	21	27	5250	3	7.9	5.1	$30,065
GMC	Split	Yukon Hybrid	RWD	NA	27	28	30	NA	NA	NA	NA	$50,920
GMC		Yukon	RWD	NA	15	19	27	NA	NA	NA	NA	Premium
GMC		Yukon	RWD	NA	17	20	27	NA	NA	NA	NA	$39,970
GMC	Split	Yukon Hybrid	4WD	NA	27	28	30	NA	3.9	9.6	5.5	$53,730
GMC		Yukon	4WD	NA	17	20	27	NA	3.4	9	5.7	$41,765
GMC	Split	Sierra Hybrid	RWD	NA	27	28	30	NA	NA	NA	NA	$38,390
GMC		Sierra	RWD	NA	17	21	27	5500	NA	NA	NA	NA
GMC		Sierra	RWD	NA	18	21	28	NA	NA	NA	NA	$26,915
GMC	Split	Sierra Hybrid	4WD	NA	27	28	30	NA	NA	NA	NA	$41,540
GMC		Sierra	4WD	NA	17	20	27	NA	NA	NA	NA	NA
GMC		Sierra	4WD	NA	18	21	27	6000	3	7.9	5.1	$30,065
Dodge	Split	Durango Hybrid	4WD	NA	25	27	30	NA	NA	NA	NA	$45,040
Dodge		Durango	4WD	NA	17	20	26	NA	2.8	7.4	5.2	NA
Chrysler	Split	Aspen Hybrid	4WD	NA	25	27	30	NA	NA	NA	NA	$45,270
Chrysler		Aspen	4WD	NA	17	20	26	NA	2.8	7.4	5.2	NA
Cadillac	Split	Escalade Hybrid	2WD	NA	27	28	30	NA	NA	NA	NA	$73,135

7

Non-Engine Technologies

INTRODUCTION

This chapter focuses on reducing fuel consumption with non-power-train technologies. These technologies affect engine performance either directly or indirectly in a manner that reduces fuel consumption. For example, a significant portion of this chapter discusses the state of readiness, cost, and impact of reducing vehicle mass. Reducing mass reduces the energy necessary to move a vehicle, and thus reduces fuel consumption. The complexity of substituting advanced, lightweight materials affects the redesign of a part or a subsystem, component manufacturing (including tooling and production costs), and joining, and raises interface issues that mixing different materials can pose. The term *material substitution* oversimplifies the complexity of introducing advanced materials, because seldom does one part change without changing others around it. Advanced lightweight materials show great promise for reducing mass throughout a vehicle's body structure and interior. Low-rolling-resistance tires and reduction of aerodynamic drag are also discussed as technologies that can lower tractive force and result in reduced fuel consumption. Improvements in energy-drawing devices such as air conditioner compressors and power steering can reduce fuel consumption either by electrification or by improving their efficiency. New transmissions with more gears or that are continuously variable improve power train efficiency. All these options either reduce the demand for power from the engine or enable operating the engine at a more efficient point to reduce fuel consumption.

NON-ENGINE TECHNOLOGIES CONSIDERED IN THIS STUDY

The committee considers car body design (aerodynamics and mass), vehicle interior materials (mass), tires, vehicle accessories (power steering and heating, ventilation, and air conditioning [HVAC] systems), and transmissions as areas of significant opportunity for achieving near-term, cost-effective reductions in fuel consumption. These will be considered in some detail below.

Aerodynamics

As discussed in Chapter 2, the force required to overcome drag is represented by the product of the drag coefficient, the frontal area, and the square of speed. The actual formula is F = ½ Cd AV2 where A is the vehicle frontal area, V is velocity, and Cd is the drag coefficient. Cd typically ranges from about 0.25 to 0.38 on production vehicles and depends on several factors with the primary influence coming from vehicle shape and smaller influences from other factors, such as external mirrors, rear spoilers, frontal inlet areas, wheel well covers, and the vehicle underside. Vehicles with higher Cd values (greater than .30) may be able to reduce the Cd by up to 10 percent at low cost without affecting the vehicle's interior volume. In trying to reduce fuel consumption, certain vehicles achieved very low drag coefficients, for example, GM's EV1 had a Cd of 0.19, and the third-generation Prius has a Cd of 0.25.[1] In the committee's judgment a Cd of less than 0.25 would require significant changes that could include the elimination of outside rear view mirrors, total enclosure of the car underbody, and other modifications that may be very costly. Vehicles that exist today with a low Cd (below 0.25) are usually specialty vehicles (e.g., sports cars and high-mileage vehicles like the Prius). The 2010 Mercedes E-class is the only production vehicle with a Cd as low as 0.25. However, this is a luxury-class vehicle and retails for $50,000 (or more). Some costs are incurred from incorporating aerodynamic features such as the integrated front spoiler, an option that may not be possible for lower-cost vehicle classes. Further reducing Cd for lower-cost vehicles is expensive and perhaps beyond a point of diminishing returns. Vehicles with higher Cd (e.g., trucks,

[1] See http://www.greencar.com/articles/20-truths-gm-ev1-electric-car.php and http://pressroom.toyota.com/pr/tms/toyota/all-new-prius-reveal.aspx, respectively.

vans, and box-like vehicles such as the Scion and Flex) can reduce Cd, although vehicle functionality is diminished. If the functionality is compromised, then the vehicle's appeal to the consumer would be reduced.

As noted above, the aerodynamic drag is the product of the drag coefficient Cd, the vehicle frontal area, and speed. Reduction in the frontal area, reducing vehicle size, and lower speed limits would also improve fuel consumption; however, exploring these options is outside the committee's statement of task.

Car Body Design and Interiors

Optimized car body design focuses on a balance between structural stiffness, noise/vibration/harshness (NVH), safety (crashworthiness), comfort (space), and mass. Today's priority of reducing fuel consumption places an emphasis on mass reduction, with the assumption that other performance criteria will not be unduly compromised. Vehicle mass can be reduced without compromising size, crashworthiness, and NVH, although countermeasures are often required to restore NVH performance when mass is reduced.

The majority of vehicle mass can be attributed to the body structure, closure panels (doors, hood, and deck lid), interior seating and trim components, glass, power train components (engine, transmission, etc.), and the chassis (axles, wheels, brakes, suspension, etc.). Steel, cast iron, fiber/reinforced composites, glass, and aluminum have been the dominant materials for these components, with steel accounting for the majority of mass. Estimates for the amount of these materials in today's average, high-volume vehicles are listed in Table 7.1 (Carpenter, 2008). The typical baseline vehicle used for comparison is described as a 3,600-lb model-year 2009 comparable to a Toyota Camry or Chevrolet Malibu.

High-volume vehicle manufacturing is generally associated with the production of more than about 100,000 vehicles per year (although some might say 50,000). Low volume might be under 25,000 vehicles per year. This is important because different materials become cost competitive at different volumes. Higher-cost materials (composites, aluminum, and magnesium) become more cost competitive at lower volumes because the forming tools in most cases have a lower investment cost offsetting the higher material cost. Steel requires high-cost forming tools but has a lower materials cost, making steel competitive at higher volumes. For example, for some non-structural applications, steel becomes cost competitive vis-à-vis plastic at around 50,000 units.

Two key strategies for achieving mass reduction are changing the design to require less material, or substituting lighter-weight materials for heavier materials. Assuming that the car size is essentially fixed, there are design techniques that can reduce mass. Several different body architectures are described below. Material substitution relies on replacing a heavier material with a lighter one while maintaining performance (safety and stiffness). For example, high-strength steel can be substituted for mild steel (and therefore a thinner gauge can be used), aluminum can be substituted for steel, plastic can be substituted for aluminum, and magnesium can be substituted for aluminum. It is often a misnomer to refer to this as material substitution. The part (or subsystem) often has to be redesigned, and the fabrication process may change and the assembly process may be different. In fact, the material cost differential may be insignificant relative to the costs associated with the changes in fabrication and assembly.

Body Design and Material Selection

The great majority of vehicles produced today are unibody design. The unibody design is a construction technique that uses the internal parts as the principal load-bearing structure. While the closure panels (doors, hood, and deck lid) provide important structural integrity to the body of the vehicle, the outer skin panels, defined as the metal outer panels on the entire automobile that are painted and visible to the consumer, do not. This design has replaced the traditional body-on-frame design primarily because it is a lighter. Body-on-frame designs, where an independent body structure (with its own structural integrity) sits on top of a separate frame (with its own structural integrity), still prevail on some heavier vehicles such as pickup trucks and larger SUVs because of its overall superior strength and stiffness. Another design, the space frame, was recently developed to accommodate aluminum. The forming and joining of aluminum cannot easily or cheaply be replicated in a steel unibody design. A typical space frame is composed of extruded metal connected at the ends, which are referred to as nodes. Both the unibody and the space frame have "hang-on" panels where the skin panels have little to no structural load. A final design architecture, the monocoque, relies on the outer skin surface as a principal load-bearing surface. The

TABLE 7.1 Distribution of Materials in Typical Vehicle (e.g., Toyota Camry and Chevrolet Malibu)

Material	Comments	Approximate Content in Cars Today, by Weight (percent)
Iron and mild steel	Under 480 Mpa	55
High-strength steel	≥ 480 Mpa (in body structure)	15
Aluminum	No aluminum closure panels; aluminum engine block and head and wheels	10
Plastic	Miscellaneous parts, mostly interior trim, light lenses, facia, instrument panel	10
Other (magnesium, titanium, rubber, etc.)	Miscellaneous parts	10

monocoque is seen in very low volumes because there are few applications where it is structurally and economically viable. Generally, these three designs are associated with the following materials:

- *Unibody*—steel-based structure (mostly steel stampings) usually with steel skin panels but sometimes plastic or aluminum skin panels. This design has high investment (engineering and tooling) costs and is designed for high volume.
- *Space frame*—usually an aluminum-based structure (aluminum castings, extrusions, and sheet). This design is less complex than the unibody and has lower investment costs, which are typically offset by higher material costs. Because of the high material costs (that are variable with volume), this is typically a low-volume design.
- *Monocoque*—reinforced resin/composite body structure using the skin to bear loads. Today, this architecture is uncommon for passenger automobiles and more common for aircraft or ships.

The space frame and monocoque structures are associated today with niche vehicle markets, whereas the unibody with its steel-based structure is common (perhaps found in more than 99 percent of today's automobiles). These design approaches differ from the body-on-frame design that is well suited for heavier "working" vehicles like trucks and SUVs. Body-on-frame readily achieves all the desired design criteria, except that it is heavy because of the large frame components.

Reducing Mass Using Alternative Materials

There are several methods to make steel structures lighter, regardless of their design construction:

- Substitute higher-strength steel for lower-strength steel. Higher-strength steel can be down-gauged (made thinner). There are, however, forming and joining issues with higher-strength steel that limit where it can be applied, and down-gauging can reduce the ability to meet stiffness criteria.
- Substitute sandwich metal material for conventional steel. Sandwich material has layers of steel or aluminum (usually three), often with the internal layer in the form of honeycomb or foam. Other layered materials can include bonded steel with plastic/polymers. This cladding material can achieve high stiffness and strength levels with low mass. Sandwich material is light, is very stiff, and can be formed for many parts. On the downside, joining it to other parts can be difficult, its availability is limited today, and it is expensive to produce.
- Introduce new steel designs that are available, such as with laser welded blanks and hydro-formed tubes or hydro-formed sheet metal. The use of tubes and laser blanks can make more optimal use of metal (steel or aluminum) and result in less mass in the structure without compromising design criteria. These methods may increase or decrease costs depending on the application.

Most steel and mixed-material vehicles (e.g., steel and aluminum) today are unibody, and aluminum-intensive vehicles tend to be space frame designs, but these are low volume due to cost. The unibody design was developed primarily for steel, and the conventional vehicle today is composed of about 65 percent steel (both mild and high strength). Various components of a unibody can have alternative lightweight materials, including high-strength steel, polymers/composites, and aluminum directly substituted on a part-by-part basis to help reduce mass on a limited basis. Sheet molding compound (SMC plastic) body panels are sometimes used for fenders or exterior closure panels to save weight, and in the case of low-volume vehicles, to save costs. The ability to substitute alternative materials, however, can be limited because of forming (part shape), joining, and interface issues between mixed materials. Steel unibody designs can accommodate polymer/composite or aluminum closure panels because these parts can be easily isolated from the remainder of the structure since they are fastened onto the structure. Many unibody steel-based vehicles made in North America have aluminum hoods and deck lids, but steel doors. Hoods and deck lids are simpler designs than doors (they are flatter and have fewer parts, and therefore are less expensive and less complex to switch over to aluminum). Steel doors could also be converted to aluminum in many cases, as is often done in Europe, but in North America their size and geometry would make this conversion relatively expensive.

The mass savings by introducing high-strength steel results from the ability to down-gauge the thickness over mild steel while maintaining the same strength as the thicker mild steel part. Down-gauging reduces stiffness, and so this is not a solution in some cases where stiffness is important. Also, as the strength of steel increases, its ability to be formed into different shapes is reduced (its allowable percent elongation is reduced). This reduced formability also limits where high-strength steel can be applied. The outside panels (skin panels) on a unibody are predominantly non-structural and subject to dents, thus also limiting the ability to down-gauge these panels. The tools that form high-strength steel parts cost more, require greater maintenance because they are subject to wear, and require greater forming pressures in production. In most cases, high-strength steel parts cost more than comparable mild steel parts. New, advanced high-strength steels are being developed to give high-strength steel greater formability and weldability. These advanced high-strength steels, expected to be available within a few years, can reduce mass on some compatible parts by around 35 percent. This is achieved by using high-strength steel to reduce part thickness by 35 percent (e.g., replacing 1.8-mm-thick mild steel with 1.2-mm-thick high-strength steel). Factors such as part geometry and subsystem stiffness can limit viable applications of high-strength steel or constrain the reduction in thickness.

An aggressive approach to introducing aluminum into the structure may dictate a totally different body design approach, such as shifting from a unibody to a space frame structure. The space frame design has been developed recently for aluminum-intensive structures. The structure is composed of aluminum castings, extrusions, and sheet. This design is lighter than a comparable steel design and is in production today, but is used only on lower-volume, higher-end vehicles because of its high cost. Introducing an aluminum-intensive structure would necessitate a complete vehicle redesign, requiring several years at extremely high development costs (see the product development process discussion in the section "Timing Considerations for Introducing New Technologies" below in this chapter).

Polymer-matrix composites (PMCs) are beginning to be introduced into higher-volume vehicles. Viable options for PMC are for it to be reinforced with glass fibers, natural fibers, or carbon fiber to give it strength. Glass- and natural-fiber-reinforced PMCs are lower cost than carbon fiber, but they have less strength. Since they incur lower cost, it is likely that these applications will be seen on higher-volume vehicles before there is significant use of carbon fiber composites. Carbon fiber is a promising lightweight material for many automotive components. Much like plastic, PMC can be molded into complex shapes, thus integrating several steel or aluminum parts into a single PMC part that reduces complexity and tooling costs. Conservative estimates are that carbon fiber PMC can reduce the mass of a steel structure by 40 to 50 percent (Powers, 2000). Both its strength and its stiffness can exceed that of steel, making it easy to substitute for steel or aluminum while offering equal or better structural performance. The greatest challenges with PMC are cost and carbon fiber availability. Also challenging is connecting composite parts with fasteners, which has delayed the introduction of the latest Boeing 787 Jet.

The price of carbon fiber is extremely volatile, with material cost typically in excess of $10/lb. Carbon fiber exceeds the cost of steel and aluminum by approximately 20-fold and 7-fold, respectively. Steel and aluminum can also be formed with high-speed stamping, which is much less costly than forming PMC, which typically involves a fairly slow autoclave process. Research at Oak Ridge National Laboratory (ORNL) is aimed at developing lignin-based carbon fiber to help reduce material cost and improve supply (Compere et al., 2001). This research in conjunction with the FreedomCar program at the United States Council for Automotive Research (USCAR) indicates that the price of carbon fiber has to fall to $5 to $7 per pound (about 50 percent) before it can be cost competitive for high-volume automobiles (Carpenter, 2008). Lignin-based carbon fiber will also help ensure a greater supply of the base material of PMC. One expert stated that carbon fiber will see wider use in the future, but primarily on lower-volume (fewer than 100,000 vehicles per year), higher-performance vehicles (Carpenter, 2008).

The cost differential (by pound) varies significantly for alternative materials. High-strength steel might cost double the price of mild steel ($0.80 versus $0.40 per pound), and aluminum might cost four or five times that of steel (per pound). Other materials such as magnesium and titanium are also expensive and have volatile price fluctuations.

It is important to recognize that the comparison of different materials is complicated by many factors, making a cost analysis difficult. Tooling costs and parts fabrication costs differ significantly for different materials.

- The amount of material (pounds) needed by the lightweight material is different from the incumbent material.
- Because of part fabrication, the optimal design with the lightweight material may be very different from the design of the original part. For example, some steel parts cannot be formed exactly the same out of aluminum because of formability constraints. Also, if you substitute a material that is cast (magnesium) instead of stamped (steel), the forming cost and the part design are different.
- The tooling to form the alternative material is likely to be different than the tooling for the incumbent material, and may cost more or less.
- The processing (part fabrication) process will likely run differently, and may operate much slower than that for the incumbent material (e.g., molding is much slower than stamping, sometimes by a factor of 10).

USCAR and the U.S. Department of Energy continue to research reducing body mass by substituting new materials, such as high-strength steel, advanced high-strength steel, aluminum, magnesium, and composites for current materials. The material industries also conduct significant research to advance new materials (for example, through the Auto-Steel Partnership, the American Iron and Steel Institute, the Aluminum Association, and the American Chemistry Council). Increased costs for lighter and stronger parts result from higher material costs and higher costs for component fabrication and joining. Estimates for the body-mass reduction that can be achieved in the near term vary from 10 percent (with mostly conventional and high-strength steels) to 50 percent (with a mostly aluminum/composite structure). Even greater reductions are feasible, but these require very expensive and aggressive use of aluminum, magnesium, and composite structures involving materials such as carbon fiber.

Non-Body Mass Reduction

Vehicle interiors also offer opportunities to reduce vehicle mass. Some opportunities can be implemented for little cost, whereas others entail significant costs. For example, composite-intensive instrument panels, recycled seating materials, and lighter-weight trim panels can reduce mass by tens of pounds at virtually no cost. However, unlike the car body for which the consumer cannot easily detect what materials are used, the interior is aesthetically critical and closely scrutinized by the consumer. Costs may be incurred by covering over the appearance of some parts. There are quality concerns, such as fit-up of panels, part texture, and appearance issues that constrain interior cockpit design alternatives. Some isolated components can have mass reduced with material substitution such as headlamps (with new resins) and wheels (with new aluminum grades) that actually enhance aesthetics but often increase cost. Non-visual parts, however, also present an opportunity, such as seat belt reinforcements, seating frames/brackets, and fire wall panels. Most non-structural applications that can be light-weighted with plastic already have been. Glass-reinforced sheet molding compound (SMC) is low cost and inexpensive to form but lacks sufficient strength to replace most structural applications responsible for much of the weight.

Isolated components on the vehicles are also candidates for aluminum, magnesium, or advanced high-strength steel substitution, such as wheels, engine cylinder heads, suspension arms, transmission cases, brake calipers, steering knuckles, and engine blocks, although many OEMs have already made these substitutions, especially in cylinder blocks and heads. Aluminum heads are more common than aluminum blocks because of performance issues in the block, but other materials including hybrid materials (both aluminum and cast iron) are being applied to the blocks. An even more aggressive approach to introducing aluminum into the structure itself will likely involve aluminum-intensive substructures (e.g., axle assemblies, engine compartment, etc.), and such components are also now starting to penetrate the new-vehicle population.

Car glass (windshield, side windows, rear window, mirrors, and sun roofs) is also heavy, and there are opportunities to reduce mass by substituting polycarbonate. Polycarbonate can be coated to provide a durable finish, and this has been applied to non-windshield glass panels where scratching is less a concern.

Rolling Resistance

Tire rolling resistance is one of many forces that must be overcome in order for a vehicle to move (see discussion in Chapter 2). When rolling, a tire is continuously deformed by the load exerted on it (from the vehicle mass). The repeated deformation during rotation causes energy loss known as rolling resistance. Rolling resistance is affected by tire design (for example, materials, shape, and tread design) and inflation. Underinflated tires increase rolling resistance. The opportunity to improve fuel economy by reducing rolling resistance is already used by OEMs to obtain better "EPA numbers," and so original equipment tires tend to have lower rolling resistance than consumer-replaced tires because typical values for the coefficient of rolling resistance (r_0) values differ between them (NRC, 2006). This represents an interesting value tradeoff. The OEMs are more interested in getting low-rolling-resistance tires to show improved fuel economy, and people buying replacement tires are more interested in low cost and durability. Therefore the total opportunity for fuel consumption reduction is defined by the fraction of the tires on the road that falls into each category. Education of the public on the subject of low-rolling-resistance tires for replacement tires and the continued introduction of tire pressure monitoring systems, which is discussed below, may help improve in-use performance of tires for fuel consumption reduction.

There are performance tradeoffs involving tires that tire manufacturers consider during design and manufacturing. These tradeoff variables include, for example, tread compound, tread and undertread design, bead/sidewall, belts, casing, and tire mass. Important tire performance criteria affected by design and manufacturing include rolling resistance, tire wear, stopping distance (stopping distance or grip can be evaluated over different surfaces, such as wet or dry), and cornering grip. Wear and grip are closely correlated to tread pattern, tread compound (e.g., softer compounds grip better but wear faster), and footprint shape.

The impact of emphasizing one performance objective (such as low rolling resistance) over other performance criteria is inconclusive. Some studies have shown that tires with low rolling resistance do not appear to compromise traction, but may wear faster than conventional tires. Another study in 2008 by Consumers Union and summarized by *Automotive News* (Automotive News, 2008) concluded that there may be a reduction in traction, because of low-rolling-resistance tires, that increases stopping distance. The study is not rigorously controlled, and other influences may confound the results. The response by one tire manufacturer, Michelin (Barrand and Bokar, 2008), argues that low-rolling-resistance tires can be achieved without sacrificing performance factors by balancing the design and manufacturing process variables. Tire makers are continuing to research how to get optimal performance (including fuel economy) without sacrificing other criteria such as safety or wear. Goodyear points out that performance tradeoffs between rolling resistance, traction, and tread wear can be made based on materials and process adjustments, which also affect cost (Goodyear Tire & Rubber Company, 2009). The incremental cost for low-resistance tires may not be significant, but the cost-benefit tradeoff with increased stopping distance, wear, and possibly noise, vibration, and harshness issues are important for the consumer.

Rolling resistance can also be affected by brakes. Low-drag brakes reduce the sliding friction of disc brake pads on rotors when the brakes are not engaged because the brake pads are pulled away from the rotating rotor. Most new vehicles have low-drag brakes. The impact over conventional brakes may be about a 1 percent reduction of fuel consumption.

Rolling resistance is also affected by tire inflation, and so any technology that affects inflation levels can also affect fuel economy. Reducing tire inflation levels increases rolling resistance, which in turn increases fuel consumption. A tire pressure monitoring system (TPMS) can be set to different pressure thresholds, and the average deviation from the recommended inflation level would be 1/2 the threshold level. For example, if the threshold is set at 10 psi, the average deviation from the recommended level would be 5 psi. Michelin believes that an accurate TPMS with an appropriately set threshold could reduce fuel consumption by up to 0.7 percent (J. Barrand, personal communication, May 12, 2009).

Vehicle Accessories

Some automakers are beginning to introduce electric devices (such as motors and actuators) that can reduce the mechanical load on the engine, reduce weight, and optimize performance, resulting in reduced fuel consumption. Of course, the electrical power used by these devices must be furnished by the engine driving the alternator. Thus the most advantageous opportunities for converting mechanical devices to electrical are devices that operate only intermittently, such as power steering and air-conditioning compressor. The benefits from electric and/or electro-hydraulic power steering and greater efficiency in air-conditioning (A/C) are not credited by current EPA fuel economy tests (since neither operates during the test), and so manufacturers are reluctant to implement them because of added costs. With the new EPA test procedures, some of the benefits will be reflected in the "sticker," and improvements in these areas are relatively "low hanging fruit."

- *Heating, ventilating, and air-conditioning (HVAC).* A more efficient system starts with (larger) heat exchangers that transfer high heat more effectively and a thermal expansion valve that controls the evaporator temperature. The compressor uses the majority of the energy of the A/C system, and variable displacement piston compressors are available and in use that significantly reduce fuel use over fixed displacement compressors. There are many other technologies, such as increased use of recirculated air, elevation of evaporator temperature, use of pulse-width modulated blower speed controllers, and internal heat exchangers, that can further reduce fuel usage.

Further reductions in fuel use can be achieved by decreasing A/C load through the use of low-transmissivity glazing (reducing both heat and ultraviolet penetration), reflective "cool" paint, and cabin ventilation while parked. Suppliers are investigating the use of directly cooling the seat either through ducting or by thermoelectric materials. Although this may increase comfort, it is not clear whether this will significantly improve fuel economy (Rugh et al., 2007).

- *Exhaust heat recovery.* Recent improvements in thermoelectric materials for HVAC and exhaust energy recovery appear promising. Research is directed primarily at new materials with higher "thermoelectric figure of merit" (Heremans et al., 2008; Hussain et al., 2009). This is accomplished by increasing the thermoelectric effect (Seebeck coefficient) and reducing the thermal conductivity. Good results have been obtained with nanomaterial processing, but at this time these are costly. Improvements in potentially low-cost bulk materials are needed for automotive applications. BMW has announced a planned introduction on production vehicles in the 2012/2013 model year.[2] It presented a model of an application at the 2006 DEER Conference[3] and in the press.[4] A DOE presentation gave more information on this vehicle and presented a rather optimistic view of energy recovery.[5] In the view of the committee significant improvements need to be made in the performance of bulk materials and in the processing of nanomaterials before thermoelectric heat recovery from the exhaust can be applied in mass production. The committee thinks that this will not happen in the 10-year horizon considered here.

Transmission Technologies

Transmission technologies can reduce fuel consumption in two ways, first by moving engine operation to more efficient regions of the engine map (cf. Figure 2.3 in Chapter 2) and second by continued reduction of the mechanical losses within transmissions. Of these two, moving engine operation to more efficient regions of the engine map (e.g., higher torque (or brake mean effective pressure; BMEP) and lower speeds) offers the largest potential gains. The major approaches to achieving this movement are by increasing the number of speeds in the transmission (whether manual, automatic, or continuously variable) and lowering final drive ratio.

Five-speed automatic transmissions are already a standard for many vehicles; 6-, 7-, and 8-speed automatic transmissions have been available on luxury cars and are penetrating into the non-luxury market. This new wave of automatic transmissions has been enabled by new power flow configurations and improved controls capability that are enabling larger numbers of speeds to be achieved at a lower cost increment over 4-speed automatics than would be the case for adding speeds to previous automatic transmission designs.

---

[2] See http://www.motorward.com/2009/02/new-details-on-next-generation-bmw-5-series/.

[3] See http://www1.eere.energy.gov/vehiclesandfuels/pdfs/deer_2006/session6/2006_deer_lagrandeur.pdf.

[4] See http://www.autobloggreen.com/2008/09/25/bmw-wins-koglobe-2008-award-for-thermoelectric-generator/.

[5] See http://www1.eere.energy.gov/vehiclesandfuels/pdfs/deer_2006/session6/2006_deer_fairbanks.pdf.

This cost improvement resulted from transmission gear train synthesis optimization studies using computational tools that uncovered gear trains requiring fewer discrete elements because some of the elements (e.g., planetary gear trains) are utilized for multiple speeds. However, increasing the number of speeds always adds some components and their associated cost. Along with higher numbers of transmission speeds, which allow operating engines in more efficient parts of their fuel consumption map, transmission internal losses are also being reduced, thus further improving power train efficiencies.

In addition to planetary-based automatic transmissions, advanced versions of manual transmissions are also being introduced that can be more efficient than automatics since torque converters are replaced by computer-controlled clutches, which slip less than torque converters. These new clutches not only are used to launch the vehicle from a stop but also enable rapid automated shifting of the manual gears since one clutch can start engagement before the other clutch has completely released. This class of manuals is called dual-clutch automated manual transmissions (DCTs).[6] With this concept, new-design manual transmissions are arranged with two parallel gear trains, one for odd-numbered speeds and the other for even-numbered speeds: for a 6-speed DCT, one gear train would contain the first, third, and fifth speed gears while the other gear train would include the second, fourth, and sixth speed gears. DCTs are then coupled to the engine through two clutches integrated into the transmission, one linking the odd-speed gear train to the engine and the other clutch linking the even-speed gear train to the engine. Finally, the clutches are actuated with electro-hydraulic systems calibrated to provide smooth launch and rapid and smooth shifting, making them automatic in their interface to the driver. In most of the current implementations of these clutches, they are immersed in transmission oil, thus providing the cooling necessary for acceptable durability. Dry-clutch versions are now also being developed for vehicles with lower torque requirements, making oil cooling unnecessary. Dry-clutch DCT designs are expected to be less costly to produce and lighter than their wet-clutch counterparts. In addition, dry-clutch DCTs will be more efficient through elimination of the hydraulic pump work to cool the wet clutches.

Both automatic and DCT transmissions feature a discrete number of gear ratios that determines the ratio of engine speed to vehicle speed. In contrast, a continuously variable transmission (CVT) offers a theoretically infinite choice of ratios between fixed limits, which allows engine operating conditions to be optimized for minimizing fuel consumption. CVT technology has tended to be used in lower-horsepower vehicles because of maximum-torque limitations with the most common metal-belt design. A few OEMs offer CVTs that utilize other drive schemes allowing usage with larger engines. CVTs have achieved some penetration into the market, but recent trends suggest that their usage may not grow further due to higher than expected costs and lower than expected internal efficiencies (EPA, 2008).

The issues discussed above generally apply to both SI and CI engines. However, the effects of moving engine operating points to lower-speed and higher-torque regions of the engine map are more beneficial for SI engines than for CI engines because intake throttling losses are reduced for SI engines, whereas CI engines are not throttled. Nonetheless, for both CI and SI engines, fuel consumption is reduced by moving to higher-torque and lower-speed regions of the engine maps because the relative effect of engine friction losses is reduced.

Another important transmission issue difference between SI and CI engines is their peak torque. As noted in Chapter 5, CI engines produce higher maximum torques than do SI engines. Maximum torque capacity is one of the most important criteria for durable transmission design, and so CI engines generally are mated with different, higher-torque-capacity transmissions than SI engines even in the same vehicle platform. Sometimes, a given transmission used for SI engines can be upgraded to higher torque capacity by more extensive and more expensive heat treating of the gears and clutch upgrading, but frequently, different transmissions originally designed for higher maximum torque capacity must be used with CI engines, thus increasing cost, weight, and to some extent internal losses.

Another transmission-related technology that is applicable to both SI and CI engines is called idle-stop. This technology is useful primarily for operation in cities and involves turning off the engine at idle. Benefits from idle-stop involve eliminating most of the idle fuel consumption during the idle-stop period. Since idle fuel consumption is relatively large for SI engines due to throttling losses and the use of ignition retard for smooth operation when accessories turn on and off, FC reductions on the Federal Test Procedure (FTP) driving cycle range from 3 to 5 percent. The real-world gain for congested city driving (e.g., New York City) could be as high as 10 percent since engines would be idled much more than on the FTP test cycle. All idle fuel consumption losses are not eliminated since some accessories may need to operated while the engine is stopped (e.g., A/C in hot climates), which not only consumes some fuel but also increases component cost by the necessity of replacing belt-driven accessories with electrically driven ones. For the CI diesel vehicle, idle-stop benefits are smaller than those attained with idle-stop for SI gasoline vehicles because diesel engines have much lower idle FC than their gasoline counterparts. The estimated gain on the U.S. cycle for CI vehicles is about 1 percent, although the real-world gain for congested city driving (e.g., in New York City) could be much higher.

Other studies of vehicle fuel consumption (e.g., NRC, 2002) have generally considered potential gains from transmission technologies in a separate category from engine efficiency technologies. In the present study, potential gains

[6] See http://www.dctfacts.com/hmStory1b.asp.

from transmission technologies are considered together with those for engines. This choice was made for the following reasons. For SI engines, the major opportunity for reducing fuel consumption (as is discussed extensively in Chapter 4) is reducing pumping losses. Many of the technology measures discussed in Chapter 4 reduce pumping losses in one way or another. As noted above, the major impact of transmission technologies toward reducing fuel consumption is to move the operation of the engine toward higher torque (or BMEP) and lower speeds at which pumping losses will be reduced. As a result, there are significant interactions between engine technologies that reduce pumping losses (e.g., valve event modulation) and transmission changes that also move engine operation to lower speeds and loads, such as increasing the number of ratios and the associated ratio spread.[7] A good example of these interactive effects is cylinder deactivation, as discussed in Chapter 4. When cylinder deactivation is used, the benefit of moving the engine operating point to lower speeds and higher torques and higher BMEP is reduced compared to engines not using cylinder deactivation, because the working cylinders are already running at higher BMEP, thereby reducing pumping losses. Thus the fuel consumption reductions possible from increasing the number of transmission ratios from 4 to 6, for example, would be lower for engines using cylinder deactivation than for those not using cylinder deactivation. This demonstrates how transmission-derived fuel reductions of fuel consumption cannot readily be separated from engine-technology-derived fuel consumption reductions. This choice is reflected in the technology paths discussed in Chapter 9.

FUEL CONSUMPTION BENEFITS OF NON-ENGINE TECHNOLOGIES

The tractive force that is needed to propel a vehicle can be written simply as the sum of three forces:

$$F_{TR} = F_m + F_r + F_a$$

where F_m accelerates the mass, F_r overcomes rolling resistance, and F_a overcomes aerodynamic drag. The integral of this force over a given driving cycle gives the amount of energy required at the wheels. Using typical values in Equation 2.1 one can calculate that for the EPA combined cycle about one-third of the tractive energy goes into each of these three components (see Table 2.7). However, as Table 2.7 shows for the urban cycle, F_m is around 60 percent of the total and for the highway cycle, F_a is about half. Before giving estimates of the benefits of fuel-saving technologies, it is necessary to make two important points.

Merely reducing tractive energy does not translate into a directly proportional reduction of fuel consumption because of (1) the accessory load and (2) the possibility that the power train may then operate at worse efficiency points. To take care of the power train efficiency it is necessary, at the same time, to downsize the engine and/or change transmission shift points, because with a lighter load, the efficiency of the power train is reduced, especially with SI engines that will then operate with more throttling. Unfortunately, many studies on the impact of reducing F_m and F_a do not change the engine operating points. For example, Barrand and Bokar (2008) do an excellent job of investigating the effect of rolling coefficient by changing tires without changing the power train. Only an OEM designing a vehicle with low-rolling-resistance tires, for example, can fully take advantage of rolling-resistance changes by reoptimizing the power train.

Theoretically reducing any one of the three components by, say, 10 percent should reduce fuel consumption by roughly 3.3 percent since, as stated above, each component accounts for roughly one-third of the total tractive energy. In fact the size of the engine is determined by acceleration performance requirements, as well as the tractive energy. Therefore all that can be said for certain is that reduction of all three components by an amount (say, *X* percent) would result in a reduction in fuel consumption by roughly the same amount (*X* percent), assuming the power train were reoptimized.

Aerodynamics

As discussed above, vehicles with higher C_d values (over .30) may be able to have the C_d reduced by 5 percent or so (up to 10 percent) at low cost. The associated impact on fuel consumption and fuel economy could be 1 to 2 percent, and this assumes that the engine operating regime is not modified. If lower acceleration can be tolerated and the engine operates at the same efficiency, the improvement with a 10 percent reduction of aerodynamic drag could be as high as 3 percent (10 percent × 0.3). Argonne calculations for the improvement in fuel consumption show that without engine modifications a 10 percent reduction in aerodynamic drag would result in about a 0.25 percent reduction in fuel consumption for the urban cycle and a 2.15 percent change for the highway cycle.

Car Body Design and Interiors

It is well established that a reduction in vehicle mass reduces fuel consumption. The specific relationship between mass reduction and fuel consumption, however, is complex and depends on many factors:

- Amount of mass reduction,
- Driving cycle,
- Type of engine, and
- Secondary benefits, such as whether or not other vehicle systems are redesigned to match the new vehicle

[7] Ratio spread is defined as the ratio of first gear divided by the ratio of the top gear. As an example, for a typical 6-speed automatic transmission, the low-gear ratio would be 4.58:1 while that of the sixth gear would be 0.75:1. The ratio spread would then be 4.58/0.75, which equals 6.1.

mass, as with, for example, engine downsizing, retuned transmission, and reduced components for crash management, braking, fuel storage, and so on.

A midsize car body structure with closure panels (no trim or glass) can weigh approximately 800 pounds (about 25 percent of the vehicle's total curb weight). Should the mass reduction be significant, a secondary benefit can accrue from reducing the size of the needed power train, braking systems, and crash management structures. These secondary benefits are difficult to estimate but can potentially approach an additional 30 percent reduction in mass, and these secondary benefits can help offset the cost of the initial effort (IBIS Associates, 2008).

A basic estimate of the relationship between fuel economy and mass is provided by the Department of Energy (Carpenter, 2008) and also by the Laboratory for Energy and Environment at the Massachusetts Institute of Technology (Cheah et al., 2007). A rule of thumb is a 6 to 8 percent improvement in fuel economy (or, equivalently, a reduction of 5.7 to 7.4 percent in fuel consumption) for every 10 percent drop in weight when secondary benefits are included that indirectly accrued from having lower mass.

In a study conducted by Ricardo, Inc. (2007), and sponsored by the Aluminum Association, this relationship was simulated for several vehicles loaded with from 2 to 5 passengers. The gasoline-powered vehicles simulated are listed in Table 7.2.

Two scenarios for these vehicles were simulated. The first scenario evaluated the impact on fuel economy when everything about the vehicle remained unchanged except for a reduction in vehicle mass. The second scenario resized the engine to reflect comparable vehicle performance (the benefits of other reductions in mass such as a smaller gas tank, smaller brakes, etc. were not included). In this scenario, the engine required less power because of the reduction in mass, and therefore, fuel economy was further improved. The vehicle type was not a major differentiator of fuel economy impact; Table 7.3 shows the range of impact on fuel economy for all types.

Table 7.3 shows the results of the Ricardo, Inc., simulation calculating the potential impact on fuel consumption from reduction of mass. The range shown in the results is due to summarizing a composite of simulation runs for different vehicle models and power trains. This discrepancy (range of fuel economy impact) in fuel economy improvement increases for different vehicle types as the reduction in mass increases from 5 to 20 percent. However, if the engine is resized to match each level of mass reduction (to maintain original vehicle performance), the range of fuel economy improvement across the vehicle classes is fairly small. This observation points to the importance of matching engine performance to vehicle mass. For small (under 5 percent) changes in mass, resizing the engine may not be justified, but as the reduction in mass increases (greater than 10 percent), it becomes more important for certain vehicles to resize the engine and seek secondary mass reduction opportunities.

Physical vehicle testing has confirmed the reductions in fuel consumption associated with reductions in vehicle mass. For an internal combustion engine, the effect of mass reduction is greater with a city driving cycle versus a highway cycle because of the frequent acceleration/deceleration of mass. For example, vehicles (combination of compact, midsize, and SUV classes) powered by internal combustion engines can reduce fuel consumption approximately as follows (Pagerit et al., 2006): 0.1 gallon per 100 miles driven can be saved with, approximately,

- 190 pounds mass reduction—city cycle, and
- 285 pounds mass reduction—highway cycle.

As discussed in Pagerit et al. (2006) and further supported by the Ricardo, Inc., study, the improvement gained from reduction of mass (expressed as fuel consumption and not miles per gallon) is the same regardless of the weight of the vehicle. Unlike changes in rolling resistance and aerodynamics, re-

TABLE 7.2 Vehicle Mass Assumptions for Ricardo, Inc. (2007) Study to Assess Effects of Mass Reduction on Fuel Economy

Type of Vehicle	Initial Weight (lb)	Load Weight (lb)	5% Reduction (lb)	10% Reduction (lb)	20% Reduction (lb)
Small car	2,875	300	3,031	2,888	2,600
Midsize car	3,625	450	3,894	3,713	3,350
Small SUV	4,250	550	4,588	4,735	3,950
Large SUV	5,250	750	5,738	5,475	4,950

NOTE: The 5 percent, 10 percent, and 20 percent mass reduction applies to the initial vehicle weight and not the load.

TABLE 7.3 Impact on Fuel Consumption Due to Reduction of Mass in Study by Ricardo, Inc. (2007)

Vehicle Mass Reduction from Baseline Vehicle	5% Mass Reduction	10% Mass Reduction	20% Mass Reduction
Mass reduction only	1-2%	3-4%	6-8%
Mass reduction and resized engine	3-3.5%	6-7%	11-13%

ducing mass not only reduces the amount of tractive energy needed but also permits a reduction in power train (engine downsized or transmission shift changes) without adversely affecting performance (acceleration). A 10 percent reduction in mass and power for the reference vehicle should reduce fuel consumption by about 5.7 to 7.4 percent (or 6 to 7 percent). In a conventional vehicle, the energy used to accelerate the mass is mostly dissipated in the brakes, whereas in a hybrid, a significant fraction of this braking energy is recovered, sent back to the battery, and reused. Thus, mass reduction in hybrid vehicles is less important than in conventional vehicles. The complexity of mass reduction increases when a conventional vehicle is compared with either a hybrid (which incurs additional battery mass) or a CI engine (which has greater power train mass). While reducing mass will always provide a fuel economy benefit, changing technology pathways (between SI, CI, or hybrid designs) has to recognize the impact that the new technology has on mass.

Rolling Resistance

A report on tires and fuel economy (NRC, 2006) estimates that a 10 percent reduction in rolling resistance will reduce fuel consumption by 1 to 2 percent. This reduction, however, is without changes in the power train. If the power train could be adjusted to give the same performance, then the benefit of a 10 percent reduction would be on the order of as much as 3 percent. Underinflated tires that are 20 percent below recommended inflation pressure (say, 35 psi) increase rolling resistance by 10 percent, and thus increase fuel consumption by 1 to 2 percent (Goodyear Tire & Rubber Company, 2009).

Again as discussed above under "Aerodynamics," if a reduction in rolling resistance is combined with a reduction in aerodynamics and mass, the power train can be significantly modified to improve efficiency. As indicated in Chapter 2, rolling resistance accounts for about a third of the energy going to the wheels for the city as well as the highway cycles. Reducing mass, aerodynamics, and rolling resistance by 10 percent reduces fuel consumption by about 10 percent with power train resizing and other drive train adjustments (e.g., changes in transmission shift points, axle ratios). As noted earlier, vehicle mass reduction for a hybrid is not as effective since some of the energy going to the brakes is recovered.

Vehicle Accessories

The opportunity may exist to decrease fuel consumption (in gallons per 100 miles driven) by about 3 to 4 percent with a variable-stroke HVAC compressor and better control of the amount of cooling and heating used to reduce humidity (Table 7.4). Estimates for further reductions that can be achieved by decreasing air conditioner load through the use of low-transmissivity glazing, reflective "cool" paint, and cabin ventilation while parked have not been determined. According to a Deutsche Bank report, electro-hydraulic power steering (EHPS) would reduce fuel consumption by 4 percent with an incremental cost of $70, while electric power steering could improve 5 percent with an incremental cost of $120, but there is little information on how this estimate was obtained (Deutsche Bank, 2008). A TRW study (Gessat, 2007) showed that while a conventional hydraulic power steering system consumed 0.35 L/100 km, the best TRW electro-hydraulic steering system consumed 0.07 and an electric power steering system 0.02. These figures are relative to a small vehicle with a 1.6-L engine. In its study of CO_2-reducing technologies for the EPA (EPA, 2008), Ricardo, Inc., found that electric power steering (EPS) reduced combined fuel consumption by about 3 percent based on FSS calculations. From this and the estimates provided in recent regulatory activities by NHTSA and EPA, the committee estimated that EPS reduces combines fuel consumption by about 1 to 3 percent on the EPA 55/45 combined cycle, which is the basis for the CAFE standard. However, the committee recognizes that the reduction of fuel consumption could be as high as 5 percent under in-use driving conditions.

Transmission Technologies

Fuel consumption reductions generally increase with additional transmission speed ratios, although interaction effects between engine technologies that reduce pumping losses and increase the number of transmission speeds are important, as noted earlier. However, since the costs also increase and the marginal gain for each additional speed gets smaller, there are diminishing returns. Table 7.5 lists the transmission technologies and estimated reductions in fuel consumption. The basis of this table is baseline engines

TABLE 7.4 Potential Reduction of Fuel Consumption with the Use of Vehicle Accessories

Vehicle Accessory	Reduction in Fuel Consumption (%)	Comments
Variable-stroke HVAC compressor	3-4	Improved cooling, heating, and humidity control
Low-transmissivity glazing, cool paint, parked-vehicle ventilation	~1	Lower heat buildup in vehicle decreases air-conditioning load
Electrohydraulic power steering	4	Combined electric and hydraulic power for midsize to larger vehicles reduces continuous load on engine
Electric power steering	1-5	Electric power steering for smaller vehicles reduces continuous load on engine—smaller benefits (1-3%) estimated for the FTP

without significant valve event modulation technologies or cylinder deactivation.

TIMING CONSIDERATIONS FOR INTRODUCING NEW TECHNOLOGIES

The timing for introducing new fuel consumption technologies can significantly influence cost and risk. The maturity of a technology affects its cost and reliability. Automobile companies have sophisticated *product and process validation procedures* that must be adhered to before products can be scaled up for mass production, or they expose themselves to large warranty or product liability concerns. Many vehicle changes are timed for implementation around the product development process to minimize cost and quality concerns. Lower-volume and higher-end vehicles often have new technologies applied first for several reasons. The lower volumes mitigate the exposure to risk, and the higher-end vehicles can bear the higher initial early cost of a new technology. During this period, competition brings the technology cost down while the supply chain develops for higher volumes in the future.

An important consideration for introducing new technologies that have broad impact concerns the product development process of new vehicles. Aggressive use of lightweight materials to obtain secondary benefits; power train modifications; and body shape modifications (to improve aerodynamics), for example, may have to be timed with future product development phases. Although material substitution for components can occur throughout the life cycle of a car in many cases, the mass saved in this way is relatively minor. Considering how to reduce mass to achieve greater energy savings requires a broad systems evaluation and reengineering of the vehicle. Once a vehicle has been validated and tooled for a specific design and production has begun, new development costs are planned for future model changes. Most significant modifications have to occur around various phases of the vehicle's production life.

Automobile manufacturers differ significantly in their approach to introducing new products. Manufacturers based in Asia, for example, are known for having shorter product life cycles but often implementing lower levels of engineering redesign at changeover. Manufacturers based in Europe and North American have traditionally had longer product cycles with a greater amount of engineering applied at changeover. There are always exceptions to these generalities even within a manufacturer, depending on the vehicle model. The strategy to implement engineering changes on a regional vehicle (e.g., North America only) versus a global platform can greatly impact timing and cost. Entire textbooks have been written around product timing for manufacturers, and so a discussion here can at best only introduce the inherent issues that affect cost and timing for any manufacturer.

Generally, 2 to 3 years is considered the quickest time frame for bringing a new vehicle to market. A significant amount of carryover technology and engineering from other models (or previous vehicle models) is usually required to launch a new vehicle this quickly. In some cases, so much of the vehicle is replicated that the new vehicle is considered a "freshened" or "re-skinned" model. The ability to significantly influence vehicle performance (e.g., through light-weighting, changing power trains, etc.) is minimal because so much of the vehicle is unchanged. More substantial changes to the vehicle occur over longer periods of time. Newly styled, engineered, and redesigned vehicles can take from 4 to 8 years, each with an increasing amount of new content.

Automobile producers generally have product development programs (PDPs) spanning at least 15 years. PDPs are extremely firm for 3 to 5 years due to the need for long-lead-time items such as tooling or supplier development requirements, and the need for extensive testing of major items such as those required for fuel economy, emissions, and safety regulations, and confirmation of reliability and durability. In general, model changeovers can be categorized into five areas (freshen, re-skin, restyle, reengineer,

TABLE 7.5 Transmission Technologies and Estimated Reductions in Fuel Consumption

Technology	Fuel Consumption Reduction[a] (%)	Comments
Five-speed automatic transmissions	2-3	Technology can also improve vehicle performance
Six-speed automatic transmissions	3-5	
Seven-speed automatic transmissions	5-7	
Eight-speed automatic transmissions	6-8	
Dual-clutch automated manual transmissions (6-speed) (DCT)	6-9	Original automatic transmissions with conventional manual transmissions supplemented with electro-hydraulic clutch and shift actuators have been replaced with DCTs
Continuously variable transmissions	1-7	Some issues related to differences in feel and engine noise; improvements depend on engine size

NOTE: Values based on EEA (2007) with adjustments to reflect range of values likely to occur.

[a]Improvements are over a 2007 naturally aspirated SI-engine vehicle with 4-speed automatic transmission of similar performance characteristics.

and redesign; see *Automotive News,* July 14, 2008, p. 28). These five categories and their potential for effecting fuel consumption improvements are described in Table 7.6. It is not accurate to say that every vehicle progresses through every one of these phases. It is possible to skip a re-skin and jump to a restyle, for example. Also, not every vehicle will be redesigned in 6 to 8 years because many factors affect this timing (market demand, finances, etc.). The potential for impacting fuel consumption is only a rough approximation, and none of these estimates consider the inclusion of hybrid or alternative power trains. The estimates for reducing fuel consumption shown in Table 7.6 are not additive (from previous changeover phases). Fuel consumption estimates also assume comparable vehicles of the same size and performance (including crash worthiness, electronic content, and other factors that are often adjusted with new vehicles).

The engine development process often follows a path separate from those of other parts of the vehicle. Engines have longer product lives, require greater capital investment, and are not as critical to the consumer in differentiating one vehicle from another as are other aspects of the car. Also, consumer-driven changes for styling change faster than the need to introduce new power train technologies. The power train development process evolves over closer to a 15-year cycle, although refinements and new technologies will be implemented throughout this period. Also, because of the complexity, costs, and resources required to launch a new power train, it is unusual to launch a new engine-related transmission simultaneously. The development of new technologies over a 15-year life cycle can be substantial, and the performance improvement for fuel consumption can be substantial with a new power train.

The estimates in Table 7.6 are based on business as usual. The "frequency" is the time from concept through prototyping, production vehicle design, tooling release, verification testing on preproduction vehicles, and start of full-scale production. Shorter time frames are possible, especially if more vehicle content is carried over between PDPs to reduce engineering, testing, etc., but this limits the degree of model changeover. Urgency to introduce new vehicles (e.g., smaller and more fuel efficient vehicles) can accelerate the nominal duration of each PDP phase, but the investment cost will grow.

Modest improvements in fuel consumption can be achieved early in the PDP cycle (e.g., freshen and re-skin stages) by introducing more aerodynamic designs and low-rolling-resistance tires. A greater impact on reducing fuel consumption can come from changes in engine, transmission, and mass reduction later in the PDP when the vehicle is redesigned or reengineered. Restyled vehicles allow for material substitution on a part-by-part basis, but without changing entire subassembly structures. Often, the substitution might be for a higher-strength metal with a thinner gage in place of the current material. Tooling and assembly processes may be altered somewhat to accommodate the new material. A reengineered vehicle allows for changing the design of major subassemblies (engine compartment, closure panels, body sides, etc.), thus allowing for entirely new approaches to reducing mass. Reengineered vehicles normally require crashworthiness testing

TABLE 7.6 Vehicle Product Development Process (non-power train) and Timing Implications to Effect Fuel Economy Changes

Type of Model Change	Frequency (Years)	Description	Fuel Consumption Reduction	Opportunities to Impact Fuel Consumption	Investment Cost
Freshen	2-3	Sheet metal untouched, may include new grille, fascia, headlights, taillights, etc.	Little to none (≤3%)	Minor impact on mass; possible impact with aerodynamics and tires	Low
Re-skin	3-5	Minor changes to sheet metal	Little to none (≤5%)	Same as above and vehicle accessories	Modest
Re-style	4-8	Extensive changes to exterior and interior	Minimal (5-8%)	Some impact on mass (mostly interior components); possible impact with aerodynamics, tires, and vehicle accessories	High
Re-engineer	4-8	Extensive makeover of vehicle's platform, chassis, and components to reduce noise, vibration, and harshness and improve qualities such as ride, handling, braking, and steering (this degree of change or the next may require recertification and crash testing), body restyling often concurrent with this phase	Moderate (7-14%)	Mass reduction opportunity with part-by-part material substitution (e.g., aluminum or high-strength steel); possible impact with aerodynamics, tires, and vehicle accessories	Very high
Redesign A	6-8	New platform, new interior and exterior styling; engine and transmission carried over; some structural subsystems possibly reengineered	Significant (13-18%)	Entire vehicle structure—opportunity to introduce lightweight materials throughout entire vehicle; impact from aerodynamics, tires, and vehicle accessories	Very high

and incur significant additional costs because of the reengineered designs. The redesigned vehicles start with a "clean sheet" affording the benefits of a reengineered vehicle, along with more optimal matching of the power train to the lighter-weight structure. In general, a redesign results in a new vehicle platform that in many cases replaces existing vehicles.

Aerodynamics

Reductions of drag coefficient Cd by 5 percent or so (up to 10 percent) have been taking place and will continue. A 5 percent reduction in aerodynamics can be achieved with minimal cost through vehicle design, and larger reductions can be achieved by sealing the undercarriage and installing covers/shields (e.g., in the wheel well areas and underbody). Elimination of outside rear view mirrors will require changes in safety regulations and improvement in vision systems. Since these changes can be costly, they are unlikely to be implemented soon except on high-end vehicles. In the longer term (about 10 years), 5 to 10 percent reductions in aerodynamic drag are plausible, but this may come with some compromise in vehicle functionality.

Car Body Design and Interiors

Reductions in weight have been taking place and will continue in the near term with reductions from 10 percent (with mostly conventional and high-strength steels) to 25 percent (with high-strength steel structures, aluminum closure panels, and body/interior components made from various lightweight materials). Table 7.7 provides an overview of the timelines for the introduction of new materials for various vehicle components. Today's new vehicles already are composed of upward of 40 to 50 percent high-strength steel (over 480 MPa yield strength), but higher-strength steels (advanced high-strength steels) are being developed (up to 1,000 MPa) that could replace even the current high-strength steel. Various vehicle components for which isolated material substitution can take place will also be the norm. For example, Ford recently indicated that aluminum calipers replaced steel ones, thus saving 7.5 pounds per vehicle. Also, aluminum wheels replaced steel wheels, resulting in 22 pounds saved per vehicle. More aggressive application of aluminum to car doors can also save another 20 pounds per door, but at a higher cost. Substitution of material in other components can also be expected, including the wiring harness. Substituting copper-clad aluminum wiring for all copper wiring can save 10 or more pounds per vehicle, but usually at a higher cost.

More aggressive reduction of mass is feasible at higher cost if aggressive targets of greater than 25 percent are set. Reduction of mass at the 50 percent level can be attained in the body with a mostly aluminum structure (probably using a space frame design), but this approach will be cost prohibitive under most conditions for high-volume vehicles.

The use of composite structures involving materials such as carbon fiber will need significant cost reduction and supply chain development over the next 15 years. The committee does not expect to see significant inroads in this time frame by this technology except in low-volume (specialized applications), high-performance vehicles. Other polymer/reinforced composites, etc. will continue to make inroads in the vehicle interior where steel or aluminum is used currently for strength. For example, all-polymer/reinforced composite instrument panels (without rear steel reinforcements) are likely to make it to production soon.

As production processes continue to be developed, broader application of both magnesium and titanium can be expected, such as for magnesium engine blocks that weigh approximately 30 pounds less than aluminum ones (see Table 7.7). Magnesium will likely make inroads for component parts such as suspension arms and interior dash panels and seating brackets. Titanium will continue to find application in suspension springs, valve springs, valves, connecting rods, and exhaust systems, resulting in 35 to 40 percent savings in mass over steel components.

Rolling Resistance

Low-rolling-resistance tires are already used by OEMs. The committee does not expect significant additional improvements without sacrificing performance. Since replacement tires are on most vehicles on the road today, a campaign to educate purchasers of replacement tires of the possibility of fuel savings is a good way to reduce fuel consumption. More vehicles today are being offered with low-tire-pressure monitors to warn the driver of underinflated tires for safety and fuel economy.

Vehicle Accessories

Variable stroke compressors and reduction of subcooling are being developed and should appear in vehicles in the next 3 to 5 years. Because the current duty cycle measuring fuel consumption does not recognize HVAC systems, there is no motivation to introduce these systems because they incur additional costs. However, the proposed new EPA test procedure may cause new interest in introducing this technology.

COSTS OF NON-ENGINE TECHNOLOGIES

Aerodynamics

A 5 percent reduction in aerodynamics can be achieved with minimal cost through vehicle design. Slightly more aggressive reductions can be achieved by sealing the undercarriage and installing covers/shields (e.g., in the wheel well areas and underbody) costing in the tens of dollars. A 10 percent reduction in aerodynamics may be aggressive, calling for wind deflectors (spoilers) and possibly elimination of rear view mirrors, which would cost a few hundred dollars.

TABLE 7.7 Estimated Timeline for Introduction of New Materials by Type of Component

Timing	High-Strength Steel	Aluminum	Magnesium	Plastics and Polymer—Composites
Current or near term (3-5 years)	Body rails, door sills, B-pillar, side roof rails, underbody, front suspension subframe, bumper beams, cross-members, brackets and reinforcements, exterior body panels, body side ring, longitudinal rails, wheels	Hood, deck lid, engine block and cylinder lining, front suspension subframe, bumper beams, rear suspension knuckles, steering hanger beam, power train components (castings), condenser/radiator wiring harness	Instrument panel, seat components Brackets Crash structures Intake manifold	Truck box Outer skin panels (doors, fenders, etc.) Instrument panel Bumpers Trim Engine parts (intake manifold, cover, etc.)
Future (5-10 years)	Same as above, only with higher-strength steels	Doors, exterior body panels (fender, roof)	Door, inner Engine block	Body side ring Roof Side pillar (B or C) Underbody Seat components Sound dampening Glass (polycarbonate)
Long term (>10 years)	New steels with greater formability allowing application to more complex part shapes and exterior panels; less steel overall in the vehicle	Increased applications (depending on material cost); subassemblies such as engine compartment, chassis, instrument panels; overall, more aluminum in the vehicle	Limited increase in applications; possibly transmission parts	New materials will be developed with higher strength, allowing them to be applied to more structural parts. Mixed-material bonding will be developed. Overall, more plastics/polymers will be in the vehicle.

Car Body Design and Interiors

The term "material substitution" often misrepresents the complexity and cost comparison when one material is substituted for another one. The cost to change materials in the vehicle, from an incumbent material to a lighter-weight material, is a function of capital and variable costs:

Fixed Costs (up-front investment costs)
- Design and engineering
- Prototype development and testing
- Tooling: fabrication, dimensional measurement, and assembly

Variable Costs (a function of the volume of production)
- Production and assembly labor cost
- Production equipment
- Material
- Joining (welding, adhesive, sealing, riveting, etc.)

An added complexity results with material substitution because part design is material dependent, and the redesigned part may provide (and often does) different functionality than the original part. For example, a molded plastic part can take on more complexity than a formed steel part, and so the direct comparison should also take the difference in functionality into account. Also, two or more parts may get integrated into a single part with one material versus that of another, and so the subsystem of parts has to be evaluated for a cost and performance comparison.

Most cost-effective materials today for reducing mass are high-strength steel and aluminum. Both materials can replace many incumbent steel parts or assemblies, and the structural components that are among the heaviest parts offering the greatest opportunity will be targeted. Plastics, composites, and other metals (magnesium and titanium) will be used on a somewhat limited basis because of cost.

In recent years, reductions in mass have been realized in the body, interior, and power train by introducing new materials such as high-strength (and advanced high-strength) steels, plastics (not including carbon fiber), and aluminum. Magnesium has also been used to reduce mass, but to a much lesser extent. In the near future (5 years), the committee expects continued mass reduction following the same pattern; through continued introduction of more and higher-strength steels, aluminum, plastics/polymers, and to a lesser extent other materials such as magnesium.

Although there are research and development costs to develop new high-strength steels and new manufacturing processes for them, once developed they have minimal net long-term incremental cost over mild steel. Tooling, fabrication, and joining costs tend to be higher for these materials because of the material strength, which has to be added to the net cost difference. Although the cost per pound of high-strength steel is higher than mild steel, less of it is needed. Hence, a 10 or 20 percent material cost premium will be offset by using 10 to 20 percent thinner steel. As high-strength steels are introduced, their net incremental cost approaches zero after a period of maturity. The DOE estimates that, on average, substituting high-strength steel for mild steel results in about a net increase in material cost of 10 percent (see Carpenter, 2008).

The cost to reduce mass (cost per pound of mass reduced) *increases* as the amount of reduced mass increases. The "low hanging fruit" of mass reduction using high-strength steel in basic applications can result in less than a 10 percent cost premium. However, increasingly aggressive reduction of mass requires more difficult parts and materials whose cost exceeds the 10 percent premium. For example, a 1 percent reduction in mass can generally be achieved at a multiplier of 1.0 to 1.1. More aggressive applications likely require more expensive materials or more expensive fabrication and joining methods, or affect the costs of other parts in the vehicle. As the aggressiveness increases (to 5 percent, 10 percent, or even 20 percent), more materials and processing options need to be considered that further increase cost. The committee believes that a 10 percent reduction in mass is achievable with a mix of materials (high-strength steels, aluminum, composites, and other metals) for approximately $2.00 per pound of mass eliminated (see Table 7.8). More aggressive reductions will cost more than $2.00 per pound.

Aluminum costs more than steel and has some forming and joining limitations that prevent its use in some applications. An incremental cost of aluminum over steel body parts in the range of 30 to 100 percent has been estimated (Carpenter, 2008; Bull, 2008). The Aluminum Association estimates that the average increment is 30 percent at the low end (premium cost per pound of mass eliminated). At the mid-point of this range, the incremental cost is $1.65/pound of mass eliminated. Higher costs will be incurred (approaching $2.00/lb cost premium) as more aggressive reduction of mass reduction is attempted.

The body of a baseline vehicle (mostly steel) weighs approximately 800 pounds. An aluminum-intensive body weighs approximately 45 percent less, or 440 pounds. The estimated cost for this savings in weight is in the range of $468 ($1.30/lb) to $594 ($1.65/lb). Mass reduction in other vehicle systems such as power train, wheels, chassis, and interior would typically come at similar or slightly higher incremental cost per pound saved. Vehicle interiors (including seats, door trim, headliners, instrument panel components, etc.) constitute approximately one-third of the vehicle mass (1,000 pounds in a 3,000-pound vehicle). By using lightweight materials, Byron Foster at Johnson Controls plans to eliminate 30 percent of the interior mass (Forbes, 2008). If the same incremental cost used for the body is assumed, approximately 300 pounds eliminated would cost $390 ($1.30/lb) to $495 ($1.65/lb).

Other opportunistic components in the vehicle include the power train, chassis, and wheel components. Many of these components have been light-weighted already with high-strength steel and aluminum where practical. One next step would be to transition to more magnesium, which comes with a cost premium of perhaps 50 percent or more over that for aluminum.

Secondary Savings Benefits

An important consideration with mass reduction is that its effects on fuel consumption can cascade. As the mass of a vehicle is reduced in, say, the body or interior, other components of the vehicle can be reduced in size as a consequence. For example, brakes, fuel system, power train, and even crash-management structures can all be downsized for a lighter vehicle. In the study conducted by Ricardo, Inc., (2007) for the Aluminum Association, the rule of thumb generated was that for every pound eliminated in the vehicle structure, an additional 0.30 lb (30 percent) of mass could be reduced in other areas of the vehicle. If this rule of thumb is applied and mass reduction comes at a cost of $1.65/lb, then at an additional 30 percent of secondary mass savings (0.3 lb) the net cost per pound becomes $1.65/1.3 lb, which becomes $1.27/lb. It is important to note that achieving secondary savings typically requires reengineering one or more systems on the vehicle, and this would likely be performed according to the product development timing plan (see above the section "Timing Considerations for Introducing New Technologies"). So the 30 percent secondary benefit is achieved in the long term and not necessarily when the initial reduction in mass is achieved.

Rolling Resistance

The incremental cost for low-rolling-resistance tires is estimated to be $2 to $5 per tire, but there is some evidence that suggests that these tires may slightly compromise stopping distance. One tire manufacturer suggested that tires that do not compromise stopping distance or tread wear could cost 10 to 20 percent more than conventional tires. (Note: The uncertainty about low-rolling-resistance tires with respect to increased tread wear and stopping distance is the reason for increasing the estimated cost beyond the $1.00 per tire cost cited in NRC (2006). The NRC (2006) study recognized that an acceptable increase in tread wear and stopping distance might occur. However, to eliminate this increase, additional costs can be expected over the $1.00 estimate.)

TABLE 7.8 Committee's Estimate of Cost to Reduce Vehicle Mass (based on 3,600-lb vehicle)

Mass Reduction (%)	Low Cost/lb ($)	High Cost/lb ($)	Average Cost/lb ($)	Mass Saved (lb)	Low Total Cost ($)	High Total Cost ($)
1	1.28	1.54	1.41	36	46.08	55.30
2	1.33	1.60	1.46	72	95.76	114.91
5	1.50	1.80	1.65	180	270.00	324.00
10	1.80	2.16	1.98	360	648.00	777.60

Vehicle Accessories

Table 7.9 shows the committee's estimates of the costs for vehicle accessories that could improve the fuel consumption of light-duty vehicles.

Transmission Technologies

The estimated retail price equivalent for each transmission technology is provided in Table 7.10. As was the case for the engine technology chapters (e.g., Chapters 4 and 5), the baseline for transmission costs is the 4-speed automatic typical of 2007 model-year vehicles. Cost estimates are from the two sources considered (EEA, 2007; Martec Group, Inc., 2008). As can be seen from Table 7.10, the cost estimates for the 5-, 6-, 7-, and 8-speed automatic transmission replacements for the baseline 4-speed automatic have a considerable numerical range. In addition to the cost estimates, Table 7.10 also includes cost estimates converted to RPE using the RPE multiplier of 1.5. Besides the estimates for 5-, 6-, 7-, and 8-speed automatic transmission replacements, estimates are also included for DCTs and CVTs. The DCT estimates reflect an even wider numerical range than those for the automatics. For example, the 6-speed automatic cost estimates range from $133 to $215, whereas the estimates for the wet-clutch, 350 N-m torque capacity range from $140 to $400.

Although DCT units have been in high-volume production for a number of years, until recently only the VW-Audi group, working closely with one supplier, has produced such a transmission. As a result, the number of cost estimates available to the committee was limited. When additional information was sought by the committee, the results reflected the still-emerging knowledge base about this transmission type. One estimate, based on a detailed teardown study conducted by FEV, Inc., for the EPA, estimated the cost of 6-speed DCTs with 350 N-m torque capacity and wet clutches at over $147 less than that for a 6-speed automatic (Kolwich, 2010). However, OEMs considering tooling up their own equivalent units had also made careful estimates of the high-volume piece cost increase of DCT6s. These OEM estimates were that high-volume DCT6s would cost nearly $200 more than 6-speed automatics. Thus, the range between estimates was approximately $350. At the present time, insufficient information is available to narrow this wide range.

SUMMARY

There is a range of non-engine technologies with varying costs and impacts to consider. Many of these technologies are continually being introduced to new vehicle models based on the timing of the product development process. Coordinating the introduction of many technologies with the product development process is critical to maximizing their impact and minimizing their cost. Relatively minor changes that do not involve reengineering the vehicle can be implemented within a 2- to 4-year time frame. This could include efforts such as aiming for minor reductions

TABLE 7.9 Estimated Incremental Costs for Vehicle Accessories That Improve Fuel Consumption

Description	Source of Cost Estimate	Estimate
HVAC—variable stroke, increased efficiency (humidity control, paint, glass, etc.)	U.S. Environmental Protection Agency[a]	$70-$90
Electric and electric-hydraulic power steering	Deutsche Bank	$70-$120
Thermoelectric energy recovery		Several hundred dollars

[a]The U.S. EPA has estimated the cost associated with improving the energy efficiency of the A/C system and reducing refrigerant leakage from the system at less than $110 to the consumer (ANPR-HQ-OAR-2008-0318; FRL 8694-2). With an RPE of 1.75 the cost to the original equipment manufacturer would be just over $60.

TABLE 7.10 Estimates of Replacement Costs for Transmission Technologies Relative to 2007 4-Speed Automatic Transmissions

Transmission Type	$Cost (EEA, 2007)	$RPE (EEA, 2007)	$Cost (Martec, 2008)	$RPE (Martec, 2008)
5-speed automatic	133	200	—	—
6-speed automatic	133	205	215	323
7-speed automatic	170	255	—	—
8-speed automatic	—	—	425	638
DCT (dry clutch, 250 N-m)	—	—	300	450
DCT (wet clutch, 350 N-m)	140	210	400	600
CVT (engine <2.8 liter)	160	240	—	—
CVT (engine >2.8 liter)	253	380	—	—

NOTE: RPE values were determined using a cost multiplier of 1.5.

in mass (material substitution), improving aerodynamics, or switching to low-rolling-resistance tires. More substantive changes require longer-term coordination with the PDP because reengineering and integration with other subsystems are necessary. This could include resizing the power train/transmission or aggressively reducing mass (e.g., changing the body structure). Substantive changes like this will take 4 to 8 years to adopt. The cost estimates provided in this chapter all assume coordination with the PDP to help contain costs and achieve maximum impact.

Two important technologies impacting fuel consumption addressed in this chapter are light-weighting and transmissions. Light-weighting has almost unlimited potential because vehicles can be made very light with exotic materials, albeit at potentially high cost. The incremental cost to reduce a pound of mass from a vehicle tends to increase as the total amount of reduced mass increases, leading to a curve with diminishing returns. About 10 percent of vehicle mass can be eliminated at a cost of roughly $700 (or about $2.00/lb; see Table 7.11). If the aggressiveness to reduce mass increases much beyond 10 percent, it is necessary to begin addressing body structure design (such as considering an aluminum-intensive car), and the cost per pound increases. A 10 percent reduction in mass over the next 5 to 10 years appears to be within reach for the typical automobile, considering the current baseline.

Transmission technology has significantly improved and, like other vehicle technologies, shows a similar curve of diminishing returns. Planetary-based automatic transmissions can have five, six, seven and eight speeds, but with incremental costs increasing faster than their impact on fuel consumption. Continuously variable transmissions have been available on the market for a number of years, but their rate of implementation seems to have flattened out, suggesting that future new implementations will be limited in number. DCTs are in production by some vehicle OEMs (e.g., VW/Audi DSG), and new DCT production capacity has been announced by other vehicle OEMs and suppliers. It is therefore expected that the predominant trend in transmission design will be conversion both to 6- to 8-speed planetary-based automatics and to DCT automated manuals, with CVTs remaining a niche application. Because of the close linkage between the effects of fuel-consumption-reducing engine technologies and those of transmission technologies, the present study has considered primarily the combined effect of engines and transmission combinations rather than potential separate effects.

TABLE 7.11 Summary of the Committee's Findings on the Costs and Impacts of Technologies for Reducing Light-Duty Vehicle Fuel Consumption

Fuel Consumption Technology	Description and Approximate Manufacturing Cost	Impact on Fuel Consumption (%)	Comments
Mass reduction (assume 3,600-pound vehicle)	1% (36 lb); $46-$55	0.25	Material substitution
	5% (180 lb); $270-$324	3-3.5[a]	Material substitution
	10% (360 lb); $648-$778	6-7[a]	Aggressive material substitution
	20% (720 lb); $1,600+	11-13[a]	Redesigned body with aluminum and composite-intensive structures
Transmission	Five-speed automatic transmissions; $133	2-3	Can also improve vehicle performance
	Six-speed automatic transmissions; $133-$215	3-5	Can also improve vehicle performance
	Seven-speed automatic transmissions; $170-$300	5-7	Can also improve vehicle performance
	Eight-speed automatic transmissions; $425	6-8	Can also improve vehicle performance
	Dual-clutch automated (DCT) manual transmissions (6/7 speed); $300 (dry clutch), –$14-$400 (wet clutch <350 N-m)	6-9	DCTs have replaced original automated manual transmissions
	Continuously variable transmissions; $150 (<2.8 L), $263 (>2.8 L)	1-7	Possible engine noise; not applicable to large engines
Aerodynamics	5 to 10% reduction in C_d (coefficient of drag); $40-$50	1-2	Wheel well and underbody covers, body shape, mirrors, etc.; bigger impact on highway drive cycle
Rolling resistance	Low-rolling-resistance tires; approximately $10 apiece ($30-$40)	1-2[b]	Stopping distance and durability can be compromised with inferior materials; optimal materials drive up costs
	Tire-inflation monitor; becoming standard equipment	0.7	Depends on monitor settings and driver behavior
	Low-drag brakes; becoming standard equipment	1	Most cars equipped already today
Electrical accessories	HVAC—variable stroke, increased efficiency (humidity control, paint, glass, etc.); $70-$90	3-4	Current FTP does not capture benefit (benefits reduced to 0.5-1.5% within Table 9.1)
	Electric and electric-hydraulic power steering; $70-$120	1-5	Electric for small cars, electric-hydraulic for bigger cars—benefits for the FTP are smaller (1-3%).

[a]With resized power train.
[b]Three percent may be feasible with resized power train.

Accessories are also being introduced to new vehicles to reduce the power load on the engine. Higher-efficiency air-conditioning systems are available that more optimally match cooling with occupant comfort. This includes, for example, humidity control, air recirculation, and increased compressor efficiency using a variable-stroke compressor. Electric and electric-hydraulic power steering also reduces the load on the engine by demanding power (electric) only when the operator turns the wheel, whereas the older technology relied on hydraulic power supplied by the engine all the time. An important motivating factor affecting the introduction of these accessories is whether or not their impact is measured during the official CAFE certification tests. The certification test currently does not take the air conditioner into account, and so there is little motivation to improve its efficiency and incur added cost; however, this situation may change with newly proposed test procedures.

Estimates for these technologies and several others are summarized in Table 7.11. The fuel consumption estimates assume ideal conditions, and there are important interaction effects among different technologies. Generally, it is not possible to apply two or more of the technologies in Table 7.11 and algebraically add the impacts on fuel consumption. The typical impact from multiple technologies will be less than the sum of their individual fuel consumption estimates.

FINDINGS

Finding 7.1: *Refresh/redesign.* With respect to reducing fuel consumption, recognition of product development process timing is important for minimizing the cost of implementing many new vehicle technologies. Only relatively modest changes can be made when vehicles are restyled, and secondary benefits from mass reduction are unlikely. The reengineering or redesign phases of product development offer the greatest opportunity for implementing new fuel-saving technologies, and this can occur from 4 to 8 years after initial introduction. Significant changes to power train and vehicle structure and materials can be made more easily at this time.

Finding 7.2: *Mass reduction.* Reduction of mass offers the greatest potential to reduce vehicle fuel consumption. To reduce mass, vehicles will continue to evolve with a broad mix of replacement materials that include high-strength steels, aluminum, magnesium, and reinforced plastics. These materials will be introduced on a component-by-component basis as companies move up the learning curve and continue to design for them. More aggressive efforts to reduce mass (by, say, 5 to 10 percent) will require system solutions (as opposed to material substitution solutions). Achieving a mass reduction of greater than 10 percent (as high as 20 percent) will require a significant change in vehicle design (such as a shift to an aluminum-intensive body or aggressive use of other higher-cost materials like carbon fiber) and will incur a significant increase in costs. The uncertainty and instability of commodity prices (e.g., for carbon fiber, resins, and aluminum versus steel) increase the risk to the vehicle manufacturer of adopting these new materials.

Finding 7.3: *Transmissions.* Another promising technology for reducing vehicle fuel consumption is transmissions with an increased ratio spread between the low and the high gears (e.g., 6-8 speeds) and dual-clutch transmissions that eliminate torque converters.

Finding 7.4: *Lower-energy-loss accessories.* A collection of relatively low-cost vehicle technologies can have a positive impact on reducing fuel consumption. Low-rolling-resistance tires, improvements to vehicle aerodynamics, and electric power steering can all cost less than $200 in total while reducing fuel consumption by about 10 percent, if HVAC is included as a component of real-world driving. Other technologies that can yield incremental reductions in fuel consumption are increased HVAC compressor efficiency, ultraviolet filtering, glazing, and cool/reflecting paints, but these technologies are not currently pursued very aggressively because they are not taken account of in the official CAFE certification tests. It would take more than the addition of HVAC in one of the five test schedules used to report fuel economy on the vehicle sticker to have a significant impact on the penetration of these technologies.

REFERENCES

Automotive News. 2008. A big fuel saver: Easy rolling tires (but watch braking). July 21.

Barrand, J., and J. Bokar. Reducing tire rolling resistance to save fuel and lower emissions. SAE Paper 2008-01-0154. SAE International, Warrendale, Pa.

Bull, M. 2008. Efficacy, cost, and applicability of lightweight structures to fuel economy. Presentation to the Committee for the Assessment of Technologies for Improving Light-Duty Vehicle Fuel Economy, June 3, Washington, D.C.

Carpenter, Jr., J., U.S. Department of Energy, Washington, D.C. 2008. Challenges to automotive composites. Presentation to American Chemistry Council, Plastics Division, Arlington, Virginia, May 20.

Cheah, L., C. Evans, A. Bandivadekar, and J.B. Heywood. 2007. Factor of Two: Halving the Fuel Consumption of New U.S. Automobiles by 2035. Laboratory for Energy and Environment, Massachusetts Institute of Technology, October.

Compere, A.L., W.L. Griffith, C.F. Leitten, and S. Petrovan. 2001. Improving the Fundamental Properties of Lignin-Based Carbon Fiber for Transportation Applications. Oak Ridge National Laboratory. Available at http://www.ornl.gov/~webworks/cppr/y2001/pres/120145.pdf.

Deutsche Bank. 2008. Electric cars: Plugged in. Deutsche Bank Global Markets Research Report. June 9.

EEA (Energy and Environmental Analysis, Inc.). 2007. Update for Advanced Technologies to Improve Fuel Economy of Light Duty Vehicles—Final Report. Prepared for U.S. Department of Energy. October.

EPA (U.S. Environmental Protection Agency). 2008. EPA Staff Technical Report: Cost and Effectiveness Estimates of Technologies Used to Reduce Light-Duty Vehicle Carbon Dioxide Emissions. EPA420-R-08-008. Ann Arbor, Mich.

Forbes. 2008. Automakers' natural appeal, Forbes Magazine, September 26.

Gessat, J. 2007. Electrically powered hydraulic steering systems for light commercial vehicles. SAE Paper 2007-01-4197. SAE International, Warrendale, Pa.

Goodyear Tire & Rubber Company. 2009. Letter from Donald Stanley, Vice President, Product Quality and Plant Technology, April 13, 2009; e-mail exchanges with Donald Stanley, January, 28, 2009, and February 25, 2009.

Heremans, J.P., V. Jovovic, E.S. Toberer, A. Saramat, K. Kurosaki, A. Charoenphakdee, S. Yamanaka, and G.J. Snyder. 2008. Enhancement of thermoelectric efficiency in PbTe by distortion of the electronic density of states. Science 321:554.

Hussain, Q., C. Maranville, and D. Brigham. 2009. Thermoelectric exhaust heat recovery for hybrid vehicles. SAE Paper 2009-01-1327. SAE International, Warrendale, Pa.

IBIS Associates. 2008. Benefit Analysis: Use of Aluminum Structures in Conjunction with Alternative Power Train Technologies in Automobiles. IBIS, Waltham, Mass.

Kolwich, G. 2010. Light-Duty Technology Cost Analysis—Report on Additional Case Studies. Prepared for U.S. Environmental Protection Agency. FEV, Inc., Auburn Hills, Mich.

Martec Group, Inc. 2008. Variable Costs of Fuel Economy Technologies. Prepared for Alliance of Automobile Manufacturers. June 1; amended September 26 and December 10.

NRC (National Research Council). 2002. Effectiveness and Impact of Corporate Average Fuel Economy (CAFE) Standards. Washington, D.C.: National Academy Press.

NRC. 2006. Tires and Passenger Vehicle Fuel Economy: Informing Consumers, Improving Performance—Special Report 286. Washington, D.C.: The National Academies Press.

Pagerit, S., A. Rousseau, and P. Sharer. 2006. Fuel economy sensitivity to vehicle mass for advanced vehicle powertrains. SAE Paper 2006-01-0665. SAE International, Warrendale, Pa.

Powers, W. 2000. Automotive materials in the 21st century. Advanced Materials and Processes, May.

Ricardo, Inc. 2007. Impact of Vehicle Weight Reduction on Fuel Economy for Various Vehicle Architectures. Prepared for the Aluminum Association, Inc.

Rugh, J.P., L. Chaney, J. Lusbader, J. Meyeer, M. Rustagi, K. Olson, and R. Kogler. 2007. Reduction in vehicle temperatures and fuel use from cabin ventilation, solar-reflective paint, and a new solar-reflective glazing. SAE Paper 2007-01-1194. SAE International, Warrendale, Pa.

8

Modeling Improvements in Vehicle Fuel Consumption

INTRODUCTION

The potential of technology to reduce fuel consumption can be estimated in three basic ways. One approach involves constructing an actual prototype vehicle with the technologies in question, performing the Environmental Protection Agency (EPA) city and highway dynamometer tests repeatedly, and then measuring the fuel consumption. Although there is some variability from test to test, this method is the most accurate but is also prohibitively expensive. A second approach is to construct a computer model that represents all of a vehicle's components and their interactions, including representations of the technologies for reducing fuel consumption, and to simulate the behavior of the vehicle over the federal test procedures. This method, which the committee refers to as full system simulation (FSS), is now the state of the art throughout the automotive industry for modeling fuel consumption. Although it is less expensive, FSS still requires very large expenditures of time and money if it is used to calibrate models to the 1,000 or so different vehicle configurations offered for sale in the United States each model-year and to test all relevant combinations of technologies. The third alternative is to construct an algorithm that adds discrete technologies to the set of base-year vehicle configurations and that then calculates their cumulative impact while attempting to account for interactions between them by means of adjustment factors. The committee refers to this third method as partial discrete approximation (PDA). The simplicity of the third approach allows fuel consumption impacts to be calculated for thousands of vehicles and tens of thousands of technology combinations. The key question is whether the third method can be executed with sufficient accuracy to support fuel economy regulation. The Volpe Model (Van Schalkwyk et al., 2009), used by the National Highway Traffic Safety Administration (NHTSA) in its rulemaking analyses, and the EPA's OMEGA model, used by the EPA in its rulemaking analysis (EPA and NHTSA, 2009), are PDA models that use data on technology costs and fuel consumption impacts from a variety of sources, including FSS models.

This chapter evaluates methods of estimating the potential to decrease automotive fuel consumption by changing vehicle design and technology. It begins with some general observations on modeling technologies' potential for reducing fuel consumption. Because of the technological complexities of vehicle systems, predicting how combinations of technologies might perform in new vehicle designs involves uncertainty. The present committee summarizes and discusses the method used by the National Research Council (NRC) Committee on the Effectiveness and Impact of Corporate Average Fuel Economy (CAFE) Standards in its 2002 report (NRC, 2002). It then goes on to compare and evaluate the two most widely used approaches to estimating the technological potential for reducing fuel consumption—PDA and FSS. Both methods are described in detail, and applications of the two methods to various types and configurations of vehicles are compared. Although it was able to make useful comparisons between modeling methods, the committee found that information comparing the results of either the FSS or the PDA method to real-world vehicles is scarce. The committee also comments briefly on the methodology used by the NHTSA in its 2011 Final Rule.

Recognizing the limitations of all modeling approaches, the committee considers the FSS method to be the state of the art and therefore the preferred method for estimating the potential of technologies to reduce fuel consumption. However, given the cost of FSS modeling at present, the committee believes that the PDA method, properly executed and supplemented with estimates of technology interaction effects developed by FSS or lumped parameter modeling, can be a reasonably accurate method for assessing the potential for reducing light-duty vehicle fuel consumption over a time horizon on the order of 10 years.

CHALLENGES IN MODELING VEHICLE FUEL CONSUMPTION

Along with the many potential benefits of using computer models to understand vehicle systems come limitations as well. In addition to enabling insight into how an overall vehicle system might operate, vehicle system modeling can also help measure the interactions between vehicular subsystems and how they affect overall vehicle performance. An understanding of the physics underlying these interactions is important when trying to estimate how future vehicles might perform with different combinations of technologies. All models are inherently simplifications of reality; the physics of real processes will always be considerably more complicated than that reflected in a model. In the end, impacts can only be known with certainty when a technological concept is realized in a real vehicle, and even then realizations of the same technological concept can differ from one vehicle to another. The meaningful question is whether any given model or methodology has sufficient fidelity to competent executions of the technological concept to achieve the goals for which the model has been developed.

With even the most complex and comprehensive models, there are challenges when modeling a *known* vehicle configuration, and even greater challenges when trying to predict the behavior of future vehicles using new combinations of technologies. When modeling a known or existing vehicle the principal problems are in capturing the desired dynamics to a sufficient level of detail or fidelity, and in collecting and inputting representative parameters or boundary conditions. The advantage of modeling a known vehicle is that data on the vehicle's actual performance are usually available to the modeler, and the data can be used to tune or validate the model's performance. Even for existing vehicles, however, experimental data sets are frequently sparse and may not include the precise performance situations of interest.

Detailed computer modeling of vehicle systems can be very expensive. Developing sufficient data on the performance of engines and other components, data that are not generally available in the open literature, is a major source of the expense of FSS modeling. An automobile company might spend many times the resources available to the committee to develop dynamic models to help answer the kinds of questions posed to the committee. On the order of 1,000 different vehicle configurations undergo fuel consumption testing each model year. FSS modeling of even the most promising combinations of advanced technologies for such a large number of vehicles would be expensive for federal agencies. PDA modeling, on the other hand, can be implemented in simplified algorithms that can estimate fuel consumption potentials for thousands of vehicles or more, considering virtually all logical combinations of technologies.

There are at least six sources of error in estimating the potential to reduce vehicle fuel consumption:

1. Differences between the attributes of the representative or typical vehicle being analyzed and the actual vehicles it represents;
2. Inaccuracies in the characterization of the base vehicle, especially its energy flows;
3. Inaccurate assessment of technology impacts, including the inability to fully represent the physics of a new technology in FSS modeling;
4. Differences in the implementation of a given technology from vehicle to vehicle;
5. Changes in the nature of a technology over time; and
6. Inaccurate estimation of the synergies among technologies and how they contribute to the overall end result of their combined application.[1]

In general, rigorous, quantitative assessments of these potential sources of error and their impacts on the potential to reduce vehicle fuel consumption are scarce.

In this chapter comparisons of the results of FSS and PDA (with lumped parameter modeling) are presented. In addition, the committee contracted with Ricardo, Inc., to perform a statistical analysis of FSS modeling. The goal was to determine whether a limited number of FSS runs could be used to generate accurate data on the main effects of technologies and their interactions, which could then be used as basic input data for PDA modeling. The results of the analysis support the feasibility of this concept. Unfortunately, scientific data about the accuracy of either modeling method in comparison to actual vehicles are very limited.

METHODOLOGY OF THE 2002 NATIONAL RESEARCH COUNCIL REPORT

The 2002 NRC report *Effectiveness and Impact of Corporate Average Fuel Economy Standards* used a type of PDA method to estimate the potential future reductions of fuel consumption by light-duty vehicles. The 2002 committee recognized the existence of synergies among technologies applied to reduce fuel consumption but did not provide explicit estimates of the effects of such interactions. Technologies were implemented in defined sequences called paths, and the impacts of technologies on fuel consumption were adjusted to account for interactions with other technologies previously adopted.

[1] In this report the committee chose to use the term *synergies* as defined in the joint EPA and NHSTA "Proposed Rulemaking to Establish Light-Duty Vehicle Greenhouse Gas Emission Standards and Corporate Average Fuel Economy Standards" (EPA and NHTSA, 2009). Two or more technologies applied together might be negatively synergistic, meaning that the sum of their effects is less than the impact of the individual technologies (contributes less to reducing fuel consumption, in this case), or might be positively synergistic, meaning that the sum of the technologies' effects is greater than the impact of the individual technologies (in this case, contributes more to reducing fuel consumption).

> Technology changes modify the system and hence have complex effects that are difficult to capture and analyze. It is usually possible, however, to estimate the impacts of specific technologies in terms of a percentage savings in fuel consumption for a typical vehicle without a full examination of all the system-level effects. (NRC, 2002, p. 33)

> For each technology assessed, the committee estimated not only the incremental percentage improvement in fuel consumption . . . but also the incremental cost that applying the technology would add to the retail price of a vehicle. (NRC, 2002, p. 35)

The 2002 NRC committee grouped technologies into three categories: engine technologies, transmission technologies, and vehicle technologies. Vehicles were grouped into 10 classes. Table 8.1 is the 2002 committee's list of technologies for passenger cars, including ranges for the estimated incremental reductions in fuel consumption and for incremental RPE impacts.

For each vehicle class three different sequences of technology implementation, called "production development paths," were mapped out. Figure 8.1 shows impacts of the technologies included in the three paths for passenger cars, as noted in Table 8.1, on the fuel consumption of a midsize car. The paths were intended to provide a logical sequence of implementation of the various technologies and to ensure that the incremental fuel consumption reductions from a given technology could be estimated conditional on the technologies that had preceded it. Paths 1 and 2 comprised proven technologies that could be introduced within the next 10 years (from 2002), with Path 2 including some more costly technologies than Path 1. Path 3 included additional emerging technologies the 2002 committee believed would become available within the next 15 years. The list of emerging technologies included several technologies that are now in use (intake valve throttling, automated manual transmission, advanced continuously variable transmissions (CVTs), integrated starter/generator, electric power steering) and several that are still not available (camless valve actuation, variable compression ratio engine). In addition, the 2002 committee judged that the potential for diesels to meet tighter emissions standards was highly uncertain and also excluded hybrids from its quantitative assessment due to uncertainty about their future potential. However, both technologies are now available on mass market vehicles in the United States.

In estimating the potential reduction in fuel consumption (gallons per 100 miles) of each technology, the 2002 committee drew on a variety of sources of information, from published reports to presentations to the committee by experts and consultations with automotive manufacturers and suppliers. Having studied the available information, the 2002 committee used its own expertise and judgment to decide on ranges of estimates for each technology. The ranges were intended to reflect uncertainties with respect to the technology of the baseline vehicle, effectiveness of the implementation, and possible tradeoffs with other vehicle attributes. Ranges were given for costs in order to reflect manufacturer-specific conditions, market uncertainties, and the potential for evolutionary costs reductions for new technologies. The 2002 committee did not specify a confidence interval for the ranges, nor did it explicitly address interdependencies or synergies of performance or cost, except via the incremental effects of sequential application in the technology paths.

The incremental fuel consumption improvement and retail price equivalent estimates in Table 8.1 are additive but only for a particular technology path. The technologies included in a path are indicated by an "X" in the columns labeled 1, 2, and 3. Technologies not contained in a path were not to be added to that path. A range of estimates is provided for both fuel consumption and cost impacts. However, only the midpoints of those estimates can be directly accumulated (as illustrated in Figure 8.1), since accumulation of all the high-end or low-end estimates without adjustment would produce misleading results.

The 2002 NRC committee's method received some criticism for being overly simplistic. One notable critique (Patton et al., 2002) cited three major issues:

1. Failure to examine system-level effects;
2. Inaccurate fuel consumption estimates for individual technologies; and
3. Overcounting of fuel consumption reductions.

The first point chiefly faulted the 2002 committee for multiplying together the impacts of individual technologies as if they were independent. It observed that when technologies address different energy-loss mechanisms, their impacts generally are independent, but when technologies address the same energy-loss mechanism (e.g., engine pumping losses), the aggregate effect may be more complex. The committee believed that it had addressed this issue by estimating the incremental effects of technologies implemented in a specified order. However, that committee neglected to quantify the energy losses addressed by each technology and did not separately quantify the interactions among technologies.

The second critique covered a variety of points including the degree of optimism in studies cited to support the committee's estimates and inadequate attention to the dependence of fuel consumption impacts on the characteristics of the vehicle to which they are applied.

> An example of this is cylinder deactivation. According to the report, cylinder deactivation is "applied to rather large engines (>4.0 L) in V8 and V12 configurations." Yet the report applies the same fuel consumption reduction factor for cylinder deactivation to vehicles with six and four cylinder engines, where the actual benefit would be smaller. (Patton et al., 2002, p. 10)

However, the 2002 committee applied cylinder deactivation only to large passenger cars, midsize and larger sport

TABLE 8.1 Fuel Consumption Technology Matrix: Passenger Cars

Baseline: overhead cams, 4-valve, fixed timing, roller finger follower	Fuel Consumption Improvement (%)	Retail Price Equivalent ($)		Subcompact			Compact			Midsize			Large		
		Low	High	1	2	3	1	2	3	1	2	3	1	2	3
Production-intent engine technology															
Engine friction reduction	1-5	35	140	X	X	X	X	X	X	X	X	X	X	X	X
Low-friction lubricants	1	8	11	X	X	X	X	X	X	X	X	X	X	X	X
Multivalve, overhead camshaft (2-V vs. 4-V)	2-5	105	140												
Variable valve timing (VVT)	2-3	35	140	X	X	X	X	X	X	X	X	X	X	X	X
Variable valve lift and timing	1-2	70	210		X	X		X	X		X	X		X	X
Cylinder deactivation	3-6	112	252											X	
Engine accessory improvement	1-2	84	112	X	X	X	X	X	X	X	X	X	X	X	X
Engine supercharging and downsizing	5-7	350	560									X			X
Production-intent transmission technology															
Five-speed automatic transmission	2-3	70	154	X			X			X	X		X	X	
Continuously variable transmission	4-8	140	350		X	X		X	X			X			X
Automatic transmission w/aggressive shift logic	1-3	—	70	X			X			X			X		
Six-speed automatic transmission	1-2	140	280								X		X	X	
Production-intent vehicle technology															
Aero drag reduction	1-2	—	140					X	X		X	X		X	X
Improved rolling resistance	1-1.5	14	56	X	X	X	X	X	X	X	X	X	X	X	X
Safety technology															
Safety weight increase	−3 to −4	0	0	X	X	X	X	X	X	X	X	X	X	X	X
Emerging engine technology															
Intake valve throttling	3-6	210	420		X			X			X			X	
Camless valve actuation	5-10	280	560			X			X			X			X
Variable compression ratio	2-6	210	490			X			X			X			X
Emerging transmission technology															
Automatic shift/manual transmission (AST/AMT)	3-5	70	280								X			X	
Advanced CVTs—allows high torque	0-2	350	840									X			X
Emerging vehicle technology															
42-V electrical system	1-2	70	280			X			X		X	X		X	X
Integrated starter/generator (idle off-restart)	4-7	210	350			X			X			X			X
Electric power steering	1.5-2.5	105	150			X			X		X	X		X	X
Vehicle weight reduction (5%)	3-4	210	350												

NOTE: An X means the technology is applicable to the particular vehicle. Safety weight added (EPA baseline + 3.5%) to initial average mileage/consumption values.
SOURCE: NRC (2002), Table 3.1.

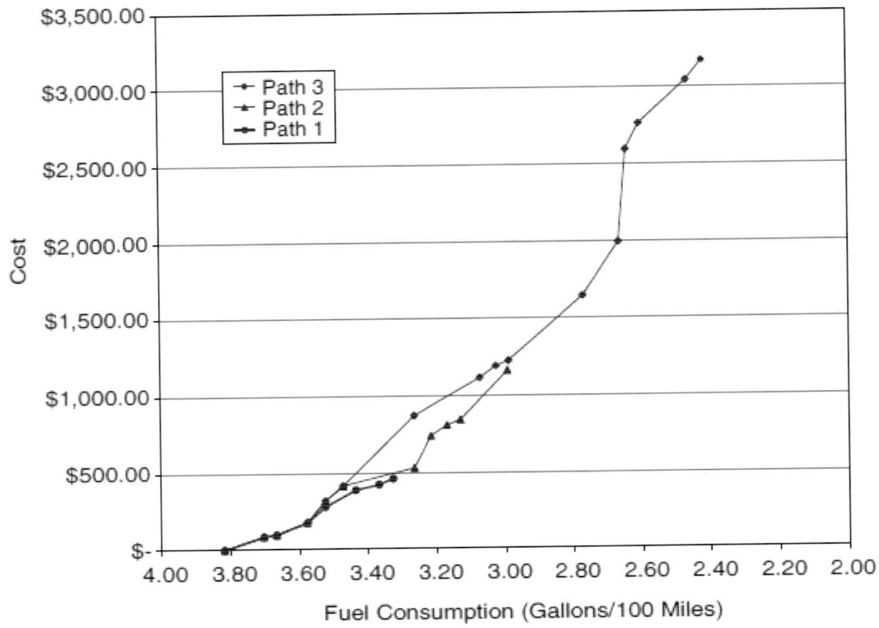

FIGURE 8.1 Estimated cost of fuel consumption reduction in model-year 1999 midsize cars. SOURCE: NRC (2002), Figure 3.6.

utility vehicles (SUVs), minivans, and pickup trucks. Nearly all of these vehicles have engines with six or more cylinders. Cylinder deactivation is applied today to six-cylinder engines. Nonetheless, the 2002 committee's characterization of baseline vehicles was based solely on the typical attributes of the 10 vehicle classes. Using the average characteristics of 10 classes of vehicles will lead to a certain degree of error if the resulting estimates are applied to the vehicles of specific manufacturers.

The criticism of inadequate attention to individual vehicle characteristics can also be leveled at the 2002 NRC committee's costs estimates. The costs of fuel consumption technologies in the 2002 NRC report were the same for all vehicle classes. In fact, the costs of many technologies scale directly with measurable vehicle attributes such as weight or cylinder count.

The third critique is that the 2002 NRC committee's estimates overstated the potential benefits of technologies that primarily addressed pumping losses because the methodology did not take into account the theoretical limits of pumping loss reduction.[2]

Using their own judgments about the allocation of the benefits of technologies to reduction of pumping losses, Patton et al. (2002) divided the 2002 committee's fuel consumption benefit estimates into six categories of energy losses. Patton et al. (2002) attributed essentially all of the 2002 committee's 4 to 8 percent benefit to reduction in pumping loss (and even added an additional 0.5 to 1.0 percent to pumping loss reduction that compensated for reduced transmission efficiency). Only a 0.0 to 0.5 percent benefit was assigned to increased thermal efficiency, presumably due to operating the engine in a more efficient portion of the engine map more of the time. Likewise, most of the benefits of 5-speed and 6-speed automatic transmissions (versus 4-speed) were attributed to reducing pumping losses with no benefits for engine thermal efficiency. Similarly, 4.0 to 6.0 percent of the committee's estimated 5.0 to 7.0 percent benefits of engine boosting and downsizing was attributed to reduced pumping losses. The 2002 committee, on the other hand, judged that the technology derives much of its benefits from increased engine efficiency at light load due to engine downsizing and, when possible, reduced friction due to reduced cylinder count at equivalent power. The 2002 committee asserted that the energy efficiency benefits of multivalve, overhead camshaft engines derived from four different sources:

> The application of single and double overhead cam designs, with two, three or four valves per cylinder, offers the potential for reduced frictional losses (reduced mass and roller followers), higher specific power (hp/liter), engine downsizing, somewhat increased compression ratios, and reduced pumping losses. (NRC, 2002, p. 36)

Patton et al. (2002) disagreed, assigning 2.0 to 5.0 percent of the committee's estimated 2.0 to 5.0 percent

[2]Patton et al. (2002) estimated the theoretical limits at between a 13 percent and 17 percent reduction in fuel consumption, depending on the vehicle in question. The U.S. EPA (2008b) estimated pumping plus friction losses at between 10 percent and 13 percent for actual vehicles, assuming a gross indicated engine efficiency of 37 percent.

improvement to reduced pumping losses, while adding a 0.5 to 1.0 percent benefit in thermal efficiency, offset by a −0.5 to −1.0 percent efficiency *loss* due to *increased* friction.

While the benefits of variable valve timing and lift (VVT + L) are largely reductions in pumping losses, they also include improved power, and the benefits of cylinder deactivation include increased engine load (operation in a more efficient region of the engine map) as well as reduced pumping losses. Estimates of the benefits of the aforementioned technologies generated by FSS models have produced results consistent with the 2002 NRC committee's estimates. Recent estimates from the DOT/NHTSA (2009) and the EPA (2008a) are compared with the 2002 NRC committee's estimates in Table 8.2. The chief area of disagreement is the benefit of cylinder deactivation applied to multivalve, overhead camshaft engines with VVT and discrete or continuous lift control. The NHTSA estimated a benefit of 0.0 to 0.5 percent, whereas the NRC and the EPA estimated benefits of 3 to 6 percent.

The critics of the 2002 NRC report's methodology make an important and valid point in calling attention to the lack of a rigorous relation between the estimates of fuel consumption reduction and the physical energy flows in a vehicle. As a consequence, the plausibility of the 2002 NRC estimates relied heavily on the expert judgment of the committee members. The 2002 NRC study's method also did not explicitly account for the current use of the identified fuel economy technologies in existing vehicles. Practitioners of the PDA method can and often do account for energy constraints using simplified modeling methods called "lumped parameter" models, based on methods developed by Sovran and Bohn (1981) and extended by Sovran and Blaser (2003, 2006) and reviewed in Chapter 2 of this report. FSS models inherently account for energy flows and ensure that physical limits will not be violated.

MODELING USING PARTIAL DISCRETE APPROXIMATION METHOD

The PDA method incrementally adds discrete fuel-consumption-reducing technologies to a baseline vehicle until certain criteria are met. The method is sometimes applied to individual vehicles but more often assumes that the fuel consumption impact and cost of a technology will be approximately the same for all vehicles within at least a subset (or class) of light-duty vehicles. In a presentation to the committee, K.G. Duleep of Energy and Environmental Analysis, Inc. (EEA) identified three important areas in which the PDA method, and especially its application in the 2002 NRC study, had come under criticism (Duleep, 2008).

1. Adequate definition of baseline vehicles;
2. Order of implementation of fuel consumption technologies; and
3. Accounting for synergies among fuel consumption technologies.

The chief disadvantage of the PDA method is that it is entirely empirically based and therefore does not explicitly represent the interactions among any set of technologies. Synergies among technologies are estimated by engineering judgment or by means of simplified analytical tools, such as lumped parameter models of vehicle energy use (Duleep, 2008; Sovran and Blaser, 2003, 2006). Computational simplicity and the ability to quickly and economically process information on thousands of individual vehicles and dozens of alternative combinations of technologies are the method's chief advantages.

The main steps in the PDA process are the following:

1. Identify discrete technologies with fuel consumption reduction potential.

TABLE 8.2 Comparison of Benefits of Valve Train Technologies as Estimated by NRC (2002), NHTSA's Final Rule for 2011, and the EPA

Technology	NRC (2002) (%)	Midpoint (%)	NHTSA[a] (%)	Midpoint (%)	EPA (%)	Midpoint (%)
Multivalve OHC	2-5	3.5	1-2.6	1.8	NA	NA
Variable valve timing	2-3	2.5	3-5	4	2-4	3
Variable valve lift and timing	1-2	1.5	1.5-3.5	2.5	3-4	3.5
Cylinder deactivation	3-6	4.5	0.0-0.5	0.25	6	6
Subtotal		**12**		**8.5**		**12.5**
Intake valve throttling[b]	3-6	4.5	1.5-3.5	2.5	1-2	1.5
Total		**16.5**		**11**		**14**
Camless valves[c]	5-10	7.5	NA	NA	5-15	10

[a]NHTSA's fuel consumption benefits are path dependent. The path shown here is for dual overhead camshaft engines.

[b]In NHTSA's terminology IVT is continuously variable valve lift (CVVL) and is a substitute for discrete variable valve lift (DVVL). NHTSA argues that cylinder deactivation applied to CVVL has little benefit since pumping losses have already been greatly reduced. Others argue that this misses the benefit of increased engine efficiency at higher load when a six-cylinder engine is operated on only three cylinders.

[c]Effect of camless valve actuation is incremental to variable valve lift and timing not to intake valve throttling. The two are mutually exclusive.

SOURCE: Based on data in NRC (2002), DOT/NHTSA (2009), and EPA (2008a).

2. Determine the applicability of each technology.
3. Estimate each technology's impact on fuel consumption and cost.
4. Determine implementation sequences based on
   a. Cost-effectiveness and
   b. Engineering and manufacturing considerations.
5. Identify and estimate synergistic effects
   a. Based on empirical data and expert judgment,
   b. Using a simplified model of vehicle energy flows (e.g., lumped-parameter model), or
   c. Using estimates from FSS models.
6. Determine the "optimal" fuel consumption level by
   a. Using a computer algorithm that sequentially applies technologies,
   b. Using fuel consumption cost curves.

Identifying Technologies That Reduce Fuel Consumption

The PDA method, like the FSS method, begins with the identification of distinct technologies that have the potential to reduce vehicle fuel consumption at a realistic cost.[3] The list of all possible technologies with some potential to reduce fuel consumption could range from lower-rolling-resistance tires and improved engine lubricants to human-powered vehicles and the compressed air engine. When the purpose is regulatory rulemaking, not all possible fuel consumption technologies should be included. The world record for automotive fuel economy is held by the Pac Car II, a fuel-cell-powered vehicle that won the 2005 Shell Ecomarathon in Ladoux, France, with a gasoline equivalent fuel economy of 12,666 miles per gallon.[4] The three-wheel vehicle accommodates one small passenger, who must drive lying down. The 0.57-m wide, 0.61-m high, 2.78-m long carbon-fiber body has no room for cup holders, not to mention air conditioning. It is a zero-emission vehicle, but meeting safety standards was not a design consideration. Clearly much of the PAC Car II's fuel economy was achieved by making unacceptable tradeoffs with other vehicle attributes. The CAFE law requires that fuel economy standards must be technologically feasible and economically practicable. This is ultimately a matter of expert judgment, yet there is remarkable agreement among diverse studies on the list of relevant technologies. Most assessments assume no reduction in size or power-to-weight ratios as a premise.

In general, studies of fuel consumption potential intended to inform the regulatory process and using the PDA method select technologies that meet all of the following three criteria:

1. Technologies already incorporated in at least one mass-produced vehicle somewhere in the world or preproduction technologies judged to have a strong likelihood of widespread adoption within the next decade;
2. Technologies having no significant negative impact, or a beneficial impact on attributes that are valued by consumers or that are necessary to meet safety and emissions regulations; and
3. Technologies whose cost does not far exceed the potential value of fuel savings and other private and social benefits.

For example, all but a few of the technologies considered by the 2002 NRC study were already in mass production. In general, PDA studies are most reliable when they are limited to technologies already in production. However, the farther one must look into the future the less tenable this constraint becomes.

Determining Applicability

Not every technology will be applicable to every vehicle. Torque limitations, for example, have so far prevented the use of CVTs in the largest, most powerful light-duty vehicles. Engine downsizing by reducing the number of cylinders with turbo-charging may be considered applicable to six-cylinder engines but less so to four-cylinder engines due to vibration and harshness considerations. Applicability appears to be largely a matter of expert judgment, determined on a case-by-case basis. The applicability step reduces the full set of technologies to only those that can be used on the baseline vehicle being considered.

Estimating Fuel Economy and Cost Impacts

Fuel consumption impacts are estimated for each technology and each class of vehicles (or each individual vehicle) to which it is applicable. Practitioners of the PDA method derive their estimates from a variety of sources. Unlike FSS, the PDA method, by itself, is not able to produce fuel consumption impact estimates for individual technologies. It is a method of aggregating the fuel consumption impacts of various technologies and must obtain the individual technology benefit estimates from other sources. In its report to the committee, EEA cited three principal sources of information on fuel economy benefit.

> First, the trade press, engineering journals and technical papers presented at engineering society meetings provide detailed information on the types of technologies available to improve fuel economy and the performance, when applied to current vehicles. Second, most of the technologies

[3] The CAFE guidance states that fuel economy standards should be set at the maximum feasible level, taking into consideration technological feasibility, economic practicability, the effect of other federal motor vehicle standards on fuel economy, and the need of the nation to conserve energy (Motor Vehicle Information and Cost Saving Act, Title V, Chapter 329, Section 32902[a]).

[4] Details about the competition, the car, and its design can be found at http://www.paccar.ethz.ch/.

> considered in this report have been introduced in at least a few vehicles sold in the marketplace, and actual test data on fuel economy can be used. Third, the world's largest automanufacturers have research and development staff with detailed knowledge of the attributes of each technology, and their inputs in an unconstrained situation can be used to estimate the benefits of technologies. (EEA, 2007, p. 9)

The EPA has provided a similar list of sources of information.

> These data sources included: vehicle fuel economy certification data; peer reviewed or publicly commented reports; peer reviewed technical journal articles and technical papers available in the literature; and confidential data submissions from vehicle manufacturers and automotive industry component suppliers. (EPA, 2008a, p. 2)

The EPA considers the vehicle certification test data to be an especially reliable source when a directly comparable vehicle is offered with and without a specific technology. In addition, the NHTSA's staff has access to proprietary data provided by vehicle manufacturers to directly support the rulemaking process.

Recently, FSS models have been extensively used to estimate the fuel economy impacts of individual technologies and combinations of technologies (e.g., Ricardo, Inc., 2008a,b; Sierra Research, Inc., 2008). A study done by Ricardo, Inc., for the committee and described below indicates that data on technologies' main and synergistic effects generated by FSS models can be used effectively in PDA analyses (Ricardo, Inc., 2009).

Sequencing Implementation

Sequences for implementing fuel economy technologies are usually determined by a combination of cost-effectiveness and engineering considerations. All else equal, it would be economically efficient to implement first the technology that offered the greatest reduction in fuel consumption per dollar of cost, followed by the technology with the second largest ratio, and so on. Engineering considerations may dictate a different sequence, however. For example, VVT for both intake and exhaust must come after VVT for intake only, regardless of cost-effectiveness.

Fuel consumption benefits must then be converted to incremental benefits, given the implementation sequence. For example, the benefit of a 6-speed transmission must be defined as incremental to that of a 5-speed transmission, even if the base vehicle has a 4-speed, assuming that the 5-speed will be implemented before the 6-speed.[5] Obvious incompatibilities (e.g., a vehicle cannot have both a 6-speed automatic transmission and a CVT at the same time) must also be taken into account.

Accounting for Synergies

Undoubtedly the most serious criticism of the PDA method is that it does not adequately account for synergies among fuel economy technologies. Whether or not the PDA approach is capable of appropriately accounting for synergies is one of the key issues addressed by the present committee.

Fuel economy technologies can have both positive and negative synergies (see footnote 1). In addition, the impacts of technologies applied to vehicle subsystems could potentially be significantly nonlinear, and therefore the effects of multiple technologies might not be accurately estimated by summing the effects of the individual technologies. Practitioners of the PDA method draw on three sources of information to estimate such synergistic effects (EEA, 2007). Because most of the technologies under consideration are in use in some mass-produced vehicle, it is occasionally possible to find models using a combination of several technologies. Comparing the actual fuel consumption performance of these vehicles to an estimate based on the sum of their individual effects can provide an estimate of the degree of synergy.

Second, simplified lumped parameter models of vehicle energy use (e.g., Sovran and Bohn, 1981) provide a means of avoiding the double counting of energy savings. Given a few key parameters, lumped parameter models allow the quantification of sources of energy loss and the components of tractive force requirements for a vehicle. By attributing the impacts of technologies to specific energy losses and tractive force requirements, analysts can check that the sequential application of technologies has plausible impacts on the factors determining energy use. A key question is whether the use of a lumped parameter model can sufficiently accurately account for synergistic effects or whether the FSS method must be used in all cases (Hancock, 2007). An analysis of this subject by Ricardo, Inc. (2009) commissioned by the committee, together with an assessment by the EPA considered below, indicates that a reasonably accurate accounting is possible.

The ability of lumped parameter models to accurately predict vehicle fuel use was first demonstrated by Sovran and Bohn (1981). In an updated version of the same methodology, Sovran and Blaser (2003) showed that despite major changes in automotive technology, lumped parameter models still predicted tractive energy requirements with a high degree of accuracy. Development of a lumped parameter model begins with the fundamental physics equations that determine the energy requirements of vehicles over fixed driving cycles, in particular the EPA urban and highway cycles (equations of the lumped parameter model are presented in Chapter 2). Any cycle can be divided into three regimes:

1. Times when tractive force (F_{TR}) is required from the engine;

[5] In the PDA method a leap from a 4-speed transmission directly to a 6-speed transmission would be calculated by combining the incremental costs and fuel consumption effects of the 4- to 5-speed transition and the 5- to 6-speed transition.

2. Times when deceleration force is greater than rolling resistance (R) and aerodynamic drag (D); and
3. Times when no tractive force is required (vehicle stationary or undergoing deceleration provided by R + D).

When tractive force is required on either cycle, it must equal the sum of forces required to overcome rolling resistance, aerodynamic drag, and inertia. The lumped parameter method simplifies the equation for tractive force and other equations for braking and idling modes by integrating over the drive cycles, as explained in detail in Chapter 2 of this report. Sovran and Blaser (2003) found that the lumped parameter model defined by Equations 2.2 and 2.3 could explain the tractive energy required at the wheels and hence indirectly the engine output of vehicles over either EPA test cycle with an $R^2 = 0.9999$.

The lumped parameter method allows changes in pumping losses, engine friction, accessory loads, and other factors to be related in a manner that can prevent double counting if done properly. It reduces the likelihood of overestimating the combined fuel consumption impacts of multiple technologies by requiring that the laws of physics controlling energy flows and tractive requirements be maintained. As such, it is a powerful tool for quantifying synergistic effects for use in the PDA method. The lumped parameter method cannot, however, predict the kind of synergistic effects that occur when two or more technologies alter each other's performance. This topic is taken up in detail in the following section.

FSS modeling more completely represents such synergistic effects and so it is useful to compare lumped parameter and FSS estimates to test the adequacy of PDA synergy estimates. The U.S. EPA (2008a) used both methods to estimate the fuel economy benefits of 26 technology packages applied to five vehicle types. For most packages they found close agreement between the two types of estimates (Figure 8.2). The EPA's general conclusion was that both methods were valuable and that the use of lumped parameter modeling in PDA estimation gave reasonable estimates of synergies.

> Based on this, EPA concludes that the synergies derived from the lumped parameter approach are generally plausible (with a few packages that garner additional investigation). (EPA, 2008b, p. 44)

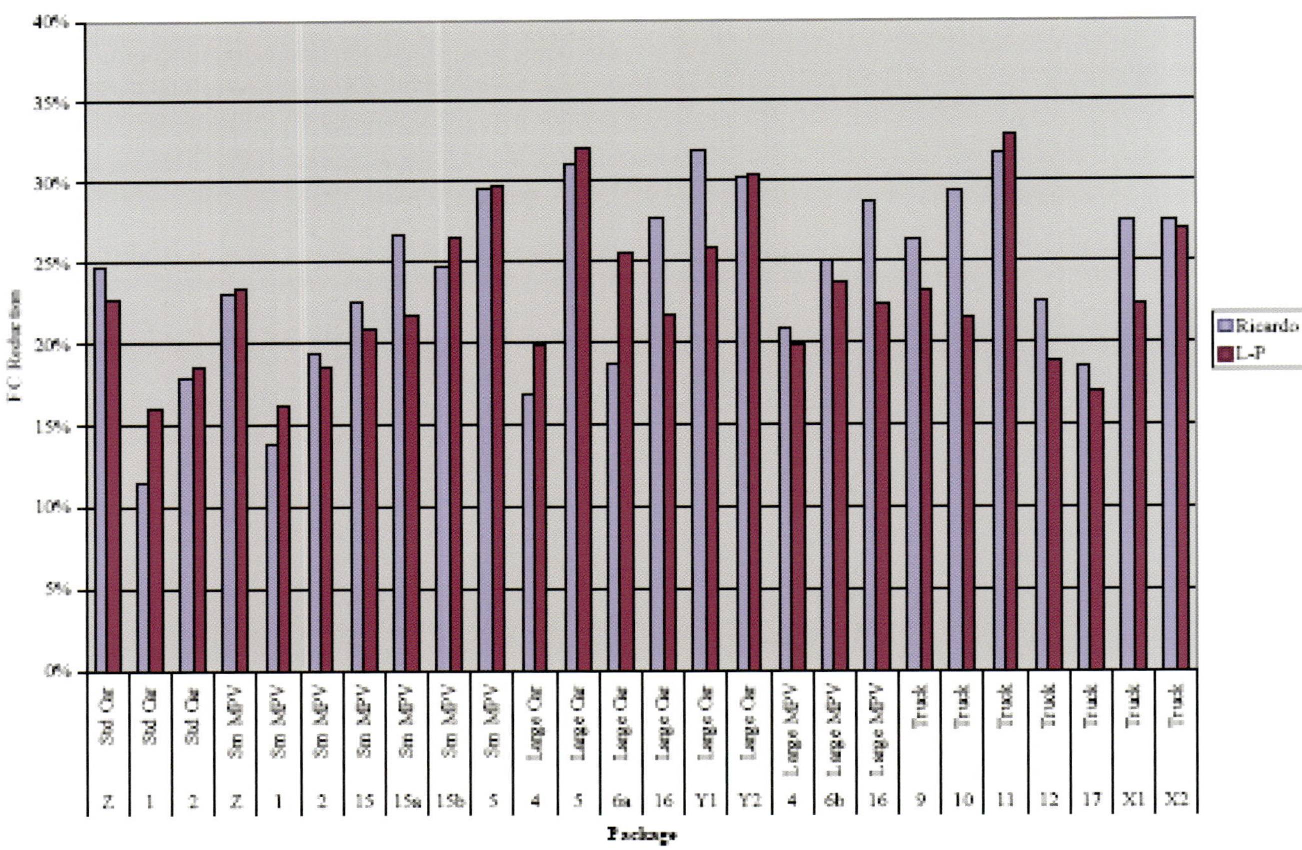

FIGURE 8.2 EPA's comparison of full vehicle simulation model (Ricardo, Inc.) and lumped parameter (L-P) PDA model results. SOURCE: EPA (2008a), Figure 3.3-1.

In 10 cases, significant differences were found (EPA, 2008b). For Standard Car Package 1 and Small MPV Package 1, the lumped parameter method estimated a larger fuel economy improvement. The difference was traced to the CVT component. The Ricardo, Inc., FSS CVT representation had a lower efficiency than assumed in the lumped parameter model. Two other cases involved turbo-charging with engine downsizing. The lumped parameter model estimate was also much higher in the case of Large Car Package 6a, involving continuously variable valve lift. In the case of Large Car Package 4, the lumped parameter model estimated a large benefit, but in the case of Truck Package 10, the FSS model produced the higher benefit estimate. For the packages including cylinder deactivation and coordinated cam phasing (Large Car 16, Large MPV, and Truck 12), the FSS modeling results were consistently higher. FSS estimates were also higher for the cases involving camless valve trains (Large Car Y1, Truck X1). The EPA staff is still investigating reasons for the differences but had identified at least some cases in which the comparison between the two methods led to the discovery of inadvertent errors in the FSS modeling. For example, EPA judged that Ricardo's modeling of cylinder deactivation and coupled cam phasing was incorrect because it did not account for cylinder deactivation's effect of approximately doubling brake mean effective pressure (BMEP) in the firing cylinders. The EPA staff suggested that conducting both FSS and lumped parameter analysis was a wise strategy since the discrepancies between the two methods had led to the discovery of correctable errors.

Twenty-three of the 26 packages evaluated by Ricardo were also estimated by EEA, Inc. (Duleep, 2008) for comparison. EEA was not able to estimate the packages including homogeneous charge compression ignition (HCCI) due to the novelty of the technology. The FSS method requires an externally provided representation of the physics of a device in order to estimate its impact on fuel consumption. While the FSS method itself cannot characterize the physics of technologies, it can produce impact estimates given such characterizations. The PDA method, on the other hand, must be given estimates of impacts for novel technologies. In 16 of the 23 comparisons the two methods produced estimates with relative differences of less than 5 percent. In two cases involving CVT transmissions the Ricardo estimate was much lower. In the committee's discussions with Ricardo and EEA, it was determined that this was due to Ricardo's estimated efficiency of the CVT being much lower than EEA's. This instance illustrates how both methods depend on assumptions about the performance of key technologies. In the remaining five cases, Ricardo's FSS estimates were higher but there appeared to be no common technology that could explain the differences. One of these cases was again the Truck Package 10 involving a turbo-charged gasoline direct injection engine: EEA's lumped parameter PDA method estimated a fuel economy benefit of 26.4 percent, whereas the Ricardo estimate was 42 percent.

Determining the "Optimal" Level of Fuel Economy

Calculation of fuel economy potential and its cost can be accomplished by algorithms that decide which technologies to apply and in what order, or by the use of fuel economy cost curves. The algorithmic approach relies on predefined technology implementation sequences (decision trees or pathways) and is the basis of the Department of Transportation's Volpe Model (Van Schalkwyk et al., 2009) and the Energy Information Administration's NEMS model's Manufacturers' Technology Choice Submodule (DOE/EIA, 2007). The decision tree methodology is described below. Cost curves developed by the NRC (2002) CAFE study and in a number of other studies have been reviewed in Greene and DeCicco (2000).

A PDA Algorithm: The NHTSA's Volpe Model

The NHTSA's Volpe model contains a compliance simulation algorithm that simulates the response of manufacturers to various forms of fuel economy standards. Data are put into the model describing a "CAFE scenario," a combination of definitions of vehicles included in the program, definitions of vehicle classes, levels of fuel economy standards that must be met each year, and the structure of the standards. The structure comprises several elements, the mathematical formulation (e.g., sales-weighted harmonic mean), the functional form (e.g., footprint metric function), the classes of vehicles to which it applies (e.g., foreign or domestic manufacture), and provisions for trading credits over time and among firms. In the description below, the focus is the determination of a manufacturer's "optimal" fuel economy level for a given CAFE scenario.

The algorithm begins with a list of vehicles expected to be available during the future period being evaluated. This is typically a narrow window of three to five model years, beginning 2 years in the future. Vehicles are distinguished by make, model, engine, and transmission, as in the EPA test car list. Many other vehicle attributes are in the vehicles data base, including sales volumes, prices, and specifications. The compliance algorithm applies technologies to each vehicle in the database individually. In the past, the technologies were largely taken from the NRC 2002 report's three technology path lists, but for the 2011 Fuel Economy Rule, the NHTSA developed a new technology list with the assistance of Ricardo, Inc. The new list adds diesel and hybrid power trains (including plug-in hybrids) and materials substitution to reduce vehicle weight. It represents other technologies at a greater level of detail. It also provides a table of estimated pair-wise synergies between technologies. However, the synergies used in the final rule appear to be the same for all vehicles classes. The analysis done for the committee by Ricardo, Inc., described below, indicates that synergy effects can vary across applications to different classes of vehicles (Ricardo, Inc., 2009).

The algorithm evaluates the applicability of each technology to each individual vehicle based on timing of availability and whether or not it is included in decision trees for that vehicle class. The Volpe model's decision trees are analogous to the 2002 NRC study's "paths" except that there are separate decision trees for internal combustion engines, transmissions, electrical accessories, material substitution, dynamic load reduction, aerodynamic drag reduction, and hybrid electric technology. The engine technology decision tree is shown in Figure 8.3. After low-friction lubricants and engine friction reduction are accomplished, the tree splits into three paths depending on camshaft configuration. This allows the NHTSA to tailor the technology sequencing to the base vehicle's engine attributes. If fuel economy is pushed to higher levels the three paths then converge on the stoichiometric, gasoline direct-injection engine. A table of notes can be used to "override" the algorithm's logic and determine applicability in special cases (e.g., as in Table 4, DOT, 2005).

In the committee's judgment, it is not necessary to have separate decision trees for engines and transmissions. This view is supported by the Ricardo, Inc. (2009) analysis, which demonstrates that the important across, or inter-decision-tree, synergies are between engines and transmissions (Ricardo, Inc., 2009). These inter-tree synergies can be transformed to incremental improvements by combining engines and transmissions into a single power train decision tree. Once this has been done, nearly all important synergies can be addressed by adjusting technology impacts to account for interactions with technologies previously implemented in the decision tree, or pathway.

In the Volpe model, the cost and fuel economy impact of each technology vary by vehicle class. Previously the 10 vehicle classes of the 2002 NRC report were used, but the 2011 rule is based on 12 vehicle classes that include 4 performance-based classes:

1. Small light truck (including SUVs and pickups),
2. Midsize light truck (including SUVs and pickups),
3. Large light truck (including SUVs and pickups and full-size vans),
4. Minivans,
5. Subcompact cars,
6. Subcompact performance cars,
7. Compact cars,
8. Compact performance cars,
9. Midsize cars,
10. Midsize performance cars,
11. Large cars, and
12. Large performance cars.

The sequence in which the technologies are applied to any given vehicle is determined by an optimization algorithm. Technologies already in use in a given vehicle are "carried over" from the previous year so that they are not duplicated.

The algorithm then begins an iterative process of determining a manufacturer's compliance with the CAFE standards. If a manufacture is not in compliance, the algorithm selects the next-best technology to add to the vehicle.[6] A technology is selected from the next steps on each of the applicable decision trees. The single technology that has the lowest "effective cost" is chosen for implementation. Effective cost is defined as the total retail price equivalent (RPE) cost of implementing the technology (the change in RPE times the number of vehicles affected), plus any change in the manufacturer's potential CAFE fine, minus the total discounted value of fuel saved by the increase in fuel economy, all divided by the number of vehicles affected. Fines are calculated so as to take account of credits for exceeding standards on some vehicles that can be transferred to other vehicles. Some manufacturers are assumed not to be willing to pay fines and so for them that option is removed. The current version of the model calculates credits or deficits (negative credits) generated by exceeding or failing to meet the standard in any given year. It does not, however, attempt to model credit trading either within a manufacturer over time or among manufacturers. The algorithm continues considering and implementing next-best technologies for all vehicle classes until a manufacturer either achieves compliance with the standard, exhausts all available technologies, or finds that paying fines is more cost-effective than increasing fuel economy (Van Schalkwyk et al., 2009, p. 2).

In a joint EPA and NHTSA (2009) notice of proposed rulemaking (NPRM) the EPA introduced its optimization model for reducing emissions of greenhouse gases from automobiles (OMEGA) model. Like the Volpe model, OMEGA is based on the PDA method and although the logic of the two models is fundamentally the same, there are some notable differences. The Volpe model operates on individual vehicle configurations (on the order of 1,000 make, model, engine, and transmission combinations), taking into account the existing or planned use of fuel economy technologies on each one. The OMEGA model deals with approximately 200 vehicle platforms broken down by engine size (EPA and NHTSA, 2009). For the purpose of estimating technology impacts the 200+ platforms are divided into 19 vehicle types that attempt to distinguish among power trains and market intent. Each of the 19 vehicle types are grouped into five vehicle classes (small car, large car, minivan, small truck, and large truck) for the purpose of scaling cost estimates. In general, the EPA's baseline vehicle is defined as one with a port-fuel-injected, naturally aspirated gasoline engine with two intake and two exhaust valves and fixed valve timing and lift, and a 4-speed automatic transmission. For the NHTSA's Volpe model the baseline is the actual configuration of each

[6]The Volpe model allows manufacturers to opt for non-compliance if paying a fine is less costly than missing the standards, and if a switch set in input data files allows such non-compliance. This option is not discussed here for the sake of brevity.

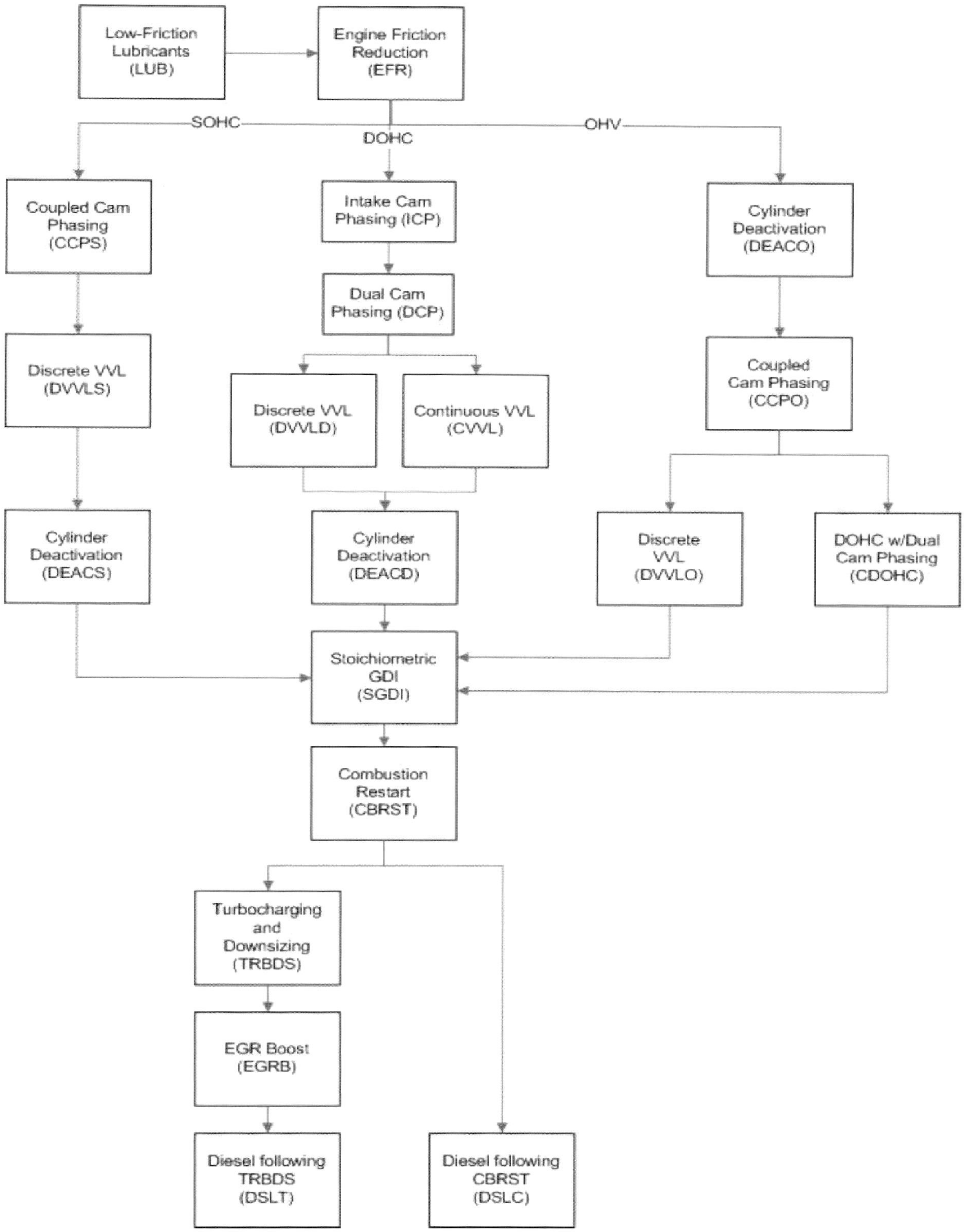

FIGURE 8.3 Volpe model engine technology decision tree.

vehicle configuration as it exists or is predicted to exist in the baseline fleet.

The Volpe model applies individual technologies one at a time in a sequential algorithm, whereas the OMEGA model applies predefined packages of technologies that have been ranked by cost-effectiveness for each vehicle type. However, the packages are assembled from individual technology impact estimates, with synergies between technologies within a package incorporated in the technology package impact estimates. The EPA used the lumped parameter method to determine the adjustment factors (EPA and DOT, 2009, p. 171).

Because neither the Volpe CAFE Compliance and Effects Modeling System nor the EPA's OMEGA model make use of cost curves but rather employ computer algorithms, neither NHTSA nor EPA require cost curves but rather a list of fuel economy technologies including cost, applicability, and synergy estimates. This committee's method is based on implementation pathways that are analogous to the Volpe model's decision trees and the OMEGA model's packages. Therefore, this committee determined that it was not necessary for this study to produce cost curves as such.

Aggregating to Estimate Manufacturers' Fleet Average Fuel Economy

Because fuel economy standards are enforced on automobile manufacturers, both the FSS and PDA methods require a means of inferring the fuel economy potential of an OEM from the fuel economy potential of individual vehicles or vehicle classes. The FSS method is sufficiently computationally intensive that it has not been practical to carry out simulations for all thousand or so vehicles in the EPA test car database for all relevant combinations of technologies. Using the PDA method, a manufacturer's fuel economy potential can be calculated using data on individual configurations (make, model, engine, transmission, i.e., a single entry in the test car database) or using data on classes of vehicles. The NHTSA's Volpe model, for example, calculates a manufacturer's fuel economy target using individual vehicle data since each vehicle has its own fuel economy target as a function of its footprint. The model also calculates each manufacturer's fuel economy potential at the test car list level of detail. Estimates based on vehicle classes can also be computed but they will only be approximately equal to estimates based on individual configurations.

Assume that the optimal level of fuel economy for a single vehicle configuration j has been determined to include technologies $k = 1$ to n_j (given a technology implementation sequence and fuel economy impacts adjusted for implementation order and synergies). The cumulative fuel economy impact is calculated by summing the fractional fuel economy (miles per gallon) improvements, adding one, and multiplying by the base fuel economy MPG_{0j}. If the sales of vehicle configuration j are s_j, then the fuel economy for manufacturer k selling configurations $j = 1$ to N_k is the following:

$$\left(\sum_{j=1}^{N_k} \frac{s_j}{\sum_{j=1}^{N_k} s_j} \left[MPG_j \left(1 + \sum_{i=1}^{n_j} \Delta_{ij} \right) \right]^{-1} \right)^{-1} \qquad \text{Equation 1}$$

If the calculation is done in terms of fuel consumption, or gallons per mile (GPM), the corresponding equation for the manufacturer's fuel consumption target is the following:

$$\sum_{j=1}^{N_k} \frac{s_j}{\sum_{j=a}^{N_k} s_j} \left[GPM_j \prod_{i-1}^{n_j} \left(1 - \delta_{ij} \right) \right] \qquad \text{Equation 2}$$

Equations 1 and 2 make two strong assumptions. First, they assume that the relative fuel consumption impact of a technology will not vary from vehicle to vehicle. Because impacts will vary depending on the initial design of each vehicle, some error will be introduced for each vehicle. In addition, it is assumed that, for a given implementation sequence, any interactions (synergies) among technologies have already been accounted for in the Δ or δ terms. Given information on technology synergies generated by FSS models, equations 1 or 2 could be modified to include synergistic effects as each technology is added. Summing relative fuel economy increases as in equation 1 produces a smaller estimate than sequentially multiplying one plus the relative fuel economy increases. Most fuel economy impact estimates have been determined with the expectation that they will be added to obtain the overall fuel economy benefit. Likewise, multiplying the terms in equation 2 will produce a smaller estimated change in fuel consumption than adding the δ_i, which could erroneously lead to negative fuel consumption. In either case, adding fuel economy impacts or multiplying fuel consumption impacts is intended to produce an approximation to the true impact in a way that reduces the chances of overestimating fuel consumption benefits.

Aggregation over Vehicles in a Class

The PDA method can be applied to an individual vehicle or to a representative vehicle (e.g., a midsize passenger car). For an individual vehicle, it is necessary to know the existing technology makeup of the vehicle so that incompatibilities are avoided and technologies are not applied twice. In the case of a representative vehicle, it is necessary to know the market shares of fuel economy technologies for vehicles in its class. In general, the exact distribution of all combinations of technologies within the vehicle class is not known. Instead, the total market shares of each technology are used, in effect assuming that their distributions are independent. This introduces a further element of approximation into the estimation.

Let $s_{ij,0}$ be the initial market share of technology i in the vehicle class j, and let $s_{ij,\max}$ be the maximum market share for technology i. The estimated change in fuel economy (MPG) by application of the full set of technologies is given by equation 3:

$$D_{j\max} = \sum_{i=1}^{n_j} \left(s_{ij,\max} - s_{ij,0} \right) \Delta_{ij} \qquad \text{Equation 3}$$

The estimated change in fuel consumption by application of the full set of technologies is given by equation 4:

$$d_{j\max} = GPM \left(\prod_{i=1}^{n_j} \left(1 - \left(s_{ij,\max} - s_{ij,0} \right) \delta_{ij} \right) \right) \qquad \text{Equation 4}$$

The cost of the above fuel economy increase is calculated similarly, where C_i is the cost of technology i in retail price equivalent:

$$C_{j\max} = \prod_{i=1}^{n_j} \left(s_{ij,\max} - s_{ij,0} \right) C_{ij} \qquad \text{Equation 5}$$

Although equation 3 approximates the share-weighted harmonic mean change in fuel economy for a class of vehicles with a mixture of technologies it does not precisely equal it. Even performing the calculations in terms of fuel consumption, as in equation 4, will not produce the exact harmonic mean fuel economy, in general.

MODELING USING FULL SYSTEM SIMULATION

The FSS approach to modeling vehicle fuel consumption involves capturing the physics or characteristics of subsystems of the vehicle in software, assembling these subsystems by passing relevant operational variables between these subsystems, and choosing preferred input variables and trajectories to simulate desired vehicle operation. The overall goal is to have the subsystem models work in a synergistic way to reflect the actual performance of the vehicle in various maneuvers. Because of the complexity and nonlinearity of these vehicle subsystems, it is often difficult to anticipate the synergistic effects, especially during transients, and this approach usually provides this useful information to some degree of accuracy. FSS modeling has been used by the automotive industry since the 1970s, and is a proven method of estimating the impacts of existing and new technologies on vehicle systems (Waters, 1972; Blumberg, 1976). More recently, regulatory agencies and other groups outside the automotive industry are undertaking efforts to develop and utilize FSS in their analysis (NESCCAF, 2004; Rousseau, 2007; EPA, 2008a).

Although modeling approaches differ, all FSS models are based on the time integration of Newton's second law (i.e., F = m·a) over some driving maneuver, in this case over the FTP and highway driving cycles. The boundary and initial conditions for this integration are based on a description of the vehicle (mass, frontal area, drag coefficient, etc.), the components that compose the driveline (engine and transmission, etc.), accessories (pumps, fans, generators, etc.), and a specification of the drive cycle, or vehicle speed trace, the vehicle is to perform. Components are represented by computer modules and may be described by performance maps represented by tables or equations. All energy flows among components are accounted for by equations linking the modules. FSS models may be backward-looking or forward-looking. Backward-looking models assume that the drive cycle's velocity and acceleration trajectory will be met, calculate the force required at the wheels, and then work backward to the resulting engine speed, and the necessary throttle and brake commands. Forward-looking models choose throttle and brake commands in order to achieve the specified trace. Some models use a combination of both strategies (see, e.g., Markel et al., 2002).

Modeling can have the potential benefit of helping one to understand these synergies and better predict future performance, either through the careful analysis of available vehicle data, or through creating dynamic models of the vehicles and analyzing the performance of these virtual vehicles. In addition to the synergies within various subsystems of the vehicle, many subsystems within the vehicle exhibit nonlinear behavior. Considering the performance of individual subsystems independently, even if this performance is well known and understood, can therefore result in misleading conclusions for the overall system. When an understanding of each subsystem can be represented by a computer model to an appropriate level of detail, and the interconnectivity or physical communication between each of these subsystems can also be adequately represented, the synergistic and nonlinear effects can be included and analyzed in the behavior of the entire system. Computer modeling of vehicle systems is widely used in the industry for this purpose, as well as to help predict future performance or performance under various conditions. Manufacturers use FSS in the product development process to optimize factors such as shift logic and final drive ratio.

For new technologies not implemented in any mass-produced vehicle, FSS model results are probably the most reliable source of estimates of synergistic effects. Historically, the PDA approach has generally not been used for estimating the fuel consumption impacts of novel vehicle systems for which there are no actual test data (Greene and DeCicco, 2000). Today FSS modeling is more widely used to estimate the potential for reducing fuel consumption than even 5 years ago. A number of studies are available that have used FSS to estimate the fuel consumption impacts of advanced technologies (e.g., Ricardo, Inc., 2008a,b, 2009; Kasseris and Heywood, 2007; Kromer and Heywood, 2007; Sierra Research, 2008). It should be noted, however, that sufficient knowledge of the technology package being investigated is necessary to allow its representation within the model to have an acceptable degree of accuracy. For an aggregate technology, this may take the form of a performance map describing its efficiency over a range of operating conditions. For a technology described by unique operation of an existing subcomponent, relevant performance insight in the corresponding new regime of operation would be necessary.

It is important to note that, although FSS models have the ability to estimate the absolute impacts of vehicle technologies due to their ability to model the physics of system components, they have limited ability to model the dynamic working of individual fuel efficiency technologies and generally rely on a limited set of input data. For novel technologies, many of the input parameters are assumptions based on engineering judgment and experience with related technologies. This emphasizes the fact stated at the outset of this chapter, that one cannot know with absolutely accuracy the impact of technologies until an actual vehicle is constructed and repeatedly tested.

Model Fidelity

An important consideration for FSS modeling is deciding what level of fidelity of the equations or look-up tables is required for the problem being addressed. No set of equations completely reflects the detailed physics of the actual process, so the choice of fidelity should be a conscious choice from a continuum of models of varying fidelity, all of which represent simplifications of the actual process. The objective is to achieve an appropriate balance of fidelity with modeling goals, modeling effort and resources, simulation speed, and available data that specifically characterizes the system being modeled. There is always a difference between the simulation and actual subsystem operation, known as the modeling error. The tolerable level of error depends upon the goals of the simulation.

Unfortunately, data on the predictive accuracy of FSS models are scarce. In part this is because some models and more often the representation of their components are proprietary to firms that use them in their own research or consulting. The committee is not aware of any rigorous study evaluating the accuracy of models for various applications. The few comparisons the committee has seen indicate that for known vehicles, simulation models can reproduce fuel consumption and performance with a high degree of accuracy. Data provided by Ricardo, Inc., based on its research for the EPA indicated a range of error in predicting fuel consumption of 1 to 3 percent for five vehicles (Figure 8.4). For this modeling, the EPA chose a specific representative vehicle for each of the five classes: the Toyota Camry for the standard car, the Saturn Vue for the small MPV, the Chrysler 300 for the full-size car, the Dodge Grand Caravan for the large MPV, and the Ford F150 for the truck. Ricardo, Inc., (2008a) attributed any discrepancies between the simulation results and the actual vehicle data to the use of generic input data for that vehicle class instead of actual data for a specific vehicle. Of course, these are known vehicles so that component representations and the overall model can be calibrated. Prediction errors for truly novel technologies for which no vehicle exists to calibrate to would presumably be larger. In any event, it is the change in fuel consumption from the implementation of a technology that is of most interest. The absolute error of a predicted change can be smaller when prediction errors similarly exist in both the "before" and "after" simulations (i.e., the modeling errors of the before and after cases are strongly correlated). Still, relative errors for a predicted change are likely to be greater. The accuracy of FSS models in predicting fuel consumption changes in actual vehicles deserves additional study. Note that such an accuracy study is made more difficult by the fact that the

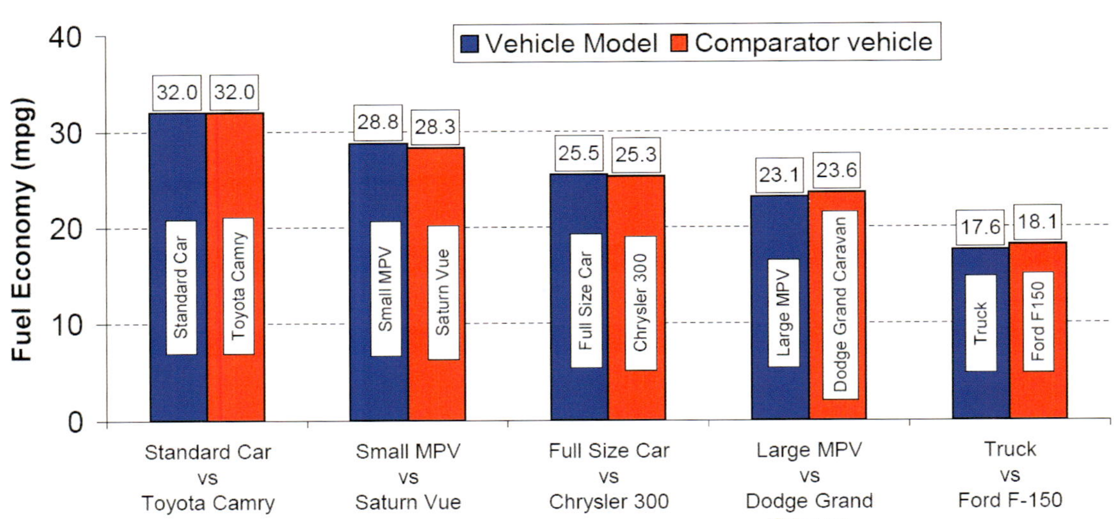

FIGURE 8.4 Comparison of actual vehicle combined fuel economies and Ricardo simulated fuel economies for five vehicles. SOURCE: Ricardo, Inc. (2008a).

accuracy of FSS estimations depends significantly on the experience and skill of the FSS practitioner.

The flexibility, rigor, and comprehensiveness of the FSS approach to vehicle modeling are significant advantages. Subsystem models may be as simple as a single parameter or table based on steady-state operation, or a detailed, nonlinear, multivariable representation of the dynamics of the subsystem, including transient operation. The choice of how to represent each subsystem model is not only based on modeling error considerations discussed above, but also on balancing fidelity between subsystem models, in order to use computational resources as effectively as possible. One way of looking at balancing fidelity between subsystems is to consider the filtering properties or bandwidth of these subsystems. If one subsystem model has a level of fidelity that generates details in an output variable that are filtered out by a subsequent system, then the effort in generating those details is mostly wasted if the intermediate variables between the subsystems are not of interest. This balance of fidelity within an overall FSS model is a judgment call that is typically developed through experience or trial and error, although the effects can be clearly seen by looking carefully at the content of the variables that are passed between subsystems to see what effects are preserved or eliminated.

An example of these considerations can be seen by examining a typical system model of a turbocharger. In many dynamic system models, the characteristics of both the turbocharger compressor and turbine are simulated based on steady-state maps. However, the rotational dynamics of the rotor is simulated based on Newton's second law (i.e., a differential equation reflecting dynamic or transient operation). The rationale for choosing and combining these two different types of models is based on the idea that the time constants for the gas dynamics in the compressor and turbine are considerably shorter (i.e., faster) than the time constant of the rotor. If much more detailed dynamic models of the gas dynamics were included in the model when the rotational speed of the rotor is the desired output variable, almost all of the gas dynamic effects would be filtered out by the rotor inertia or rotational bandwidth. This combination of steady-state and dynamic models to represent the turbocharger usually provides an effective dynamic model of its rotational dynamics and transient operation in relation to the rest of the engine. However, if the goal is to capture the pulsed gas dynamics in the turbine or compressor, this choice of subsystem models may not be appropriate (for that specific goal). The important point is that more detail is not necessarily better, but fidelity and balance should be conscious decisions reflecting modeling goals.

Model Validation

An effective way of carrying out model validation, given available data on the system operation, is to subdivide the data into at least two sets covering different operating conditions. One set of data can be used to determine parameters or tune the subsystem models, and a separate and distinct set of data can be used to test the predictive capabilities of the model in different situations after it has been tuned or calibrated. The model should not be tested using the same set of data that was used to calibrate the model.

FSS Model Example

An example of an FSS compression-engine model is illustrated in Figure 8.5 in order to give the reader a better visual idea of a possible subdivision of subsystems within the overall system model, as well as possible choices of fidelity within each subsystem. The overall goal of this model is to represent engine transient performance within the vehicle power train, including cylinder-by-cylinder rotational dynamic effects as well as first order intake and exhaust dynamics that affect turbocharger transient effects on the engine. Some simple emission transient predictive capability is included but is not comprehensive for all constituents.

This model was developed using the MATLAB/Simulink modeling software, and its overall structure is presented by the block diagram structure of MATLAB/Simulink in hierarchical form. Most of the subsystem models are identified for the reader. The core of the model is the engine map that provides brake-specific fuel consumption as a function of engine speed and load. Numerous other modules are necessary to represent the many interacting components of the engine system. Most of these components must be calibrated to the specific engine system of interest.

AN ANALYSIS OF SYNERGISTIC EFFECTS AMONG TECHNOLOGIES USING FULL SYSTEM SIMULATION

At the request of the committee, Ricardo, Inc. (2009) undertook a study to quantify the synergistic effects captured by FSS models. It is important to note that the study is based solely on the predictions of Ricardo, Inc.'s FSS models and therefore can quantify only the synergies those models can represent. In its report, Ricardo estimated the accuracy of its models for predicting fuel economy at 1 percent for well-characterized vehicle systems (systems for which nearly all model subsystems have been calibrated to the actual components) and 3 percent for novel vehicle systems. However, each estimate of accuracy was based on a single data point and so cannot be considered definitive.

Ricardo's approach was to simulate the technologies contained in five different packages of technologies it had previously modeled for the EPA (2008a) as applied to five different types of vehicles. The technologies were applied one at a time and in combinations according to a rigorously defined design of experiments. The results were then fitted by a response surface model using a neural network method. The response surface model fit the data with maximum errors of 1 percent using terms no higher than second order (Figure 8.6).

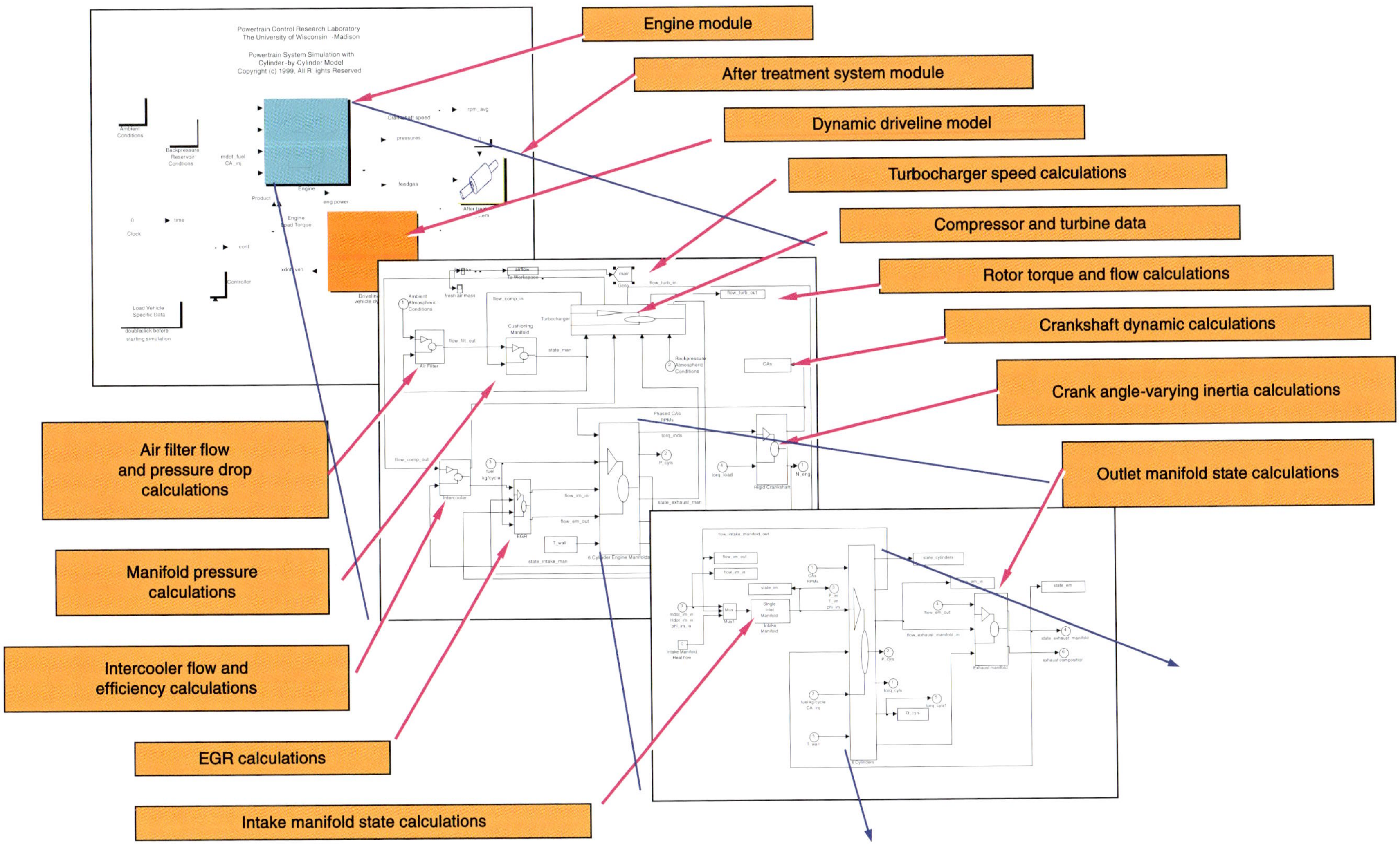

FIGURE 8.5 An example structure for a full system simulation diesel engine dynamic model. SOURCE: Reprinted with permission from John J. Moskwa, Powertrain Control Research Laboratory, University of Wisconsin, Madison, Wisc.

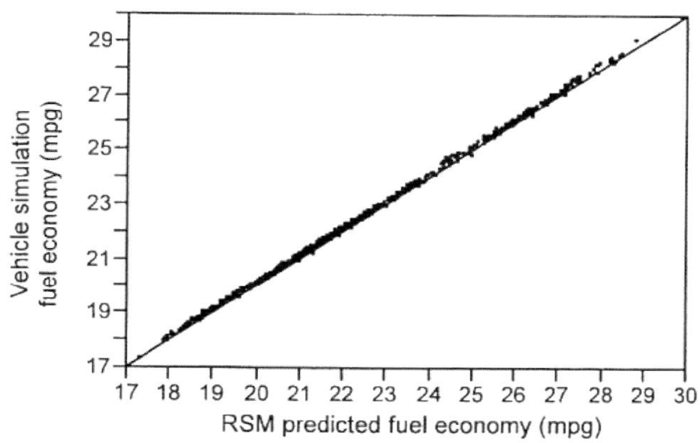

FIGURE 8.6 Ricardo, Inc., statistical (response surface model [RSM]) predictions versus full system simulation model predictions. SOURCE: Ricardo, Inc. (2009), Figure 3-2.

> This shows that a relatively simple 2nd order regression model provides a very suitable representation of the more complex vehicle simulation output with maximum RSM (Response Surface Model, ed.) residual errors of about 1 percent, or that higher order effects (3rd order and above) account for less than 1 percent of the vehicle simulation output characteristics. (Ricardo, Inc., 2009, p. 13)

This finding is significant in that it indicates that important synergistic effects (as represented by the FSS models) are of no higher order than two-way interactions. It is also generally consistent with the ability of a much simpler lumped parameter model to accurately estimate fuel economy over the federal test cycles with Sovran and Blaser (2006).

The next step was to carry out an analysis of variance (ANOVA) to quantify the first-order (main) and second-order effects. The ANOVA estimated that main effects of technologies accounted for 80 to 86 percent of the fuel economy increase. Interaction effects, taken together, accounted for 14 to 20 percent. Ricardo, Inc., concluded that simplified models that did not properly account for interaction effects could have estimation errors of up to 20 percent. However, 20 percent not only is the upper bound on estimation error but also assumes that the error in estimating interaction effects is 100 percent (for example, if they were all estimated to be 0). Interaction effects estimated using lumped parameter models, for example, are likely to be much smaller.

Even more importantly, the interaction effects include second-order main effects and incremental effects. Second-order main effects represent the interaction of a technology with itself and are introduced to account for nonlinear effects in the linear ANOVA model. Thus, they do not depend on the presence or absence of other technologies and so are not synergies in the sense that is of interest. Incremental effects include some true synergistic effects and some purely incremental effects. Purely incremental effects reflect the fact that when technologies are applied in sequence the fuel consumption impact of a technology depends on which technologies have been previously applied. For example, given a base vehicle with a 4-speed transmission, the impact of a 6-speed transmission will be smaller if a 5-speed transmission has been previously applied. The PDA method explicitly recognizes this kind of interaction by ordering technologies for interaction and using only incremental impacts, given that ordering, to estimate the total fuel consumption impact. But incremental effects, as defined in the Ricardo ANOVA, also include true synergistic effects, such as when a 42-volt electrical system is implemented together with electric accessories (e.g., electric power steering). Most PDA modelers attempt to take such interactions into account, but the accuracy with which they do so will depend on the available data sources and the engineering judgment of the analyst.

There are additional synergies of interest that Ricardo terms "inter-tree" or "true" synergies. These are the interactions among technologies that are neither second-order main effects nor incremental effects. PDA modeling cannot, in general, account for this type of synergy. According to the results of Ricardo's study, these effects are quite small. For example, adding up the synergy (inter-tree) effects for Small MPV Package 5 (allowing positive and negative effects to cancel) results in a total synergy effect of −1.3 percent of the total fuel economy impact of the technology package. Adding up the inter-tree synergies produces a positive synergy of 4.6 percent for Small MPV Package 15, a positive synergy of 2.8 percent for Large MPV Package 16, and a positive synergy of 10.3 percent of the total fuel economy impact for Truck Package 11. These are percentages of the total fuel economy change and so suggest that errors due to completely ignoring inter-tree synergies are on the order of 10 percent or less for the total fuel economy impact. The size of these effects is roughly consistent with the discrepancies EPA (2008b) found in its comparison of lumped parameter and FSS modeling.

Ricardo, Inc. (2009) concluded that PDA modeling, such as that used in the NHTSA's Volpe model, if informed by rigorously designed FSS modeling of the kind represented in its study, can produce accurate estimates of fuel consumption reduction potential. This conclusion, however, is conditional on the accuracy of FSS models for predicting EPA test cycle fuel economy. Given the scarcity of evidence on this subject and its importance to validating Ricardo's conclusion, it merits further investigation.

FINDINGS

Finding 8.1: The state of the art in estimating the impacts of fuel economy technologies on vehicle fuel consumption is full system simulation (FSS) because it is based on integration of the equations of motion for the vehicle carried out over the speed-time representation of the appropriate driving or test cycle. Done well, FSS can provide an accurate assessment (within +/–5 percent or less) of the impacts on fuel consumption of implementing one or more technologies. The validity of FSS modeling depends on the accuracy of representations of system components (e.g., engine maps). Expert judgment is also required at many points (e.g., determining engine warm-up rates or engine control strategies) and is critical to obtaining accurate results.

Finding 8.2: The partial discrete approximation (PDA) method relies on other sources of data for estimates of the impacts of fuel economy technologies. Unlike FSS, the PDA method cannot be used to generate estimates of the impacts of individual technologies on vehicle fuel consumption. Thus, the PDA method by itself, unlike FSS, is not suitable for estimating the impacts on fuel consumption of technologies that have not already been tested in actual vehicles or whose fuel consumption benefits have not been estimated by means of FSS. Likewise, the effects of technology interactions must be determined from external estimates or approximated by a method such as lumped parameter modeling. Even FSS, however, depends directly on externally generated information on the performance of individual technology components.

Finding 8.3: Comparisons of FSS modeling and PDA estimation (within the range of cases where the PDA method is applicable) supported by lumped parameter modeling to eliminate double counting of energy efficiency improvements have shown that the two methods produce similar results when similar assumptions are used. In some instances, comparing the estimates made by the two methods has enhanced the overall validity of estimated fuel consumption impacts by uncovering inadvertent errors in one or the other method. In the committee's judgment both methods are valuable, especially when used together, one providing a check on the other. However, more work needs to be done to establish the accuracy of both methods relative to actual motor vehicles. In particular, the accuracy of applying class-specific estimates of fuel consumption impacts to individual vehicle configurations needs to be investigated. The magnitude of the errors produced when such estimates are aggregated to calculate the potential of individual automobile manufacturers to reduce fuel consumption should also be analyzed.

Finding 8.4: The U.S. Department of Transportation's Volpe National Transportation Systems Center has developed a model for the NHTSA to estimate how manufacturers can comply with fuel economy regulations by applying additional fuel savings technologies to the vehicles they plan to produce. The model employs a PDA algorithm that includes estimates of the effects of technology synergies. The validity of the Volpe model, and probably also the OMEGA model, could be improved by making use of main effects and interaction effects produced by the FSS methodology described in this chapter. In particular, research done for the committee has demonstrated a practical method for using data generated by FSS models to accurately assess the fuel consumption potentials of combinations of dozens of technologies on thousands of vehicle configurations. A design-of-experiments statistical analysis of FSS model runs demonstrated that main effects and first-order interaction effects alone could predict FSS model outputs with an R^2 of better than 0.99. Using such an approach could appropriately combine the strengths of both the FSS and the PDA modeling methods. However, in Chapter 9 the committee recommends an alternate approach that would use FSS to better assess the contributory effects of technologies applied for the reduction of vehicle energy losses and to better couple the modeling of fuel economy technologies to the testing of such technologies on production vehicles.

REFERENCES

Blumberg, P.N. 1976. Powertrain simulation: A tool for the design and evaluation of engine control strategies in vehicles, SAE Technical Paper Series 760158. SAE International, Warrendale, Pa. February 23.

DOE/EIA (U.S. Department of Energy/Energy Information Administration). 2007. Transportation sector module of the National Energy Modeling System: Model documentation 2007, DOE/EIA-M070(2007). Office of Integrated Analysis and Forecasting, Washington, D.C.

DOT (U.S. Department of Transportation). 2005. CAFE compliance and effects modeling system. Volpe Systems Center, Cambridge, Mass., July 19.

DOT/NHTSA (U.S. Department of Transportation/National Highway Traffic Safety Administration). 2009. Average fuel economy standards, passenger cars and light trucks, model year 2011: Final rule, RIN 2127 AK-29, Docket No. NHTSA 2009-0062, Washington, D.C., March 23.

Duleep, K.G. 2008. EEA-ICF analysis update, Presentation to the Committee on Technologies for Improving Light-Duty Vehicle Fuel Economy, April 1, Washington, D.C.

EEA (Energy and Environmental Analysis, Inc.). 2007. Technologies to improve light-duty vehicle fuel economy, Draft report to the National Research Council Committee on Fuel Economy of Light-Duty Vehicles, Arlington, Va., September.

EPA (U.S. Environmental Protection Agency). 2008a. EPA Staff Technical Report: Cost and Effectiveness Estimates of Technologies Used to Reduce Light-Duty Vehicle Carbon Dioxide Emissions. EPA420-R-08-008, Ann Arbor, Mich.

EPA. 2008b. EPA's technical review of Ricardo simulations, Presentation to the Committee on Technologies for Improving Light-Duty Vehicle Fuel Economy, March 31, 2008, Detroit, Mich.

EPA and DOT (U.S. Environmental Protection Agency and U.S. Department of Transportation). 2009. Proposed Rulemaking to Establish Light-Duty Vehicle Greenhouse Gas Emission Standards and Corporate Average Fuel Economy Standards. August 24. Washington, D.C.

EPA and NHTSA (U.S. Environmental Protection Agency and National Highway Traffic Safety Administration). 2009. Draft Joint Technical Support Document: Proposed Rulemaking to Establish Light-Duty Vehicle Greenhouse Gas Emission Standards and Corporate Average Fuel Economy Standards, EPA-420-D-09-901, September.

Greene, D.L., and J. DeCicco. 2000. Engineering-economic analysis of automotive fuel economy potential in the United States. Annual Review of Energy and the Environment 25:477-536.

Hancock, D. 2007. Assessing fuel economy. Presentation to the Committee on Fuel Economy of Light-Duty Vehicles, National Research Council, September 10, Washington, D.C.

Kasseris, E.P., and J.B. Heywood. 2007. Comparative analysis of automotive powertrain choices for the next 25 years. SAE Technical Paper Series No. 2007-01-1605. SAE International, Warrendale, Pa.

Kromer, M.A., and J.B. Heywood. 2007. Electric powertrains: opportunities and challenges in the U.S. light-duty vehicle fleet. Publication No. LFEE 2007-02 RP. Sloan Automotive Laboratory, Massachusetts Institute of Technology, Cambridge, Mass.

Markel, T., A. Brooker, T. Hendricks, V. Johnson, K. Kelly, B. Kramer, M. O'Keefe, S. Sprik, and K. Wipke. 2002. ADVISOR: A systems analysis tool for advanced vehicle modeling. Journal of Power Sources 110(2):255-266.

Moskwa, J. 2008. Powertrain Control Research Laboratory, University of Wisconsin, Madison, Wisc.

NESCCAF (Northeast States Center for a Clean Air Future). 2004. Reducing Greenhouse Gas Emissions from Light-Duty Motor Vehicles. March.

NRC (National Research Council). 2002. Effectiveness and Impact of Corporate Average Fuel Economy Standards. National Academy Press, Washington, D.C.

Patton, K.J., A.M. Sullivan, R.B. Rask, and M.A. Theobald. 2002. Aggregating technologies for reduced fuel consumption: A review of the technical content in the 2002 National Research Council Report on CAFE. SAE Paper 2002-01-0628. SAE International, Warrendale, Pa.

Ricardo, Inc. 2008a. A Study of Potential Effectiveness of Carbon Dioxide Reducing Vehicle Technologies. Prepared for the U.S. Environmental Protection Agency, EPA420-R-08-004, Contract No. EP-C-06-003, Work Assignment No. 1-14, Ann Arbor, Mich.

Ricardo, Inc. 2008b. A study of potential effectiveness of carbon dioxide reducing vehicle technologies. Presentation to the Committee on Fuel Economy of Light-Duty Vehicles, National Research Council, January 24.

Ricardo, Inc. 2009. A Study of Interaction Effects Between Light Duty Vehicle Technologies. Prepared for the NRC Committee on Assessment of Technologies for Improving Light-Duty Vehicle Fuel Economy by Ricardo Inc., Van Buren, Mich., February 27.

Rousseau, A. 2007. Designing advanced vehicle powertrains using PSAT. Presentation to the Committee on Fuel Economy of Light-Duty Vehicles, National Research Council, September 10.

Sierra Research, Inc. 2008. Technology and retail price implications of HR6 CAFE standards based on vehicle simulation modeling (preliminary results). Presentation to the committee by James M. Lyons, January 24.

Sovran, G., and D. Blaser. 2003. A contribution to understanding automotive fuel economy and its limits. SAE Paper 2003-01-2070. SAE International, Warrendale, Pa.

Sovran, G., and D. Blaser. 2006. Quantifying the potential impacts of regenerative braking on a vehicle's tractive-fuel consumption for the U.S., European, and Japanese driving schedules. SAE Paper 2006-01-0664. SAE International, Warrendale, Pa.

Sovran, G., and M.S. Bohn. 1981. Formulae for the tractive-energy requirements of vehicles driving the EPA schedules. SAE Paper 810184. SAE International, Warrendale, Pa.

Van Schalkwyk, J., W. Gazda, K. Green, D. Pickrell, and M. Shaulov. 2009. Corporate average fuel economy compliance and effects modeling system documentation. DOT HS 811 012. U.S. Department of Transportation, Research and Innovative Technology Administration, Energy Technology Division, John A. Volpe National Transportation Systems Center, Cambridge, Mass., April.

Waters, W.C. 1972. General purpose automotive vehicle performance and economy simulator. SAE Paper 720043. January 10. SAE International, Warrendale, Pa.

9

Application of Vehicle Technologies to Vehicle Classes

INTRODUCTION

In conducting its assessment of technology applicability to different vehicle classes, the committee was guided by the following question included in the statement of task: "What are the estimated cost and potential fuel economy benefits of technology that could be applied to improve fuel economy of future passenger vehicles, given the constraints imposed by vehicle performance, functionality, and safety and emission regulations?" Note that applying technology to improve fuel economy and reduce fuel consumption should not be interpreted to mean simply attaching a component or subsystem that then achieves a subsequent reduction in fuel consumption. Such reductions in fuel consumption typically evolve through an incremental, evolutionary application of components, subsystems, and new power train or vehicle technologies.

Previous chapters of this report have provided technical summaries of current and advanced technologies that are currently being applied to vehicles, or developed for future vehicle applications. Other reports from the National Research Council (NRC) have also looked at the impacts of technologies for reducing fuel consumption—Appendix H provides a summary of other recent NRC studies related to light-duty vehicle technologies. Many of these technologies could, in principle, be applied to almost any vehicle. However, the intended use of the vehicle, its price range, consumer characteristics, emissions and safety standards compliance, and packaging constraints influence which technologies will see market penetration on different vehicle types.

Many of the technologies have already seen significant penetration into European or Asian markets where regulatory and market pressure, including significant taxation that results in high fuel prices for consumers, have encouraged early adoption. Others, such as turbocharged, direct-injection gasoline engines, have gained significant attention in the United States because fuel consumption can be reduced with minimal redesign of the total vehicle system.

DEVELOPING BASELINE VEHICLE CLASSES

The concept of dividing U.S. passenger vehicles into so-called classes is both an outcome of regulatory segmentation for the purpose of varying standards and a means whereby vehicle sales categories are differentiated by vehicle size, geometry, and intended use. The NRC CAFE report segmented U.S. passenger vehicle sales into 10 classes that were a subset of the larger number of type and weight classes that the U.S. EPA uses as part of its vehicle certification process (NRC, 2002). These 10 classes are as follows:

1. Small SUV,
2. Medium SUV,
3. Large SUV,
4. Minivans,
5. Small pickups,
6. Large pickups,
7. Subcompact cars,
8. Compact cars,
9. Midsize cars, and
10. Large cars.

The statement of task directs the current committee to evaluate these vehicle classes and update the technology outlook for future model introduction. However, shifts in consumer preference and vehicle sales have been significantly influenced by the recent instability in fuel prices, vehicle financing costs, U.S. and global economic conditions, and regulatory uncertainty. Significant shifts in vehicle sales between 2002 and 2007 showed a continuing increase in SUV sales, with sales of small pickups essentially disappearing (EPA, 2008a). However, in 2008, large increases in fuel price (above $4 per gallon of gasoline) resulted in a greater than 50 percent reduction in the sale of midsize and large SUVs. Subsequent U.S. and global instability in the financial markets, followed by a period of recession, has resulted in an overall reduction of vehicle sales in the United States of more than 20 percent from 2008 to 2009.

Therefore, the choice of vehicle classes for future consideration as part of this assessment of potential fuel economy technologies should focus on vehicle size, weight, interior passenger volume, intended use, and the potential for implementation of next-generation power trains, including hybrid electrics. Based on various factors outlined below, the following classification of light-duty vehicles in the United States was determined by the committee to be an appropriate basis for future technology development and introduction into production.

1. *Two-seater convertibles and coupes*—Small vehicles by interior volume whose function is high-performance and handling. The average 2007 model-year vehicle for this class was developed from EPA (2008a) and has the following characteristics: a six-cylinder, four-valve, dual overhead cam engine with intake cam phasing and a 6-speed automatic transmission. The average vehicle for this class is used as the base vehicle in the estimation of fuel consumption reductions for multiple technologies as discussed later in this chapter.
2. *Small cars*—Mini-, sub-, and compact cars, standard performance, mostly four-cylinder, mostly front-wheel drive (FWD), including small station wagons. The average 2007 model-year vehicle for this class was developed from EPA (2008a) and has the following characteristics: a four-cylinder, four-valve, dual overhead cam engine with intake cam phasing and a 6-speed automatic transmission. The average vehicle for this class is used as the base vehicle in the estimation of fuel consumption reductions for multiple technologies as discussed later in this chapter.
3. *Intermediate and large cars*—Standard performance, mostly FWD, mostly six-cylinder, including large station wagons with less than 0.07 hp/lb of vehicle weight. The average 2007 model-year vehicle for this class was developed from EPA (2008a) and has the following characteristics: a six-cylinder, four-valve, dual overhead cam engine with intake cam phasing and a 4-speed automatic transmission. The average vehicle for this class is used as the base vehicle in the estimation of fuel consumption reductions for multiple technologies as discussed later in this chapter.
4. *High-performance sedans*—Passenger cars with greater than or equal to 0.07 hp/lb of vehicle weight that are not two-seaters. The average 2007 model-year vehicle for this class was developed from EPA (2008a) and has the following characteristics: a six-cylinder, four-valve, dual overhead cam engine with intake cam phasing and a 6-speed automatic transmission. The average vehicle for this class is used as the base vehicle in the estimation of fuel consumption reductions for multiple technologies as discussed later in this chapter.
5. *Unit-body standard trucks*—Non-pickup trucks with unibody construction and hp/lb of vehicle weight ratios of under 0.055 including crossover vehicles, SUVs, and minivans. Most vehicles employ front-wheel drive. The average 2007 model-year vehicle for this class was developed from EPA (2008a) and has the following characteristics: a six-cylinder, four-valve, dual overhead cam engine with intake cam phasing and a 6-speed automatic transmission. The average vehicle for this class is used as the base vehicle in the estimation of fuel consumption reductions for multiple technologies as discussed later in this chapter.
6. *Unit-body high-performance trucks*—Crossover vehicles, SUVs, and minivans with hp/lb of vehicle weight ratios of 0.055 or greater. Most have rear-wheel drive (RWD) or all-wheel drive (AWD) and unibody construction, and most are luxury vehicles. The average 2007 model-year vehicle for this class was developed from EPA (2008a) and has the following characteristics: a six-cylinder, four-valve, dual overhead cam engine with intake cam phasing and a 6-speed automatic transmission. The average vehicle for this class is used as the base vehicle in the estimation of fuel consumption reductions for multiple technologies as discussed later in this chapter.
7. *Body-on-frame small and midsize trucks*—Pickups less than or equal to 1,500 lb payload capacity (CEC class 14) and SUVs of up to 175 cubic feet of passenger volume plus cargo volume with RWD or AWD. The average 2007 model-year vehicle for this class was developed from EPA (2008a) and has the following characteristics: a six-cylinder, two-valve, single overhead cam engine with a 5-speed automatic transmission. The average vehicle for this class is used as the base vehicle in the estimation of fuel consumption reductions for multiple technologies as discussed later in this chapter.
8. *Body-on-frame large trucks*—Pickups of greater than 1,500 lb payload but less than 10,000 lb GVW, and SUVs with 175 cubic feet or greater of passenger plus cargo volume with RWD or AWD, including all standard vans, cargo and passenger. The average 2007 model-year vehicle for this class was developed from EPA (2008a) and has the following characteristics: an eight-cylinder, two-valve, overhead valve engine with a 4-speed automatic transmission. The average vehicle for this class is used as the base vehicle in the estimation of fuel consumption reductions for multiple technologies as discussed later in this chapter.

These eight classes allow an evaluation of similar base vehicles designs, where the vehicle size, baseline chassis configuration, aerodynamic characteristics, vehicle weight and type of drive train (FWD, RWD, and AWD) are similar. This grouping should result in vehicle classes where similar calibration criteria are associated with similar vehicle performance characteristics. A greater number of classes would also be possible if there was a desire to narrow the variability in vehicle characteristics in each class.

ESTIMATION OF FUEL CONSUMPTION BENEFITS

Incremental reductions in fuel consumption through the application of technologies were estimated by the committee. As discussed earlier in this report, input came from many sources including component suppliers, vehicle manufacturers, and the review of many published analyses conducted by, or for, the U.S. Department of Transportation National Highway Traffic Safety Administration (NHTSA), U.S. Environmental Protection Agency (EPA), California Air Resources Board (CARB), and other agencies or trade associations. The committee also contracted with several consultants to provide input.

Relative reductions in fuel consumption can result from several factors, many of which are interrelated:

- Reduction in the tractive force needed to propel the vehicle (reduced rolling resistance, aerodynamic drag, vehicle weight, etc.);
- Improvement in the energy conversion efficiency of the fuel in the engine into maximum usable energy through increased thermal efficiency (compression ratio increase for gasoline engines, lean combustion, diesel, etc.);
- Reductions in the engine and power train energy losses that consume portions of the available energy before and after combustion (gas exchange losses, power train friction, accessory losses, etc.);
- Optimization of operational parameters that allow the engine to run in regions of highest efficiency (increased number of transmission gears, CVTs, improved lugging characteristics, aggressive shift logic, etc.); and
- Some form of hybridization that allows other forms of energy capture, storage, and management to reduce the total energy consumed over the driving cycle.

The committee thinks that the most accurate method of analyzing potential reductions in fuel consumption, which considers the extent to which any of the efficiency improvements or energy loss reductions identified above can be realized while maintaining energy balance criteria, utilizes full system simulation (FSS). This analysis technique, as described in Chapter 8, represents the state of the art in predicting vehicle performance, fuel consumption, direct CO_2 emissions, and other regulated and non-regulated emissions. However, FSS analyses require detailed vehicle, engine, transmission, accessory, and other subsystem data, typically expressed in the form of data maps that quantify power, torque, fuel consumption, and exhaust emissions over the complete range of operation. Historically, such data (which may not yet exist for the most advanced technologies) have been considered proprietary by automobile manufacturers (referred to as original equipment manufacturers; OEMs) and suppliers, such that only those companies associated with the design, development, and production of such systems have had the data to conduct such analyses. However, partnerships currently exist between the automotive industry and the U.S. government such that more complete experimental data will be made available in the future.

Another factor in successfully modeling full vehicle systems is the need to understand and capture the tradeoffs that OEMs must make in developing final production calibrations of vehicles and their power trains. Calibration is the process of power train and vehicle performance optimization that focuses on achieving predetermined performance, drivability, fuel consumption, durability, fuel octane sensitivity, and many other parameters while still complying with statutory requirements such as those for levels of emissions, onboard diagnostics (OBD), and safety standards. In particular, many potential technologies that can be applied for improving fuel consumption could influence performance parameters such as 0-60 mph acceleration times, vehicle passing capability, towing capability, transmission shift quality, or noise and vibration characteristics. Different manufacturers must thus determine their customer-preferred compromises and calibrate the vehicle control algorithms accordingly. Based on the number of potential parameters that may be varied in modern passenger car engines, tens of thousands of combinations are possible. Therefore, manufacturers and calibration service companies have developed optimization strategies and algorithms to fine-tune these variables while achieving an OEM's criteria for performance and drivability within the constraints of emissions, fuel economy, and other standards. Calibration logic is normally a highly confidential process that requires the experience of companies involved in the production release of vehicles (OEMs, Tier 1 suppliers, production engineering services companies, etc.) to accurately assess the necessary performance, fuel consumption and exhaust emission, and drivability tradeoffs for accurate modeling.

Partial discrete approximation (PDA) and lumped parameter modeling techniques, as described in Chapter 8, examine and estimate incremental reduction in fuel consumption associated with applications of discrete technologies or subsystems and their effect on reducing energy losses. They represent a more time- and cost-effective method of estimating potential reductions in fuel consumption and may incorporate routines that attempt to tabulate or account for aggregation of energy-loss reductions that focus on fluid mechanical losses, frictional losses, and heat transfer losses. However, the ultimate accuracy of such analyses relies on a sufficiently broad set of empirical or system-simulation data that do not necessarily provide enough detail to understand the base test vehicle distribution of energy losses. Calibration of such models against actual test vehicles provides a benchmark of the modeler's attempt to match performance data, but does not provide the same level of explanation of the subsystem contribution to total vehicle energy losses that is accomplished in the FSS cases. Furthermore, the influences of variations in calibration strategies owing to such factors as driver comfort; noise, vibration, and harshness (NVH)-related issues; and performance/emissions tradeoffs are typically not considered in such analyses.

With either modeling approach, it is imperative to understand the role that any previously applied technologies play in reducing energy losses and/or improving the thermodynamic efficiency of the power train.

APPLICABILITY OF TECHNOLOGIES TO VEHICLE CLASSES

Not all of the technologies identified in Chapters 4 through 7 of this report can be justifiably applied to all vehicle types. Applicability of the technologies to the various vehicle classes requires an analysis of parametric tradeoffs which considers functionality, intended use, impact on warranty, ease of implementation, product cycle timing, market demand, cost-effectiveness, and many other factors. Some technologies may be discounted for technical reasons, for example, the limitations of continuously variable transmissions (CVTs) in transmitting high torque on vehicle classes with larger engines where towing or off-road capability is a primary product feature. Others may be excluded based on the intended purpose of the vehicle. For example, low-rolling-resistance tires appear to be a cost-effective means of reducing fuel consumption, potentially justifying their use on all vehicles. However, in higher performance classes of vehicles, where tire grip is an important product feature or for SUV applications where the vehicle may travel off-road, the use of such tires is likely restricted.

Table 9.1 shows the committee's estimation of incremental reductions in fuel consumption that may be expected from the application of different technologies and ranges associated with the reductions. In general, the committee estimated what it considered to be the average fuel consumption reduction for a technology before it attempted to estimate the range. These data, shown in the form of ranges, are in some cases dependent upon the level of technology applied to a vehicle before the next increment is taken. As identified above, these data represent estimates by the committee developed from evaluating published data, and analyses conducted by the NHTSA, the EPA, and others. Appendix I contains results from some of these other studies, although the reader should refer to the original source for the assumptions and study approaches used in these other studies. The expert judgment of members of the committee whose careers have focused on vehicle and power train design and development were also incorporated in the estimates. Examination of the data in Table 9.1 suggests that significant variations in estimates of the potential for reducing fuel consumption are due to the lack of detailed simulation data on actual or theoretically modeled vehicles or power trains against which to refine the estimates. This variability in estimates for fuel consumption reductions also reflects the fact that different OEMs may obtain different benefits from the same technology due to differences in implementation and calibration. Also, positive benefits may vary depending on the particular engine/transmission/vehicle architecture. These factors have been considered by the committee in its range of estimates or its decision to include or exclude the potential application of technologies into different technology paths. Note that the ranges associated with these technologies do not reflect the possibility that, over time, the average fuel consumption benefit could tend toward the high end of the range as the lessons learned from the best examples of the technology spread across the industry and as the impacts of higher CAFE standards increase. Although the committee recognizes that the implementation of these technologies with fuel consumption benefits at the higher end of the ranges could occur, it is difficult to assert that this will occur or to what degree this would impact the average consumption benefit over time.

The issue of how multiple technologies might interact when used to reduce fuel consumption is critical. FSS analyses conducted by Ricardo, Inc., for the EPA and for the committee shed some light on the issue of synergistic interaction of multiple technologies that may attempt to reduce energy losses of a similar type, such as pumping losses (Ricardo, Inc., 2008, 2009). These analyses show the need to carefully understand the contribution of technologies in reducing losses whose impact may be only a 1 to 2 percent reduction in fuel consumption. The Ricardo, Inc., analyses also show that the type of vehicle and power train influences the extent to which different technologies reduce fuel consumption, especially between vehicles of different classes with different intended uses. This effect is discussed in Chapter 8, where the primary effect of synergies was shown to dominate the potential for improvement. Accordingly, secondary effects of influences that interact across technology improvement paths were found to be minor.

ESTIMATING INCREMENTAL COSTS ASSOCIATED WITH TECHNOLOGY EVOLUTION

Chapter 3 describes the methodologies used for the estimation of incremental costs associated with the introduction of advanced technology for reducing fuel consumption. A range of estimated costs was also prepared and is outlined in the technology sections presented in Chapters 4 through 7. Table 9.2 shows the collection of these cost estimates for all technologies included in this report. The cost estimates represent estimates for the current (2009/2010) time period to about 5 years in the future. As with the data on fuel consumption reductions, incremental cost information was provided to the committee by OEMs, Tier 1 suppliers, and studies published by trade associations, governmental agencies, manufacturing consultants, and earlier NRC reports. Appendix I

TABLE 9.1 Committee's Estimates of Effectiveness (shown as a percentage) of Near-Term Technologies in Reducing Vehicle Fuel Consumption

Technologies		Incremental values - A preceding technology must be included								
		I4			V6			V8		
Spark Ignition Techs	Abbreviation	Low	High	AVG	Low	High	AVG	Low	High	AVG
Low Friction Lubricants	LUB	0.5	0.5	**0.5**	0.5	0.5	**0.5**	0.5	0.5	**0.5**
Engine Friction Reduction	EFR	0.5	2.0	**1.3**	0.5	2.0	**1.3**	1.0	2.0	**1.5**
VVT- Coupled Cam Phasing (CCP), SOHC	CCP	1.5	3.0	**2.3**	1.5	3.5	**2.5**	2.0	4.0	**3.0**
Discrete Variable Valve Lift (DVVL), SOHC	DVVL	1.5	3.0	**2.3**	1.5	3.0	**2.3**	2.0	3.0	**2.5**
Cylinder Deactivation, SOHC	DEAC	NA	NA	NA	4.0	6.0	**5.0**	5.0	10.0	**7.5**
VVT - In take Cam Phasing (ICP)	ICP	1.0	2.0	**1.5**	1.0	2.0	**1.5**	1.5	2.0	**1.8**
VVT - Dual Cam Phasing (DCP)	DCP	1.5	2.5	**2.0**	1.5	3.0	**2.3**	1.5	3.0	**2.3**
Discrete Variable Valve Lift (DVVL), DOHC	DVVL	1.5	3.0	**2.3**	1.5	3.5	**2.5**	2.0	4.0	**3.0**
Continuously Variable Valve Lift (CVVL)	CVVL	3.5	6.0	**4.8**	3.5	6.5	**5.0**	4.0	6.5	**5.3**
Cylinder Deactivation, OHV	DEAC	NA	NA	NA	4.0	6.0	**5.0**	5.0	10.0	**7.5**
VVT - Coupled Cam Phasing (CCP), OHV	CCP	1.5	3.0	**2.3**	1.5	3.5	**2.5**	2.0	4.0	**3.0**
Discrete Variable Valve Lift (DVVL), OHV	DVVL	1.5	2.5	**2.0**	1.5	3.0	**2.3**	2.0	3.0	**2.5**
Stoichiometric Gasoline Direct Injection (GDI)	SGDI	1.5	3.0	**2.3**	1.5	3.0	**2.3**	1.5	3.0	**2.3**
Turbocharging and Downsizing	TRBDS	2.0	5.0	**3.5**	4.0	6.0	**5.0**	4.0	6.0	**5.0**
Diesel Techs										
Conversion to Diesel	DSL	15.0	35.0	**25.0**	15.0	35.0	**25.0**	NA	NA	NA
Conversion to Advanced Diesel	ADSL	7.0	13.0	**10.0**	7.0	13.0	**10.0**	22.0	38.0	**30.0**
Electrification/Accessory Techs										
Electric Power Steering (EPS)	EPS	1.0	3.0	**2.0**	1.0	3.0	**2.0**	1.0	3.0	**2.0**
Improved Accessories	IACC	0.5	1.5	**1.0**	0.5	1.5	**1.0**	0.5	1.5	**1.0**
Higher Voltage/Improved Alternato	HVIA	0.0	0.5	**0.3**	0.0	0.5	**0.3**	0.0	0.5	**0.3**
Transmission Techs										
Continuously Variable Transmission (CVT)	CVT	1.0	7.0	**4.0**	1.0	7.0	**4.0**	1.0	7.0	**4.0**
5-spd Auto. Trans. w/ Improved Internals		2.0	3.0	**2.5**	2.0	3.0	**2.5**	2.0	3.0	**2.5**
6-spd Auto. Trans. w/ Improved Internals		1.0	2.0	**1.5**	1.0	2.0	**1.5**	1.0	2.0	**1.5**
7-spd Auto. Trans. w/ Improved Internals		2.0		**2.0**	2.0		**2.0**	2.0		**2.0**
8-spd Auto. Trans. w/ Improved Internals		1.0		**1.0**	1.0		**1.0**	1.0		**1.0**
6/7/8-spd Auto. Trans. w/ Improved Internals	NAUTO	3.0	8.0	**5.5**	3.0	8.0	**5.5**	3.0	8.0	**5.5**
6/7-spd DCT from 4-spd AT	DCT	6.0	9.0	**7.5**	6.0	9.0	**7.5**	6.0	9.0	**7.5**
6/7-spd DCT from 6-spd AT	DCT	3.0	4.0	**3.5**	3.0	4.0	**3.5**	3.0	4.0	**3.5**
Hybrid Techs										
12V BAS Micro-Hybrid	MHEV	2.0	4.0	**3.0**	2.0	4.0	**3.0**	2.0	4.0	**3.0**
Integrated Starter Generator	ISG	29.0	39.0	**34.0**	29.0	39.0	**34.0**	29.0	39.0	**34.0**
Power Split Hybrid	PSHEV	24.0	50.0	**37.0**	24.0	50.0	**37.0**	24.0	50.0	**37.0**
2-Mode Hybrid	2MHEV	25.0	45.0	**35.0**	25.0	45.0	**35.0**	25.0	45.0	**35.0**
Plug-in hybric	PHEV	NA	NA	NA	NA	NA	NA	NA	NA	NA
Vehicle Techs										
Mass Reduction - 1%	MR1	0.3		**0.3**	0.3		**0.3**	0.3		**0.3**
Mass Reduction - 2%	MR2	1.4		**1.4**	1.4		**1.4**	1.4		**1.4**
Mass Reduction - 5%	MR5	3.0	3.5	**3.3**	3.0	3.5	**3.3**	3.0	3.5	**3.3**
Mass Reduction - 10%	MR10	6.0	7.0	**6.5**	6.0	7.0	**6.5**	6.0	7.0	**6.5**
Mass Reduction - 20%	MR20	11.0	13.0	**12.0**	11.0	13.0	**12.0**	11.0	13.0	**12.0**
Low Rolling Resistance Tires	ROLL	1.0	3.0	**2.0**	1.0	3.0	**2.0**	1.0	3.0	**2.0**
Low Drag Brakes	LDB	1.0		**1.0**	1.0		**1.0**	1.0		**1.0**
Aero Drag Reduction 10%	AERO	1.0	2.0	**1.5**	1.0	2.0	**1.5**	1.0	2.0	**1.5**

NOTE: Some of the benefits (highlighted in green) are incremental to those obtained with preceding technologies shown in the technology pathways described in Chapter 9.

contains results from some of these other studies, although, again, the reader should refer to the original source for the assumptions and study approaches used in these other studies. During the data gathering process, it became clear that the estimated incremental cost ranges were, in many cases, very large, depending on the boundary conditions identified by the organization offering the information. Generally, the committee notes that cost estimates are always more uncertain than the fuel consumption impact estimates, and the estimates presented here should be considered very uncertain until more detailed studies are completed. A boundary condition in the cost estimations is an assumption of long-term, high-volume production, whereby analysts attempt to normalize all incremental costs into a scenario where the capitalized development cost becomes a small portion of the final unit production cost. This is accomplished by assuming that production volumes are several hundred thousand units per year and remain in production for many years.

Although this assumption may be quite appropriate to normalize overall annual societal costs, it does not necessarily recognize the initial development-based costs and quality hurdles that may prevent a manufacturer from pursuing new product or technology areas. For example, such analyses would not consider factors that may inhibit or prevent the

TABLE 9.2 Committee's Estimates of Technology Costs in U.S. Dollars (2008)

NRC 2009 Costs													
Technologies		Incremental Values - A preceding technology must be included											
		I4				V6				V8			
Spark Ignition Techs	Abbreviation	Low	High	AVG	AVG w/1.5 RPE	Low	High	AVG	AVG w/1.5 RPE	Low	High	AVG	AVG w/1.5 RPE
Low Friction Lubricants	LUB	3	5	**4**	**6**	3	5	**4**	**6**	3	5	**4**	**6**
Engine Friction Reduction	EFR	32.0	52.0	**42**	**63**	48	78	**63**	**94.5**	64	104	**84**	**126**
VVT- Coupled Cam Phasing (CCP), SOHC	CCP	35		**35**	**52.5**	70		**70**	**105**	70		**70**	**105**
Discrete Variable Valve Lift (DVVL), SOHC	DVVL	130	160	**145**	**217.5**	180	210	**195**	**292.5**	280	320	**300**	**450**
Cylinder Deactivation, SOHC	DEAC	NA	NA	**NA**	**NA**	340	400	**370**	**555**	357	420	**388.5**	**582.75**
VVT - In take Cam Phasing (ICP)	ICP	35		**35**	**52.5**	70		**70**	**105**	70		**70**	**105**
VVT - Dual Cam Phasing (DCP)	DCP	35		**35**	**52.5**	70		**70**	**105**	70		**70**	**105**
Discrete Variable Valve Lift (DVVL), DOHC	DVVL	130	160	**145**	**217.5**	180	220	**200**	**300**	260	300	**280**	**420**
Continuously Variable Valve Lift (CVVL)	CVVL	159	205	**182**	**273**	290	310	**300**	**450**	350	390	**370**	**555**
Cylinder Deactivation, OHV	DEAC	NA	NA	**NA**	**NA**	220	250	**235**	**352.5**	255		**255**	**382.5**
VVT - Coupled Cam Phasing (CCP), OHV	CCP	35		**35**	**52.5**	35		**35**	**52.5**	35		**35**	**52.5**
Discrete Variable Valve Lift (DVVL), OHV	DVVL	130	160	**145**	**218**	210	240	**225**	**338**	280	320	**300**	**450**
Stoichiometric Gasoline Direct Injection (GDI)	SGDI	117	195	**156**	**234**	169	256	**213**	**319**	295	351	**323**	**485**
Turbocharging and Downsizing	TRBDS	370	490	**430**	**645**	-144	205	**31**	**46**	525	790	**658**	**986**
Diesel Techs													
Conversion to Diesel	DSL	2154	2632	**2393**	**3590**	2857	3491	**3174**	**4761**	NA	NA	**NA**	**NA**
Conversion to Advanced Diese	ADSL	520	520	**520**	**780**	683	683	**683**	**1025**	3513	4293	**3903**	**5855**
Electrification/Accessory Techs													
Electric Power Steering (EPS)	EPS	70	120	**95**	**143**	70	120	**95**	**143**	70	120	**95**	**143**
Improved Accessories	IACC	70	90	**80**	**120**	70	90	**80**	**120**	70	90	**80**	**120**
Higher Voltage/Improved Alternator	HVIA	15	55	**35**	**53**	15	55	**35**	**53**	15	55	**35**	**53**
Transmission Techs													
Continuously Variable Transmission (CVT)	CVT	150	170	**160**	**240**	243	263	**253**	**380**	243	263	**253**	**380**
5-spd Auto. Trans. w/ Improved Internals		133		**133**	**200**	133		**133**	**200**	133		**133**	**200**
6-spd Auto. Trans. w/ Improved Internals		133	215	**174**	**262**	133	215	**174**	**262**	133	215	**174**	**262**
7-spd Auto. Trans. w/ Improved Internals		170	300	**235**	**353**	170	300	**235**	**353**	170	300	**235**	**353**
8-spd Auto. Trans. w/ Improved Internals		425		**425**	**638**	425		**425**	**638**	425		**425**	**638**
6/7/8-Speed Auto. Trans. with Improved Internals	NAUTO	137	425	**281**	**422**	137	425	**281**	**422**	137	425	**281**	**422**
6/7-spd DCT from 6-spd AT	DCT	-147	185	**19**	**29**	-147	185	**19**	**29**	-147	185	**19**	**29**
6/7-spd DCT from 4-spd AT	DCT	-14	400	**193**	**290**	-14	400	**193**	**290**	-14	400	**193**	**290**
Hybrid Techs													
12V BAS Micro-Hybrid	MHEV	450	550	**500**	**665**	585	715	**650**	**865**	720	880	**800**	**1064**
Integrated Starter Generator	ISG	1760	2640	**2200**	**2926**	2000	3000	**2500**	**3325**	3200	4800	**4000**	**5320**
Power Split Hybrid	PSHEV	2708	4062	**3385**	**4502**	3120	4680	**3900**	**5187**	4000	6000	**5000**	**6650**
2-Mode Hybrid	2MHEV	5200	7800	**6500**	**8645**	5200	7800	**6500**	**8645**	5200	7800	**6500**	**8645**
Series PHEV 40	PHEV	8000	12000	**10000**	**13300**	9600	14400	**12000**	**15960**	13600	20400	**17000**	**22610**
Vehicle Techs													
Mass Reduction - 1%	MR1	37	45	**41**	**61**	48	58	**53**	**80**	68	82	**75**	**113**
Mass Reduction - 2%	MR2	77	93	**85**	**127**	100	121	**111**	**166**	142	170	**156**	**234**
Mass Reduction - 5%	MR5	217	260	**239**	**358**	283	339	**311**	**467**	399	479	**439**	**659**
Mass Reduction - 10%	MR10	520	624	**572**	**859**	679	815	**747**	**1120**	958	1150	**1054**	**1581**
Mass Reduction - 20%	MR20	1600	1700	**1650**	**2475**	1600	1800	**1700**	**2550**	1600	1900	**1750**	**2625**
Low Rolling Resistance Tires	ROLL	30	40	**35**	**53**	30	40	**35**	**53**	30	40	**35**	**53**
Aero Drag Reduction 10%	AERO	40	50	**45**	**68**	40	50	**45**	**68**	40	50	**45**	**68**

introduction of diesel technology into passenger vehicles due to the significant investment and general lack of familiarity of North American automotive OEMs and suppliers in the production of small, light-duty diesels and the durability of necessary exhaust aftertreatment systems. This serves as a reminder that, while overall costs to the industry of new technologies is an important consideration, it is the individual manufacturers that bear the risk in adapting a technology to a specific vehicle and this risk may not be fully captured in a metric of overall industry costs.

The committee was briefed on the very detailed and transparent teardown cost assessment methodology being utilized by the EPA as part of the process for estimating the cost of fuel economy technologies. Cost estimation using the teardown approach is discussed in Chapter 3. The committee finds this approach an improvement over one where cost estimates are developed through expert knowledge and surveys of suppliers and OEMs, which have been the basis for most published studies and the majority of this report. Furthermore, the committee recommends that the use of teardown studies be expanded for future assessments when cost-effectiveness is an important evaluation criterion.

ASSESSING POTENTIAL TECHNOLOGY SEQUENCING PATHS

When manufacturers consider a strategy for implementing technologies that reduce fuel consumption, a normal business decision process must tradeoff many different parameters, including cost-effectiveness (fuel consumption reduction versus production cost), the ability to be integrated into product planning cycles, intended product use, reliability, impact on vehicle performance characteristics, and customer acceptance. To conduct the current assessment, the committee employed a method whereby cost-effectiveness (fuel consumption reduction divided by high-volume production incremental cost), vehicle intended use, base power train configuration, and technology state of readiness were considered in estimating potential technology paths for the eight vehicle classes described earlier.

As previously stated, an attempt to perform FSS on every vehicle model with all combinations of technologies is not practicable. Such a process would necessitate the analysis of (at least) tens of thousands of vehicle and power train technology combinations. It would require potentially confidential engine, transmission, accessory, and hybrid power train system maps; vehicle data such as friction as a function of vehicle speed; aerodynamic parameters; and many others parameters that are either proprietary or would require significant vehicle testing to generate for all of the combinations that are possible.

In some published studies, OEMs have supported such analyses for a limited number of vehicles that were chosen as sufficiently representative for discussion of the technology benefits and associated costs. For example, Sierra Research, in its report for the Alliance of Automobile Manufacturers (Auto Alliance), used the DOE-supported VEHSIM model to estimate fuel consumption reduction for various technologies using a composite of engine maps provided by manufacturers. Although the committee recommends a more practical approach to apply FSS for future regulatory actions, which is discussed later in this chapter, the exclusive use of FSS simulation for the assessment of all technologies considered under this study was not possible. The committee believes that sufficient experimental data can be gathered by the government to support future analysis and regulatory activities through consortia that include both regulatory agencies and automotive manufacturers and suppliers.

With this background, the committee evaluated potential technology paths that could be considered by a manufacturer, depending upon the manufacturer's actual state of technology and production capability. Rather than creating decision trees from which an extremely large number of possible technology combinations could be created for each vehicle class, the committee estimated possible technology evolution paths for each class that develop from the average baseline vehicle. These pathways are summarized in Figures 9.1 to 9.5.

The baseline attributes were determined on a class-by-class basis using the 2007 EPA test list. If 51 percent of the vehicles in a given class had variable valve timing (VVT), then the baseline, class-representative vehicle was given VVT, and this technology would not be added in the path. Because the characteristic vehicle in each class represents the average attributes for that class, there will be some vehicles in that class that have more or less technology content. The below-average vehicles may require additional technologies and associated costs to address future standards while the above-average vehicles may not. Using the average attributes should provide a good overall representation of technology benefits relative to the baseline fleet within a class of vehicles. The technologies of the baseline vehicles are listed in the title bar of each technology path.

In the absence of a very large number of FSS analyses, but guided by a limited number of FSS runs performed for the committee by Ricardo, Inc. (2009), the committee evaluated possible sequences of technology implementation for different classes of vehicles. The development of the technology sequences shown in Figures 9.1 to 9.5 also was done with input from OEMs, Tier 1 suppliers, other published analyses, and the expert judgment of committee members. In developing the ranges of fuel consumption reduction, the committee recognized that the potential reduction for each incremental step is highly dependent on the extent to which system losses could have been reduced by previous technologies. These pathways attempt to include such factors as cost-effectiveness (percent fuel consumption reduction/incremental cost), logical sequencing based upon preexisting technology, technical limitations, and ease of implementation (requirements for major or minor manufacturing changes, including production line considerations). Subjective judgment by the committee also played a role in the pathway definition process.

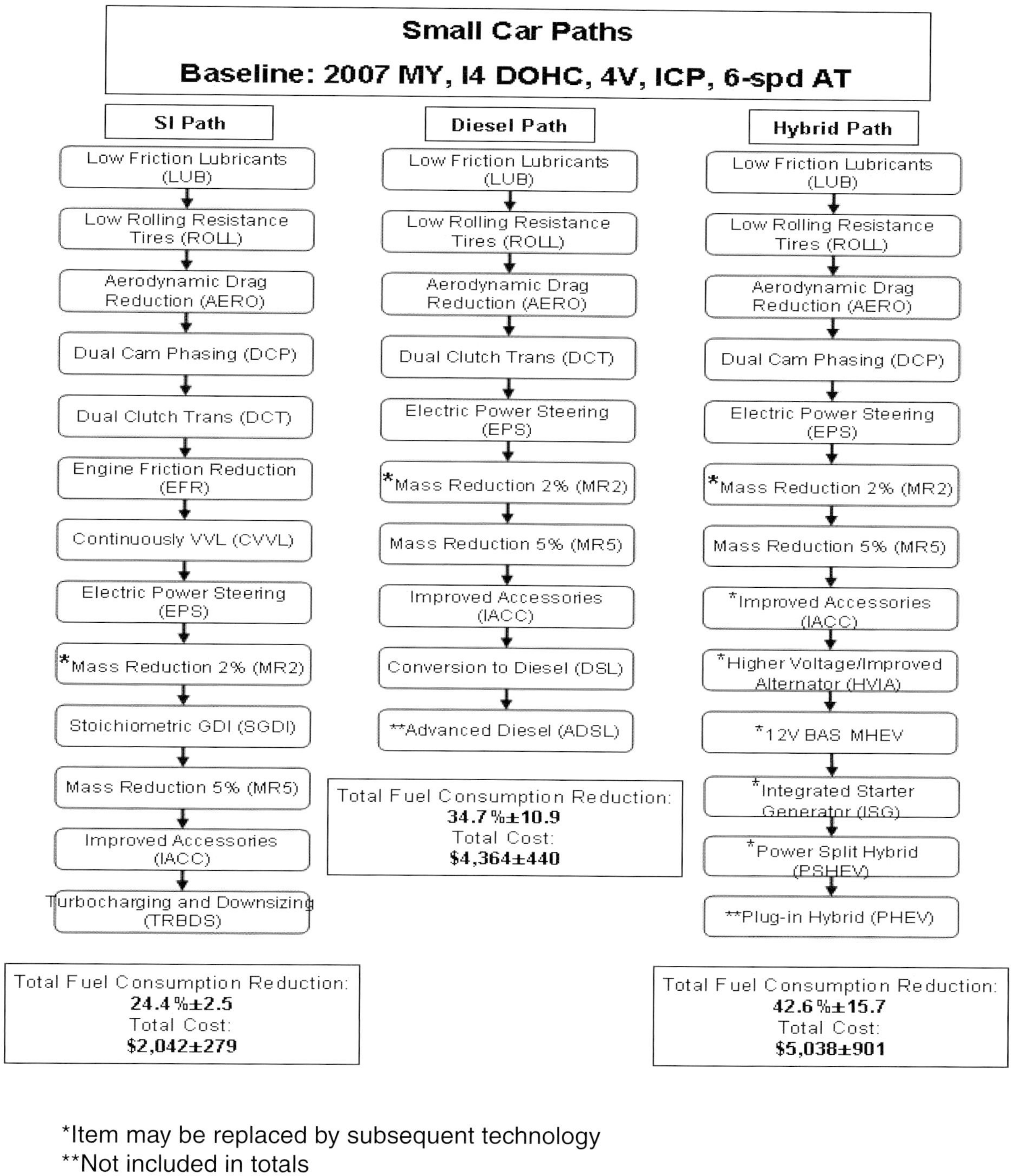

FIGURE 9.1 Small-car pathways with estimated total fuel consumption reduction and cost shown.

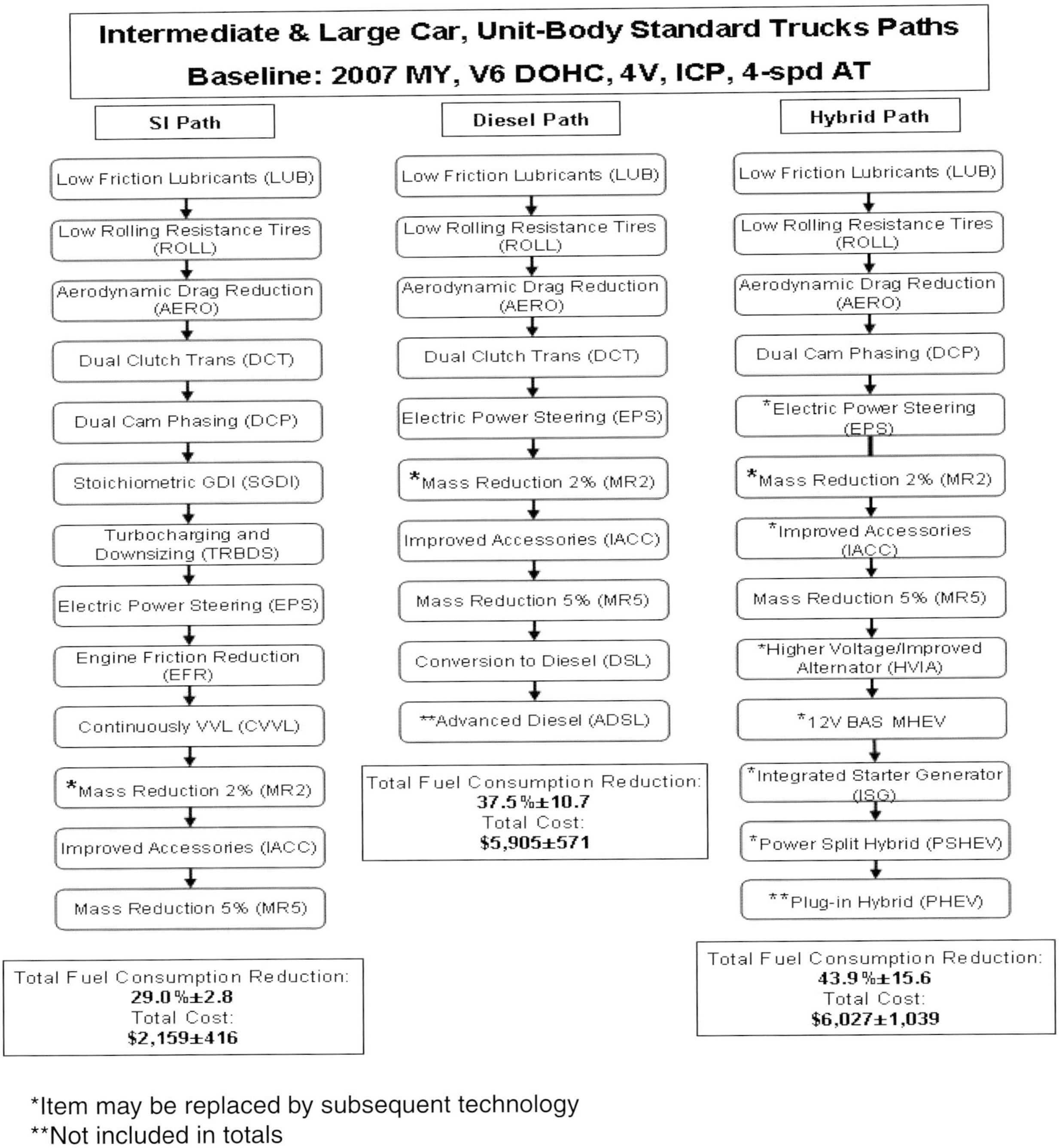

FIGURE 9.2 Intermediate- and large-car and unit-body standard truck pathways with estimated total fuel consumption reduction and cost shown.

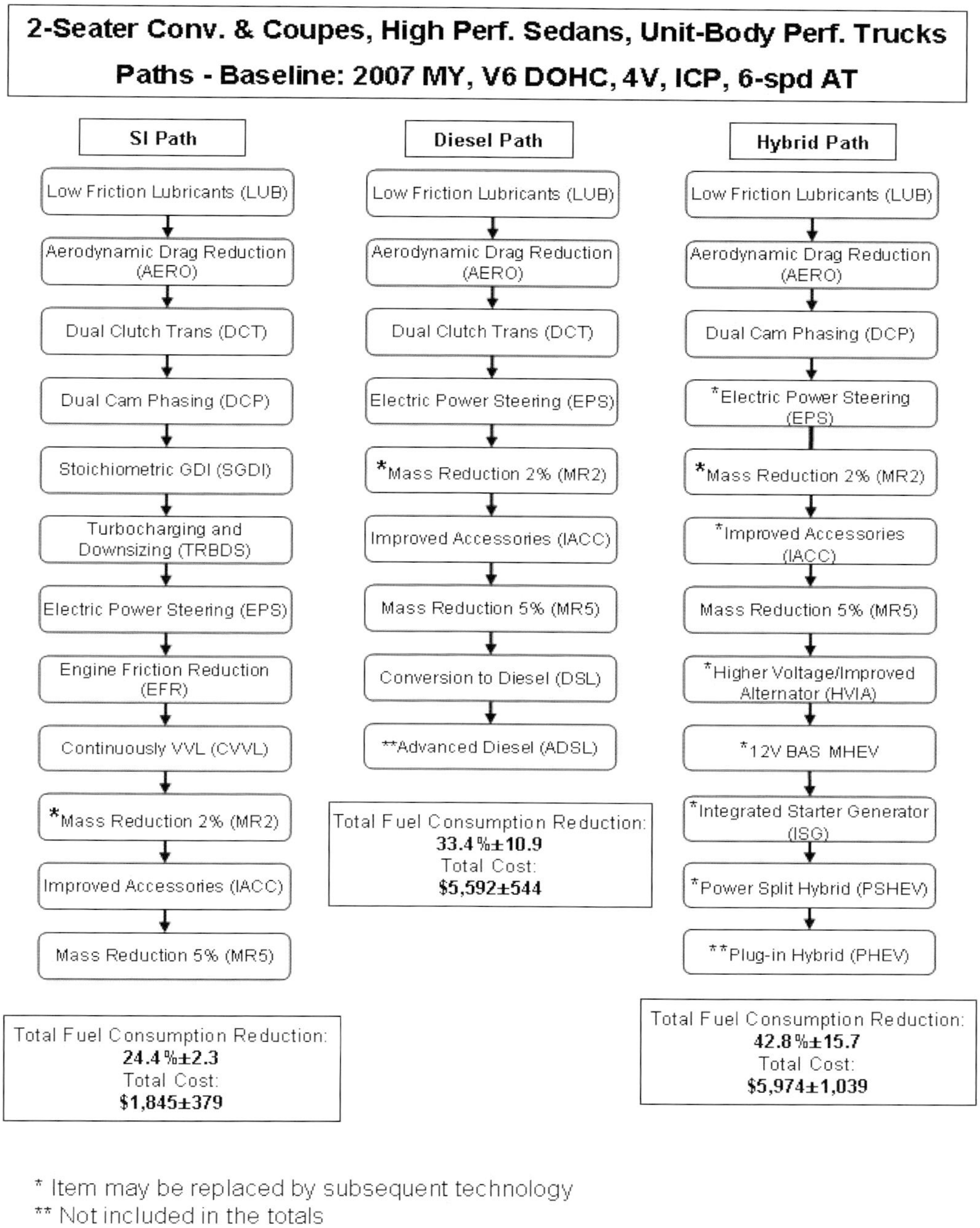

FIGURE 9.3 Two-seater convertible and coupe, high-performance sedan, and unit-body performance truck pathways with estimated total fuel consumption reduction and cost shown.

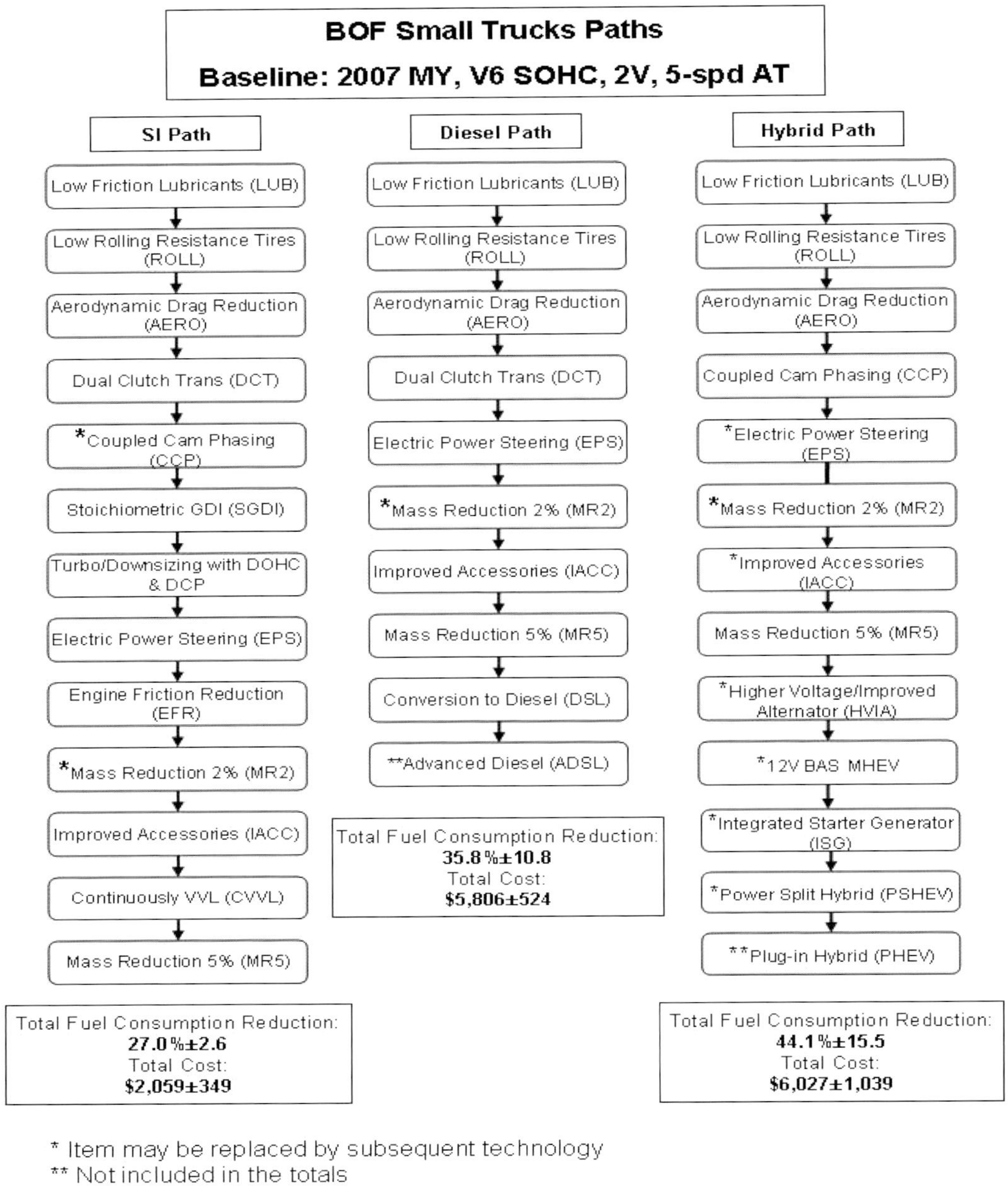

FIGURE 9.4 Body-on-frame small-truck pathways with estimated total fuel consumption reduction and cost shown.

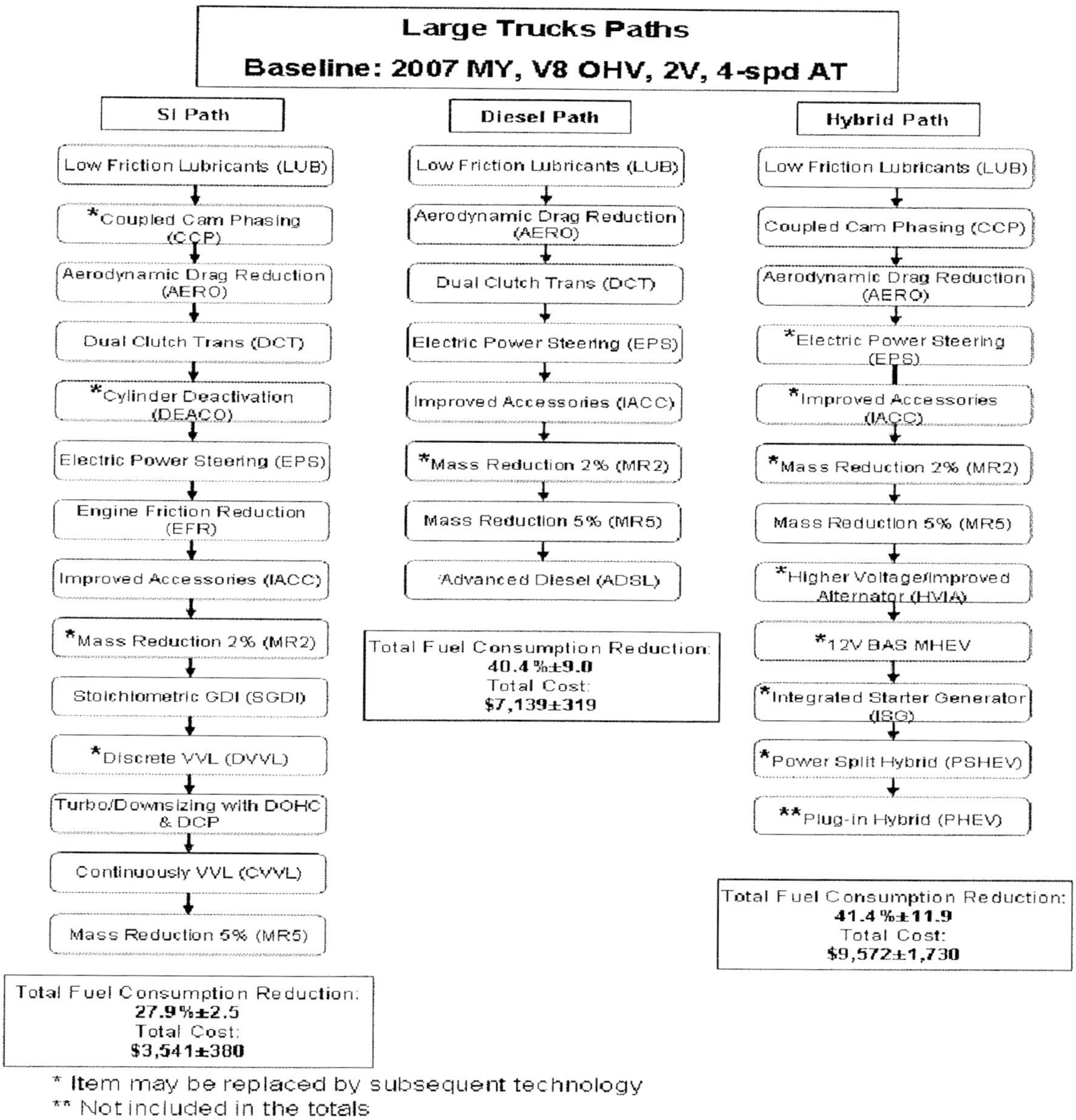

FIGURE 9.5 Large-truck pathways with estimated total fuel consumption reduction and cost shown.

Although the committee believes that some potential reduction is possible with each of the technologies considered, the extent to which a system energy loss can be reduced is highly dependent on all of the system interactive effects, the extent to which the baseline technology package has already reduced different categories of energy losses, and the production calibration parameters chosen by each manufacturer for the final release of each vehicle. Evaluating the energy losses associated with these technology pathways is discussed later in this chapter.

Review of Figures 9.1 through 9.5 shows that in certain cases (intermediate and large cars; unit-body standard trucks; two-seater convertibles, coupes, and high-performance sedans; unit-body performance trucks) the technology pathways are the same because of the similar base vehicle power train. However, the tradeoffs made as a result of varying performance metrics as these vehicle types go through their product evolution would result in different levels of fuel consumption improvement depending on the specific vehicle application.

Each range in potential fuel consumption reduction is an attempt by the committee to estimate the potential variation in energy loss reduction that might be possible when applying the technology to different power train and vehicle packages, taking into consideration known system features that will likely be optimized for different classes of vehicles with different intended uses. An example would be the bias inherent in production calibration of light-duty trucks or SUVs where reasonable towing capability is required.

A simple, multiplicative aggregation of the potential fuel consumption reduction is presented below each path in Figures 9.1 through 9.5 as a means to roughly estimate the total potential that might be possible. A probabilistic methodology based on the mean square rule was applied to estimate the confidence intervals for the aggregation of fuel consumption improvements and costs. Appendix J provides the mathematical explanation for this methodology. It assumes that the confidence intervals on each individual estimate of technology effectiveness or cost are the same. It also assumes that ranges in estimates are independent of each other and that errors are normally distributed. The approach then maintains a confidence interval for the aggregation of the low or high ends of the estimates that is equal to the confidence intervals estimated for the individual technologies. The committee assumes that the ranges for the individual costs and effectiveness represent a 90 percent confidence level. As such, the ranges were increased in technical areas where, in the opinion of the committee, more uncertainty existed with initial estimates.

It should also be noted that when the combination of fuel consumption improvement predictions and associated incremental costs is considered, the probability drops to 81 percent that any actual production technology introduction would fall within the ranges bounded by both the fuel consumption and cost ranges. This reduction is due to the (multiplicative) product of two 90 percent probabilities. *Although prepared in response to the committee's statement of task, these data are approximate in nature and as such should not be used as input to analyses where modeling accuracy is important. They are provided here as rough estimates that can be used in a qualitative comparative sense when comparing the relative cost-benefits of spark-ignition (SI)-related technologies that are potential candidates for FSS analyses. The committee's estimates can also be used for a qualitative comparison of SI-related technologies to other candidates such as light-duty diesel or hybrid vehicles.*

The results show that significant reductions in fuel consumption are possible with technologies that are already in production in U.S., European, or Asian markets. For example, for the intermediate car, large car, and unibody standard truck classes, the average reduction in fuel consumption for the SI path is 29 percent at a cost of approximately $2,200; the average reduction for the compression-ignition (CI) engine path is 38 percent at a cost of approximately $5,900; and the average reduction for the hybrids path is about 44 percent at an average cost of approximately $6,000. In general, diesel engine and hybrid vehicle technology options offer greater improvement potential compared to the SI pathway, but at a higher incremental cost. However, as evidenced by the increasingly wide range in estimated fuel consumption reduction and incremental cost, actual fuel consumption improvement can vary significantly depending on an individual manufacturer's product strategy. Further, it may be that the needs to reduce vehicle fuel consumption as mandated by recent legislation will result in OEMs implementing these technologies in such a way that the benefits fall toward the high end of the range. It should be noted that among its provisions related to fuel economy, the Energy Independence and Security Act (EISA) of 2007 required periodic assessments by the NRC of automobile vehicle fuel economy technologies and their costs. Thus, follow-on NRC committees will be responsible for responding to the EISA mandates, including the periodic evaluation of costs and fuel consumption benefits of individual technologies and the combined impacts of multiple technologies.

When developing the effectiveness numbers, attempts were made by the committee to incrementally adjust the effectiveness numbers of certain technologies that would normally be preceded by another technology. This process allowed the committee to approximate the inclusion of the synergistic effects resulting from the combination of certain technologies that were deemed to usually be packaged together. In an attempt to evaluate the incremental effectiveness numbers for the SI pathway derived by the committee, comparisons were conducted using the FSS data from the Ricardo report prepared for the committee (Ricardo, Inc., 2009), the EPA-provided lumped parameter model, and various other SAE papers where combinations of technologies were assessed. A comparison to the Ricardo data is shown in Figure 9.6. Packages involving CVTs were excluded because the committee agrees with the EPA (EPA, 2008b) that

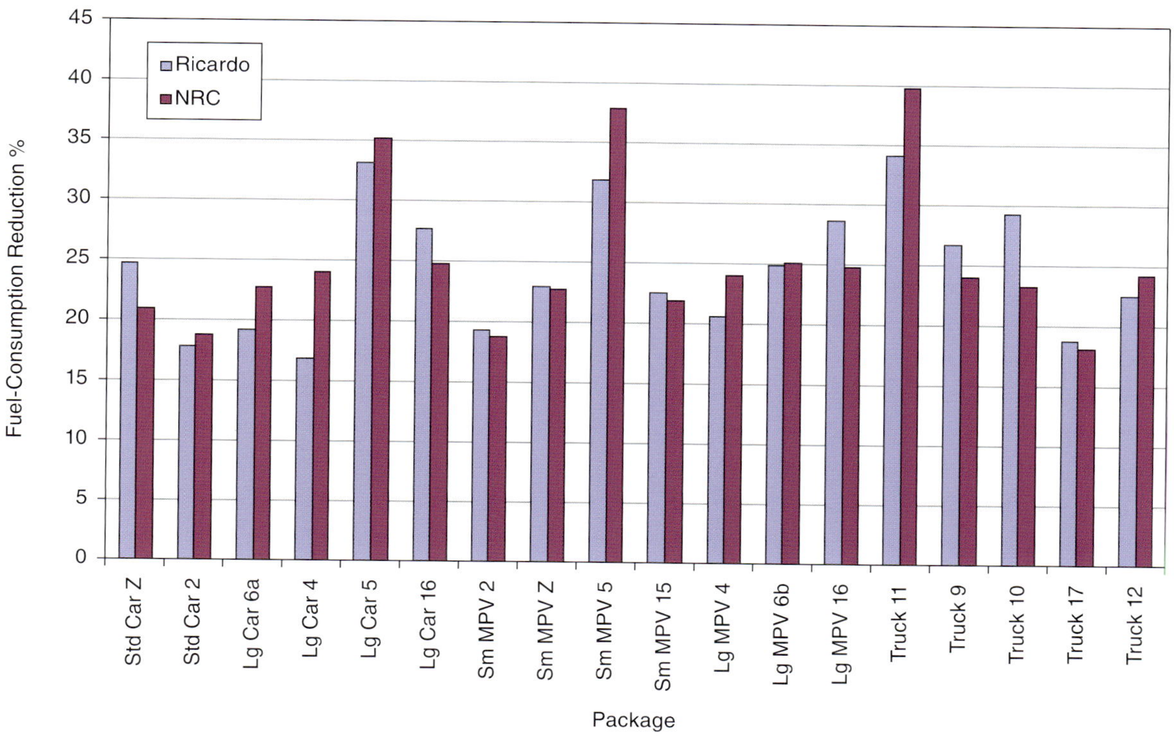

FIGURE 9.6 NRC estimates of effectiveness in reducing the fuel consumption of various light-duty vehicles compared with Ricardo, Inc. (2009) estimates based on data obtained with full system simulation.

Ricardo, Inc., used an abnormally small fuel consumption effectiveness value for this type of transmission.

As can be seen in Figure 9.6, the packages' fuel consumption reduction results generally follow the relative comparisons between the packages analyzed by Ricardo, Inc. This is likely due to the engineering judgment of the members of the committee whose experience in power train engineering could be applied to the assessment. However, the absolute levels of potential improvement can vary significantly between the committee estimates and Ricardo analyses. Furthermore, a comparison of the step-by-step incremental estimates that would result from the application of single technologies was not conducted. Therefore, it is not possible to determine whether the demonstrated correlations were a result of accurate incremental estimates, or whether a combination of over- and underestimates resulted in a rough approximation, where such occurs.

In any case, the Ricardo, Inc., packages represent only a subset of the greater number of technology combinations that would result from proceeding down the entire pathway evaluated by the committee. This underscores the importance of using FSS to account for the larger number of technology synergies and ensure that system loss reduction is not overstated.

Due to the approximate nature of the estimates of incremental improvements in fuel consumption, the committee recognizes the potential to overestimate the potential reduction in energy losses, despite consideration given to the total system energy consumers. Therefore, as another check on the predicted aggregation of potential technologies, the committee contracted with EEA to apply its lumped parameter modeling approach to evaluate the committee's estimates. Simplified lumped parameter models of vehicle energy use (e.g., Sovran and Bohn, 1981) provide a means of evaluating whether the fuel consumption benefits estimate for combinations of technologies by the multiplication methods result from forcing categories of energy losses (pumping and friction) to physically impossible levels. Appendix K provides a description of the EEA lumped parameter model as well as a description of the results in terms of the tractive energy requirements and the engine efficiency for the SI and diesel test cycles. These results indicate that the results from the multiplication method used here likely do not greatly overstate the benefits because this method does not explicitly take into account the theoretical limits of pumping loss reduction.

Figures 9.7 and 9.8 show the model results versus the committee estimates for eight cases (four for SI paths, and four for diesel paths). The model estimates for incremental improvements are relatively close to those of the committee, with the committee's estimates generally exceeding those of the EEA model by a small amount. These comparisons are made between the average level of the committee's estimates and the EEA data with no range presented. It should also be

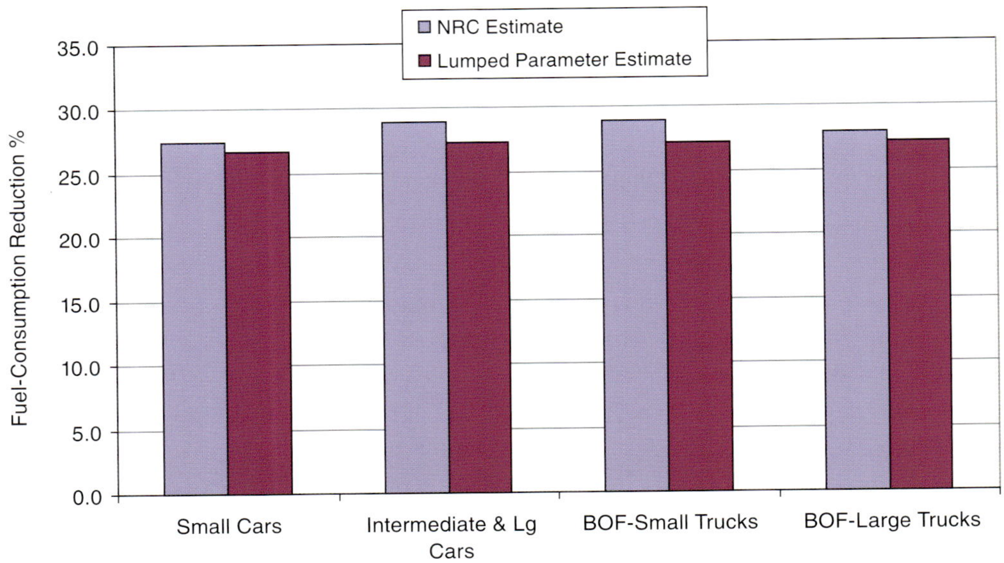

FIGURE 9.7 NRC estimates of effectiveness in reducing fuel consumption in spark-ignition engine pathways compared to EEA model outputs.

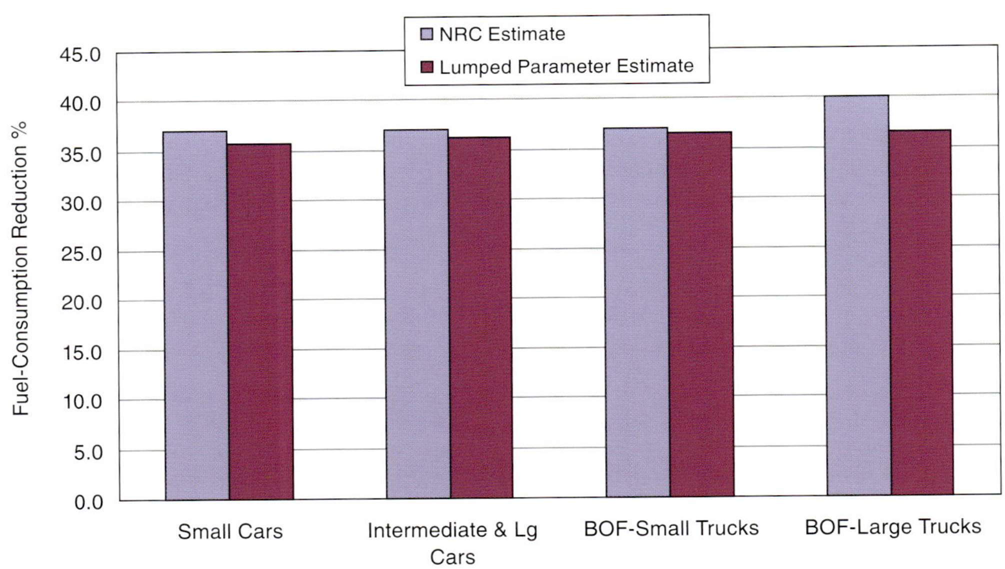

FIGURE 9.8 NRC estimates of effectiveness in reducing fuel consumption in diesel engine pathways compared to EEA model outputs.

noted that a baseline 4-speed automatic was used for both the committee's and EEA estimates because these comparisons were conducted prior to the committee's decision to utilize the average class transmission from the 2007 EPA test data in the technology paths.

One might conclude that the EEA modeling does, in fact, suggest that the committee's estimates slightly overpredict the estimate. However, the same general method of comparison with known production vehicles and estimating the potential levels of energy loss reduction are employed in both the EEA lumped parameter approach and the expert opinion of the committee members. The EEA model does employ an algorithm to account for incremental reductions of energy losses, as predicted by an industry-derived set of equations (see Chapter 8). Therefore, it is not surprising that the estimations are relatively close.

However, the applications of the EEA's or the committee's estimation approach is done without a detailed understanding of the actual levels of thermal efficiency of a subject vehicle's engine, the influence of combustion chamber design on the fuel conversion efficiency, the actual levels of gas exchange or frictional losses, and all of the other parameters for which additional technologies can be applied to reduce fuel consumption. This is only possible through a combination of experimental and analytical analyses, which are necessary to predict the absolute level of fuel consumption.

Stated another way, in the opinion of the committee, neither the lumped parameter approaches evaluated by the committee nor the committee's aggregated estimates define the actual level of energy efficiencies and/or losses of a randomly chosen vehicle with sufficient accuracy to allow accurate predictions of future technology introductions. Furthermore, this inaccuracy further degrades as an increasing number of technologies is employed. Therefore, the committee believes that a physics-based, FSS, in combination with experimentally generated data, is required for such predictions, especially if technology that is not currently in production is considered.

IMPROVEMENTS TO MODELING OF MULTIPLE FUEL ECONOMY TECHNOLOGIES

The application of FSS, in which the engine load, thermodynamic efficiency, operational losses of energy, and accessory loads are varied as a function of vehicle operational performance, offers the best opportunity to evaluate the effectiveness of incremental application of vehicle systems in reducing vehicle energy losses, thereby improving overall operational cycle efficiency and reducing fuel consumption. However, since different technologies may be attempting to reduce the same type of loss, for instance, pumping loss, it is necessary to evaluate the contribution of each incremental technology in reducing the different losses in each step along a potential product improvement path. Through the application of incremental technologies, one at a time, and then optimizing the predicted overall vehicle performance and fuel economy tradeoffs, it is possible to understand and quantify, at least for the vehicle model being evaluated, the interactive or synergistic effects that result. These may be positive or negative synergies, as outlined in the Ricardo report prepared for this committee (Ricardo, Inc., 2009) and discussed in Chapter 8 of this report. An example of these synergistic effects is presented in Table 9.3.

Table 9.3 shows that the total improvement in fuel consumption is gained from a combination of primary benefits attributed to a technology pair and a synergistic benefit (or detriment) as a result of the energy losses that are targeted for reduction. If one considers the engine and transmission combination, benefits in reduced pumping losses occur if a downsized, higher-specific-power engine is applied. Additional benefits can be gained from a more efficient transmission with reduced hydraulic losses or reduced friction. However, when these two are applied, there are additional benefits that arise from the ability to run the engine at a lower operating speed for a given power level, thereby increasing the brake mean effective pressure in the cylinders and further reducing the pumping losses. This contributes to the 2.17 percent improvement outlined in Table 9.3. However, it is important to note that the absolute level and relative levels of improvement outlined in Table 9.3 may vary significantly, depending on the application of the same technology sequence to another vehicle application.

As evidenced by the Ricardo, Inc., FSS analyses conducted for the committee, different vehicle types, with differing intended uses, demonstrate different optimization-of-performance characteristics. Therefore, when attempting to estimate the incremental benefits from the application of discrete technologies, the vehicle class, intended use, and associated performance metrics must be considered. Furthermore, the positive or negative synergistic effects of multiple vehicle energy-loss-reducing technologies will vary depending on the vehicle class and intended performance.

As outlined in Chapter 8 of this report, the current NHTSA method of applying technologies to vehicles applies them incrementally and individually to each vehicle in the NHTSA database, starting from the experimentally deter-

TABLE 9.3 Fuel Consumption Synergy Values for Inter-tree Technology Pairs—Results for Truck Package 11

Inter-tree Technology Pair	Fraction of Total Fuel Consumption Impact Attributed to Inter-tree Technology Pair (%)	Synergy Value—Impact on Total Fuel Consumption Reduction from Synergy of Technology Pair (%)
Engine–transmission	6.62	2.17
Final drive ratio–engine	2.81	0.88
Aggressive shift logic–engine	−1.28	−0.39
Electric accessories–engine	0.88	0.27
Aerodynamic drag–engine	0.44	0.13
Aerodynamic drag–transmission	0.36	0.11
Aerodynamic drag–final drive ratio	0.23	0.07
Aerodynamic drag–electric accessories	0.21	0.06
Aggressive shift logic–aerodynamic drag	0.14	0.04

NOTE: The modeling included three decision trees or families of technologies, one for engine technologies, one for transmission technologies, and a third for vehicle technologies. The results shown are for Truck Package 11 (Ricardo, Inc., 2009)

mined value for combined fuel economy as demonstrated in the EPA vehicle exhaust emission certification process. One potential flaw in this methodology results from the process in which the lumped parameter model is used to predict the magnitude of energy loss reduction through the application of discrete technologies on an actual vehicle-by-vehicle basis. Without knowing the starting point in terms of how much the energy losses have been already been reduced, the ability to accurately project further reductions in such system energy losses, and therefore fuel consumption, can be highly erroneous.

Stated another way, it appears most logical to begin any predictive analysis with actual vehicle experimental data, if they are available, as is the case with all vehicles certified under the EPA Test Car List. However, without knowing how successful each manufacturer has been on a vehicle-by-vehicle basis in an ongoing attempt to reduce such energy losses, it is not possible, without detailed vehicle and power train experimental methods, to determine the extent to which any such loss can be further reduced, with a reasonable level of accuracy, on an actual vehicle model.

With an understanding of the potential errors that will result from the approximation method presented above, or other lumped parameter approaches where insufficient information is known about the level of energy loss reduction that has previously occurred on a particular vehicle, the committee proposes an alternative method whereby the potential for fuel consumption reduction and its associated costs can be assessed. This proposed method would determine a characteristic vehicle that would be defined as a reasonable average representative of a class of vehicles. This representative vehicle, whether real or theoretical, would undergo sufficient FSS, combined with experimentally determined and vehicle-class-specific system mapping, to allow a reasonable understanding of the contributory effects of the applied technologies in the reduction of energy losses. The reference to a "theoretical" vehicle suggests that if, during the regulatory process, the NHTSA and the EPA conclude that a vehicle may be characterized to represent a class that may not be in production, FSS models may still be created using physics-based vehicle models combined with experimentally generated engine maps.

In any full system simulation, the engine/power train/vehicle system is defined by input data that are generated by other physics-based analyses, engineering judgment, or experimentally or empirically derived tests. Experimentally measured data for engine maps can incorporate manufacturer-predetermined calibration parameters that have taken into consideration production operational factors such as knock-preventing spark timing or air/fuel ratio adjustments, which are used to protect from component temperature extremes. Physics-based engine maps, generated from engine combustion models, may also be used, but calibration-specific parameters must also be incorporated into such models to achieve best possible predictive results. The use of such models may be necessary when evaluating advanced technologies, such as variable compression ratio, that may not be readily available from production vehicles.

Vehicle-related data, such as data on frontal area, rolling resistance, and weight also are required input for modeling of vehicle performance and fuel economy. However, these data are more readily approximated based upon simplified physics-based calculations or are published in accordance with vehicle certification testing. Therefore, although physics-based engine simulation models are available, the use of experimental engine data, as described above, greatly improves the accuracy of the modeling.

Experimental methods used to understand the effects of different technologies in an attempt to reduce system energy losses have been developed under the United States Council for Automotive Research (USCAR) Benchmarking Consortium. Actual production vehicles are subjected to a battery of vehicle, engine, and transmission tests in sufficient detail to understand how each is applied and how they contribute to the overall performance and fuel consumption factors in light-duty vehicles. Combining such experimental methods with FSS modeling, wherein all simulation variables and subsystem maps would be transparent to all interested parties (both the regulatory agencies and automotive manufacturers, for example), would allow, in the opinion of the committee, the best opportunity to define a technical baseline against which potential improvements could be more accurately and openly analyzed than the current methods employed.

The advantages of such a method include the ability to explicitly account for all energy loss categories, the ability to directly estimate fuel consumption (as opposed to the percent change in fuel consumption), and the ability to represent new technologies and combinations of technologies. It also recognizes the increasingly common utilization of FSS models by regulatory agencies and other entities outside the automotive industry. Finally, the method proposes a procedure whereby engine and vehicle experimental data can be obtained without reliance on proprietary data, such as engine maps, that have posed a barrier to effective utilization of FSS models by non-OEMs in the past.

The steps in the recommended process are as follows:

1. Develop a set of baseline vehicle classes from which a characteristic vehicle can be chosen to represent each class. The vehicle may be either real or theoretical and will possess the average attributes of that class as determined by sales-weighted averages.
2. Identify technologies with a potential to reduce fuel consumption.
3. Determine the applicability of each technology to the various vehicle classes.
4. Estimate the technology's preliminary impact on fuel consumption and cost.

5. Determine the optimum implementation sequence (technology pathway) based on cost-effectiveness and engineering considerations.
6. Document the cost-effectiveness and engineering judgment assumptions used in step 5 and make this information part of a widely accessible database.
7. Utilize modeling software (FSS) to progress through each technology pathway for each vehicle class to obtain the final incremental effects of adding each technology.

If such a process were adopted as part of a regulatory rule-making procedure, it could be completed on 3-year cycles to allow regulatory agencies sufficient lead time to integrate the results into future proposed and enacted rules.

Based on the eight new vehicle classes proposed by the committee, an average vehicle, either real or theoretical, would be chosen that possessed the average attributes of the vehicles in that class. It would be of average weight, footprint, engine displacement, and other characteristics. The resulting vehicle would serve as the baseline for FSS analysis. This would also allow a very important starting point for the vehicle systems from which potential improvements could be evaluated. Using detailed benchmark data, defined levels of energy losses would be used as input into the simulation model. The data used to choose the vehicle consists of the following specifications available from the EPA test car list:

- Footprint,
- Weight,
  Engine (displacement, cylinder count, horsepower, torque),
- Valve train configuration (OHV, SOHC, DOHC),
- Valve event modulation technology (VVT, VVL),
- Combustion technology (SI, CI, HCCI),
- Fuel injection method and fuel type (SEQ, GDI, DFI, gasoline, diesel),
- Aspiration method (natural, supercharged, turbocharged),
- Number of occupants,
- Power/vehicle weight ratio,
- Transmission type and gear ratio spread,
- Driveline (FWD, RWD, AWD), and
- EPA vehicle class.

FINDINGS AND RECOMMENDATION

Finding 9.1: Many vehicle and power train technologies that reduce fuel consumption are currently in or entering production or are in advanced stages of development in European or Asian markets where high consumer prices for fuel have justified their commercialization. Depending on the intended vehicle use or current state of energy-loss minimization, the application of incremental technologies will produce varying levels of fuel consumption reduction.

Finding 9.2: Data made available to the committee from original equipment manufacturers and Tier 1 suppliers and found in various published studies suggest a very wide range in estimated incremental cost that makes assessments of cost-effectiveness very approximate. Generally, the committee notes that estimates of cost are always more uncertain than estimates of impact on fuel consumption, and the estimates presented here should be considered very uncertain until more detailed studies are completed. As noted in Chapter 3, estimates based on teardown cost analysis, currently being utilized by the EPA in its regulatory analysis for light-duty vehicle greenhouse gas emissions standards, should be expanded for developing cost impact analyses.

Finding 9.3: In response to the statement of task, the committee estimated possible technology evolution paths for each vehicle class that arise from the average baseline vehicle. A very simple, multiplicative aggregation of potential for reducing fuel consumption is presented as a means to roughly estimate the total potential that might be possible. The results from this analysis show that, for the intermediate car, large car, and unibody standard truck classes, the average reduction in fuel consumption for the SI path is about 29 percent at a cost of approximately $2,200; the average reduction for the CI path is about 38 percent at a cost of approximately $5,900; and the average reduction for the hybrids path is about 44 percent at a cost of $6,000. However, unless calibrated methods are used to accurately consider the synergistic effects of applying several technologies—effects that may reduce the same sources of power train and vehicle energy losses—these results are extremely approximate in nature and, in the committee's opinion, should not be used as input to analyses for which modeling accuracy is important. In general, the technology tables that present incremental data for percent reduction in fuel consumption and estimated incremental cost cannot be used in their current form as input into lumped parameter-type models without methods to accurately consider the synergistic effects of applying several technologies and without significant expertise in vehicle technologies to fully understand integration issues.

Recommendation 9.1: As noted in Chapter 8, full system simulation (FSS), based on empirically derived power train and vehicle performance and fuel consumption data maps, offers what the committee believes is the best available method to fully account for system energy losses and synergies and to analyze potential reductions in fuel consumption as technologies are introduced into the market. FSS analyses conducted for the committee show that synergy effects between differing types of energy-loss-reducing technologies vary greatly from vehicle application to vehicle application.

The committee proposes a method whereby FSS analyses are used on class-characterizing vehicles, so that synergies and effectiveness in implementing multiple fuel economy technologies can be evaluated with what should be greater accuracy. This proposed method would determine a characteristic vehicle that would be defined as a reasonable average representative of a class of vehicles. This representative vehicle, whether real or theoretical, would undergo sufficient FSS, combined with experimentally determined and vehicle-class-specific system mapping, to allow a reasonable understanding of the contributory effects of the technologies applied to reduce vehicle energy losses. Data developed under the United States Council for Automotive Research (USCAR) Benchmarking Consortium should be considered as a source for such analysis and potentially expanded. Under the USCAR program, actual production vehicles are subjected to a battery of vehicle, engine, and transmission tests in sufficient detail to understand how each candidate technology is applied and how it contributes to the overall performance and fuel consumption of light-duty vehicles. Combining the results of such testing with FSS modeling, and thereby making all simulation variables and subsystem maps transparent to all interested parties, would allow the best opportunity to define a technical baseline against which potential improvements could be analyzed more accurately and openly than is the case with the current methods employed.

The steps in the recommended process are as follows:

1. Develop a set of baseline vehicle classes from which a characteristic vehicle can be chosen to represent each class. The vehicle may be either real or theoretical and will possess the average attributes of that class as determined by sales-weighted averages.
2. Identify technologies with a potential to reduce fuel consumption.
3. Determine the applicability of each technology to the various vehicle classes.
4. Estimate each technology's preliminary impact on fuel consumption and cost.
5. Determine the optimum implementation sequence (technology pathway) based on cost-effectiveness and engineering considerations.
6. Document the cost-effectiveness and engineering judgment assumptions used in step 5 and make this information part of a widely accessible database.
7. Utilize modeling software (FSS) to progress through each technology pathway for each vehicle class to obtain the final incremental effects of adding each technology.

If such a process were adopted as part of a regulatory rulemaking procedure, it could be completed on 3-year cycles to allow regulatory agencies sufficient lead time to integrate the results into future proposed and enacted rules.

BIBLIOGRAPHY

EPA (U.S. Environmental Protection Agency). 2008a. Light-Duty Automotive Technology and Fuel Economy Trends: 1975 Through 2008. EPA420-R-08-015. September.

EPA. 2008b. EPA Staff Technical Report: Cost and Effectiveness Estimates of Technologies Used to Reduce Light-Duty Vehicle Carbon Dioxide Emissions. EPA420-R-08-008. Ann Arbor, Mich.

NHTSA (National Highway Traffic Safety Administration). 2009. Average fuel economy standards, passenger cars and light trucks, model-year 2011: final rule, RIN 2127 AK-29, Docket No. NHTSA 2009-0062. Washington, D.C., March 23.

NRC (National Research Council). 2002. Effectiveness and Impact of Corporate Average Fuel Economy (CAFE) Standards. National Academy Press, Washington, D.C.

Ricardo, Inc. 2008. A Study of Potential Effectiveness of Carbon Dioxide Reducing Vehicle Technologies. EPA420-R-08-004. Prepared for the U.S. Environmental Protection Agency, Contract No. EP-C-06-003, Work Assignment No. 1-14. Ann Arbor, Mich.

Ricardo, Inc. 2009. A Study of Interaction Effects Between Light Duty Vehicle Technologies. Prepared for the NRC Committee on Assessment of Technologies for Improving Light-Duty Vehicle Fuel Economy by Ricardo Inc., Van Buren, Mich., February 27.

Sovran, G., and M.S. Bohn. 1981. Formulae for the tractive energy requirements of vehicles driving the EPA schedule, SAE Paper 810184. February.

Appendixes

A

Committee Biographies

Trevor O. Jones (NAE), *Chair* is founder, chairman, and chief executive officer (CEO) of ElectroSonics Medical, Inc. Before that, he was founder, chairman, and CEO of Biomec, Incorporated, a biomedical device company. He was formerly chairman of the board of Echlin, Incorporated, a supplier of automotive components, primarily to the aftermarket. Dr. Jones is also chairman and CEO of the International Development Corporation, a private management consulting company that advises automotive supplier companies on strategy and technology. He was chair, president, and CEO (retired) of Libbey-Owens-Ford Company, a large manufacturer of glass for automotive and construction applications. Previously, he served as vice president of engineering in the Automotive Worldwide Sector of TRW, Incorporated, and as group vice president, Transportation Electronics Group. Before joining TRW, he was employed by General Motors (GM) in many aerospace and automotive executive positions, including director of GM Proving Grounds; of the Delco Electronics Division, Automotive Electronic, and Safety Systems; and director of the GM Advanced Product Engineering Group. Dr. Jones is a life fellow of the Institute of Electrical and Electronics Engineers (IEEE) and has been cited for leadership in the application of electronics to the automobile. He is also a fellow of the Society of Automotive Engineers (SAE), a fellow of the British Institution of Electrical Engineers, a fellow of the Engineering Society of Detroit, a registered professional engineer in Wisconsin, and a chartered engineer in the United Kingdom. He holds many patents and has lectured and written on automotive safety and electronics. He is a member of the National Academy of Engineering (NAE) and a former commissioner of the National Research Council (NRC) Commission on Engineering and Technical Systems. Dr. Jones has served on several other NRC study committees, including the Committee for a Strategic Transportation Research Study on Highway Safety. He chaired the NAE Steering Committee on the Impact of Products Liability Law on Innovation and the Committee on Review of the Research Program of the Partnership for a New Generation of Vehicles for six reviews. He holds a higher national certificate in electrical engineering from Aston Technical College and an ordinary national certificate in mechanical engineering from Liverpool Technical College. Cleveland State University awarded Dr. Jones an honorary doctorate of science and cited him for contributions in the development of fuel cells and biomedical devices.

Thomas W. Asmus (NAE) is a retired senior research executive of DaimlerChrysler Corporation. He has also held positions at Mead Corporation, as an adjunct faculty member of mechanical engineering at the University of Michigan, and as a professor of physical chemistry at the University of Guadalajara, in Mexico. He has more than 30 years of experience and has played a leadership role in nearly all aspects of internal combustion engine and fuels research and development, focusing mainly on fuel consumption and exhaust emissions reduction. His entry into the field was initially based on his background in combustion and emissions formation mechanisms for both gasoline and diesel engines, but with time and circumstances his activities expanded to include gas exchange processes, controls, lubrication, many types of fault diagnoses, and heat management. New-concept analysis has become routine for Dr. Asmus. Besides having been a member of the NAE, he is a fellow of the SAE and was a recipient of the Soichiro Honda Lecture Award recipient in 1999. He has a B.S. in paper science and engineering from Western Michigan University and an M.S. and a Ph.D. in physical chemistry from Western Michigan University.

Rodica Baranescu (NAE) is a professor in the College of Engineering, Department of Mechanical and Industrial Engineering, University of Illinois at Chicago. Before that, she was manager of the fuels, lubricants, and engine group of the International Truck and Engine Corporation, at Melrose Park, Illinois. She is an internationally sought after public speaker on technical issues related to mobility technology, environmental control, fuels, and energy. She has extensive expertise in diesel engine technology and was elected to the NAE in 2001 for research leading to effective and environmentally sensitive diesel and alternative-fuel engines and leadership in automotive engineering. She is a fellow of SAE International and was its president in 2000. In 2003 she received the Internal Combustion Engine Award of the American Society of Mechanical Engineering (ASME). Dr. Baranescu received her M.S. and Ph.D. degrees in mechanical engineering in 1961 and 1970, respectively, from the Politechnica University in Bucharest, Romania, where she served as assistant professor (1964-1968), lecturer (1970-1974), and associate professor (1974-1978).

Jay Baron is president of the Center for Automotive Research (CAR) and the director of its Manufacturing, Engineering and Technology Group. Dr. Baron's recent research has focused on developing new methods for the analysis and validation of sheet metal processes, including die making, tool and die tryout, and sheet metal assembly processes. He also developed functional build procedures that result in lower tooling costs and shorter development lead times, while improving quality—particularly with sheet metal assemblies. He also has been researching new technologies in the auto industry, including looking at body shop design and flexibility and evaluating the manufacturing capability of evolving technologies. He recently completed investigations on state-of-the-art tailor-welded blank technologies, the economics of weld-bond adhesives, and the analysis of car door quality and construction methods. Before becoming first the director of manufacturing systems at CAR and then president, Dr. Baron was the manager of manufacturing systems at the Office for the Study of Automotive Transportation at the University of Michigan Transportation Research Institute. He also worked for Volkswagen of America in quality assurance and as staff engineer and project manager at the Industrial Technology Institute in Ann Arbor and at the Rensselaer Polytechnic Institute's Center for Manufacturing Productivity in Troy, New York. Dr. Baron holds a Ph.D. and a master's degree in industrial and operations engineering from the University of Michigan and an M.B.A. from Rensselaer Polytechnic Institute.

David Friedman is the research director of the Clean Vehicles Program of the Union of Concerned Scientists (UCS), Washington, D.C. He is the author or coauthor of more than 30 technical papers and reports on advancements in conventional, fuel cell, and hybrid electric vehicles and alternative energy sources with an emphasis on clean and efficient technologies. Before joining UCS in 2001, he worked for the University of California, Davis, in the fuel cell vehicle modeling program, developing simulation tools to evaluate fuel cell technology for automotive applications. He worked there on University of California's FutureCar team to build a hybrid electric family car that doubled its fuel economy. He also once worked at Arthur D. Little researching fuel cell, battery electric, and hybrid electric vehicle technologies, as well as photovoltaics. He served as a member of the NRC Panel on the Benefits of Fuel Cell R&D of the Committee on Prospective Benefits of DOE's Energy Efficiency and Fossil Energy R&D Programs, Phase 1, and is currently a member of the NRC Committee on National Tire Efficiency. He earned a bachelor's degree in mechanical engineering from Worcester Polytechnic Institute and is a doctoral candidate in transportation technology and policy at the University of California, Davis.

David Greene is a corporate fellow at the Oak Ridge National Laboratory (ORNL). He has spent more than 20 years researching transportation and energy policy issues. His research interests include energy demand modeling, economic analysis of petroleum dependence, modeling market responses to advanced transportation technologies and alternative—fuels, economic analysis of policies to mitigate greenhouse gas emissions from transportation, and developing theory and methods for measuring the sustainability of transportation systems. After joining ORNL in 1977, he founded the Transportation Energy Group in 1980 and in 1987 established the Transportation Research Section. Dr. Greene spent 1988 to 1989 in Washington, D.C., as a senior research analyst in the Office of Domestic and International Energy Policy, at the Department of Energy (DOE). He has published more than 150 articles in professional journals, written contributions to books and technical reports, and given congressional testimony on transportation and energy issues. From 1997 to 2000 Dr. Greene served as the first editor-in-chief of the *Journal of Transportation and Statistics*, the only scholarly periodical published by the U.S. Department of Transportation. He currently serves on the editorial boards of *Transportation Research D*, *Energy Policy*, *Transportation Quarterly*, and the *Journal of Transportation and Statistics*. Active in the Transportation Research Board (TRB) and the NRC, Dr. Greene has served on several standing and ad hoc committees. He is past chairman and member emeritus of TRB's Energy Committee, was past chair of the Section on Environmental and Energy Concerns, and was a recipient of TRB's Pyke Johnson Award. Dr. Greene received a B.A. degree from Columbia University in 1971, an M.A. from the University of Oregon in 1973, and a Ph.D. in geography and environmental engineering from the Johns Hopkins University in 1978.

Linos Jacovides (NAE) recently retired as director, Delphi Research Labs, a position he held from 1998 to 2007. Dr. Jacovides joined General Motors Research and Development in 1967 and became department head of electrical engineering in 1985. He is a fellow of the IEEE. His areas of research were the interactions between power electronics and electrical machines in electric vehicles and locomotives. He later transitioned to Delphi with a group of researchers from GM to set up the Delphi Research Laboratories. He received a B.S. in electrical engineering and a master's in machine theory from the University of Glasgow, Scotland. He received a Ph.D. in generator control systems from the Imperial College, University of London, in 1965.

John H. Johnson is a presidential professor emeritus in the Department of Mechanical Engineering-Engineering Mechanics at Michigan Technological University (MTU) and a fellow of the SAE and the ASME. His experience spans a wide range of analysis and experimental work on advanced engine concepts, diesel and other internal engine emissions studies, fuel systems, and engine simulation. He was previously project engineer at the U.S. Army Tank Automotive Center, and chief engineer in applied engine research at the International Harvester Company before joining the MTU mechanical engineering faculty. He served as chairman of the MTU mechanical engineering and engineering mechanics department from 1986 to 1993. He has served on many committees related to engine technology, engine emissions, and health effects—for example, committees of the SAE, the NRC, the Combustion Institute, the Health Effects Institute, and the Environmental Protection Agency—and consults to a number of government and private sector institutions. In particular, he served on many NRC committees, including the Committee on Fuel Economy of Automobiles and Light Trucks, the Committee on Advanced Automotive Technologies Plan, the Committee on the Impact and Effectiveness of Corporate Average Fuel Economy (CAFE) Standards, and the Committee to Assess Fuel Economy for Medium and Heavy-Duty Vehicles. He chaired the NRC Committee on Review of DOE's Office of Heavy Vehicle Technologies and the NRC Committee on Review of the 21st Century Truck partnership. He received his Ph.D. in mechanical engineering from the University of Wisconsin.

John G. Kassakian (NAE) is professor of electrical engineering and director of the Massachusetts Institute of Technology's (MIT's) Laboratory for Electromagnetic and Electronic Systems. His expertise is in the use of electronics for the control and conversion of electrical energy, industrial and utility applications of power electronics, electronic manufacturing technologies, and automotive electrical and electronic systems. Before joining the MIT faculty, he served in the U.S. Navy. Dr. Kassakian is on the boards of directors of a number of companies and has held numerous positions with the IEEE, including founding president of the IEEE Power Electronics Society. He is a member of the NAE, a fellow of the IEEE, and a recipient of the IEEE's William E. Newell Award for Outstanding Achievements in Power Electronics (1987), the IEEE Centennial Medal (1984), and the IEEE Power Electronics Society's Distinguished Service Award (1998). He has served on a number of NRC committees, including the Committee on Review of the Research Program of the Partnership for a New Generation of Vehicles and the Review of the FreedomCAR and Fuel Research Program. He has an Sc.D. in electrical engineering from MIT.

Roger B. Krieger is currently an adjunct professor at the engine research center of the University of Wisconsin, Madison. Before that, he was laboratory group manager, Compression Ignition Engine Systems Group at the Powertrain Systems Research Laboratory. He also held a position at the Institut Francais du Petrôle, Applications Division, Rueil-Malmaison, in France. Dr. Krieger has approximately 35 years of research and development experience in internal combustion engines, especially diesel engines and combustion. He holds approximately 10 patents related to engine and emissions control technologies. He served as vice-chair and chair of the Diesel Engine Committee, SAE. He has a B.S. and a Ph.D. in mechanical engineering from the University of Wisconsin-Madison.

Gary W. Rogers is president, chief executive officer, and sole director, FEV, Inc. His previous positions included director, Power Plant Engineering Services Division, and senior analytical engineer, Failure Analysis Associates, Inc.; design development engineer, Garrett Turbine Engine Company; and Exploration Geophysicist, Shell Oil Company. He has extensive experience in research, design, and development of advanced engine and powertrain systems, including homogeneous and direct-injected gasoline engines, high-speed direction injection passenger car diesel engines, heavy-duty diesel engines, hybrid vehicle systems, gas turbines, pumps, and compressors. He provides corporate leadership for a multinational research, design, and development organization specializing in engines and energy systems. He is a member of the SAE, is an advisor to the Defense Advanced Research Projects Agency on heavy-fuel engines, and sits on the advisory board to the College of Engineering and Computer Science, Oakland University, Rochester, Michigan. He served as a member of the NRC Committee on Review of DOE's Office of Heavy Vehicle Technologies Program, the NRC Committee on the Effectiveness and Impact of Corporate Average Fuel Economy (CAFE) Standards, and the NRC Panel on Benefits of DOE's Light-Duty Hybrid Vehicle R&D Program. He also recently supported the Department of Transportation's National Highway Traffic Safety Administration by conducting a peer review of the NHTSA CAFE Model. He has a B.S.M.E. from Northern Arizona University.

Robert F. Sawyer (NAE) is the Class of 1935 Professor of Energy Emeritus at the University of California, Berkeley. He is a member of the NAE and recently served as chair of the California Air Resources Board. His previous positions include research engineer and chief, Liquid Systems Analysis, U.S. Air Force Rocket Propulsion Laboratory; member of the research staff, Princeton University; member, California Air Resources Board; and chair, Energy and Resources Group, University of California, Berkeley. He is a past president of the Combustion Institute. His research includes combustion chemistry, pollutant formation and control, engine emissions, toxic waste incineration, alternative fuels, and regulatory policy. Dr. Sawyer served on numerous National Research Council committees, including the Committee for the Evaluation of the Congestion Mitigation and Air Quality Improvement Program, the Committee to Review EPA's Mobile Source Emissions Factor (MOBILE) Model, and the Committee on Adiabatic Diesel Technology, among others. He holds a B.S. and an M.S. (mechanical engineering) from Stanford University and an M.A. (aeronautical engineering) and a Ph.D. (aerospace science) from Princeton University.

B

Statement of Task

The committee formed to carry out this study will provide updated estimates of the cost and potential efficiency improvements of technologies that might be employed over the next 15 years to increase the fuel economy of various light-duty vehicle classes. Specifically, the committee shall:

1. Reassess the technologies analyzed in Chapter 3 of the NRC report, *Impact and Effectiveness of Corporate Average Fuel Economy (CAFE) Standards* (2002), for efficacy, cost, and applicability to the classes of vehicles considered in that report. In addition, technologies that were noted but not analyzed in depth in that report, including direct injection engines, diesel engines, and hybrid electric vehicles, shall be assessed for efficacy, cost and applicability. Weight and power reductions also shall be included, though consideration of weight reductions should be limited to advances in structural design and lightweight materials. The assessments shall include the effects of "technology sequencing"—in what order manufacturers might conceivably incorporate fuel economy technologies, and how such ordering affects technology cost and applicability.
2. Estimate the efficacy, cost, and applicability of emerging fuel economy technologies that might be employed over the next 15 years. The assessments shall include the effects of technology sequencing as defined in (1) above.
3. Identify and assess leading computer models for projecting vehicle fuel economy as a function of additional technology. These models would include both:

   - Lumped parameter (or Partial Discrete Approximation) type models, where interactions among technologies are represented using energy partitioning and/or scalar adjustment factors (also known as synergy factors), and
   - Full vehicle simulation, in which such interactions are analyzed using explicit drive cycle and engine cycle simulation, based on detailed vehicle engineering characteristics (*e.g.*, including engine maps, transmission shift points, etc.).

   Check the models against current, known fuel economy examples and select one of each type to perform the analyses of the effects of the technologies in 1 and 2 above.

4. Develop a set of cost/potential efficiency improvement curves, as in Chapter 3 of the 2002 NRC report, that is guided by the following question: "What is the estimated cost and potential fuel economy benefit of technologies that could be applied to improve the fuel economy of future passenger vehicles, given the constraints imposed by vehicle performance, functionality, safety and emission regulations?" The ten vehicle classes considered in the 2002 report shall be analyzed, including important variants such as different engine sizes (e.g., 6 and 8 cylinders). Most analyses shall be performed with the lumped parameter model, but sufficient cases to ensure overall accuracy shall be checked with the engine mapping model.
5. Define and document the specific methodology(ies) and inputs used to estimate the incremental costs and benefits of the fuel economy technologies chosen by the committee, including the methods used to account for variations in vehicle characteristics (e.g., size, weight, engine characteristics) and to account for the sequential application of technologies. Use flow charts or similar methods to document sequencing upon which the committee's estimates of incremental costs and benefits are based. Although methodologies vary, the committee's report should detail all of its calculation methodology(ies), even those as basic as simple mathematical relationships (if used) and as complex as structural representations, such as decision trees (if used). It should do so to levels of specificity, clarity and completeness sufficient for implementation and inte-

gration into models that project the fuel economy capability of vehicles, fleets and manufacturers, including fleets specified at the level of individual vehicle models, engines, and transmissions. The report should also provide and document estimates of all input data required for implementation of these methodologies.

6. Assess how ongoing changes to manufacturers' refresh and redesign cycles for vehicle models affect the incorporation of new fuel-economy technologies.

The committee's analysis and methodologies will be documented in two NRC-approved reports. An interim report will discuss the technologies to be analyzed, the classes of vehicles which may employ them, the estimated improvement in fuel economy that may result, and the models that will be used for analysis. The final report will include the detailed specifications for the methodologies used and the results of the modeling, and will make use of the input from the interim report and any new information that is available.

C

List of Presentations at Public Committee Meetings

WASHINGTON, D.C., SEPTEMBER 10-11, 2007

Julie Abraham, National Highway Traffic Safety Administration, *Fuel Economy Technology Study*
William Charmley, EPA Office of Transportation and Air Quality representative, *Greenhouse Gases and Light-Duty Vehicles*
Coralie Cooper, Northeast States for Coordinated Air Use Management, *Technical Feasibility and Costs Associated with Reducing Passenger Car GHG Emissions*
John German, USA Honda, *Advanced Technologies, Diesels, and Hybrids*
Dan Hancock, GM Powertrain, *Assessing Powertrain Fuel Economy*
John Heywood, Massachusetts Institute of Technology, *Challenges in Estimating Future Vehicle Fuel Consumption*
Aymeric Rousseau, Argonne National Laboratory, *Designing Advanced Vehicle Powertrains Using PSAT*
Wolfgang Stütz, BMW of North America, *Fuel Economy of BMW Diesel Vehicles*

WASHINGTON, D.C., SEPTEMBER 27, 2007

K.G. Duleep, Energy and Environmental Analysis, Inc., *Approaches to Modeling Vehicle Fuel Economy*
Kevin Green, The Volpe Center, *CAFE Compliance and Effects Modeling System*
Marc Wiseman, Ricardo, Inc., *Potential Approaches to Modeling Fuel Economy Technologies: Engine Simulation Modeling Capabilities and Cost Analysis Capabilities*

WASHINGTON, D.C., OCTOBER 25-26, 2007

Manahem Anderman, Advanced Automotive Batteries, *Lithium-Ion Batteries for Hybrid Electric Vehicles: Opportunities and Challenges*
Mark Daroux, Stratum Technologies, Inc., *Lithium Ion Phosphate Batteries for Traction Application*
Tien Duong, U.S. Department of Energy, *Status of Electrical Energy Storage Technologies*
Michel Forissier, Valeo, *Fuel Economy Solutions*
Bart Riley, A123 Systems, *A123 Systems Battery Technologies*

WASHINGTON, D.C., NOVEMBER 27-28, 2007

Khalil Amine, Argonne National Laboratory, *Advanced High Power Chemistries for HEV Applications*
Paul Blumberg, Ethanol Boosting Systems, LLC, *Ethanol Turbo Boost for Gasoline Engines: Diesel and Hybrid Equivalent Efficiency at an Affordable Cost*
Frank Fodal, Chrysler LLC, *Fuel Economy/Fuels*
David Geanacopoulos, Volkswagen of America, Inc., *Diesel Technology for VW*
Johannes Ruger, Bosch, *Increasing Fuel Economy: Contribution of Bosch to Reach Future Goals*
Robert Wimmer and Shunsuke Fushiki, Toyota, *Toyota Hybrid Program*

WASHINGTON, D.C., JANUARY 24-25, 2008

Steve Albu, California Air Resources Board, *ARB Perspective on Vehicle Technology Costs for Reducing Greenhouse Gases*
Wynn Bussman, Consultant, *Study of Industry-Average Mark-up Factors Used to Estimate Retail Price Equivalents (RPE)*
K.G. Duleep, Energy and Environmental Analysis, Inc., *Analysis of Technology Cost and Retail Price*
Kevin McMahon, Martec Group, *Variable Costs of Fuel Economy Technologies*
James Lyons, Sierra Research, Inc., *Technology and Retail Price Implications of More Stringent CAFE Standards Based on Vehicle Simulation Modeling*

WASHINGTON, D.C., FEBRUARY 25-26, 2008

Julie Abraham, National Highway Traffic Safety Administration, *Update from NHTSA on Regulatory Activities and Other Analysis*

WASHINGTON, D.C., MARCH 31-APRIL 1, 2008

David Haugen and Matt Brustar, EPA Office of Transportation and Air Quality, *Discussion of EPA's Modeling of Fuel Economy*

K.G. Duleep, Energy and Environmental Analysis, Inc., *Assessment of Costs and Fuel Economy Benefits*

WASHINGTON, D.C., JUNE 3-4, 2008

Michael Bull, Aluminum Association, *Opportunities for Reducing Vehicle Mass*

Bruce Moor, Delphi Electronics and Safety, *Power Electronics Systems Solutions for HEV Architectures*

Huang-Yee Iu, Hymotion, *Hymotion Plug-in Hybrid Vehicle*

WASHINGTON, D.C., SEPTEMBER 9-10, 2008

Susan Yester, Chrysler, *Opportunities for Reducing Vehicle Mass*

Joseph Kubish, Manufacturers of Emissions, Control Equipment Association, *Aftertreatment Technologies and Strategies for Light Duty Vehicles with Emphasis on NO_x and Particulates*

Frank Fronczak, University of Wisconsin, *Hydraulic Hybrid Vehicle*

John Kargul, EPA Clean Automotive Technology Program, *EPA's Hydraulic Hybrid Program*

WASHINGTON, D.C., MARCH 16-18, 2009

EPA Office of Transportation and Air Quality, *Update from EPA on Analysis of RPE and Separate Ongoing Work on Estimates of Analysis of Direct Manufacturing Costs of Technologies*

D

Select Acronyms

AWD	all-wheel drive
BMEP	brake mean effective pressure
BOM	bill of materials
BSFC	brake specific fuel consumption
CAFE	corporate average fuel economy
CDPF	catalyzed diesel particulate filter
CI	compression ignition
CO_2	carbon dioxide
CR	compression ratio
CVVL	continuously variable valve lift
DCP	dual cam phasing
DCT	dual-clutch transmission
DI	direct injection
DISI	direct injection spark ignition
DOC	diesel oxidation catalyst
DOHC	dual overhead cam
DOT	U.S. Department of Transportation
DPF	diesel particulate filter
DVVL	discrete variable valve lift
E85	85 percent ethanol
EACC	electric accessories
ECU	engine control unit
EEA	Energy and Environmental Analysis, Inc.
EGR	exhaust gas recirculation
EPA	U.S. Environmental Protection Agency
EU	European Union
EVO	exhaust valve opening
FAME	fatty acid methyl ester
FC	fuel consumption
FE	fuel economy
FSS	full system simulation
FTP	Federal Test Procedure
FWD	four-wheel drive
GDI	gasoline direct injection
GHG	greenhouse gas
GM	General Motors Company
HC	hydrocarbon
HCCI	homogeneous-charge compression ignition
HEV	hybrid-electric vehicle
HWFET	highway fuel economy test schedule (or highway cycle)
I4	inline four-cylinder engine
IC	internal combustion
ICP	intake-cam phasing
IVC	intake-valve closing
LBL	low-viscosity lubricants
LDV	light-duty vehicle
LEV	low-emissions vehicle
LNT	lean NO_x traps
LP	low pressure
LTC	low-temperature combustion
LVL	low-viscosity lubricant
MBT	maximum brake torque
MPFI	multipoint fuel injection
mpg	miles per gallon
MSRP	manufacturer's suggested retail price
NA	North American
NESCCAF	Northeast States Center for a Clean Air Future
NHTSA	National Highway and Traffic Safety Administration
NO_x	nitrous oxides
NSC	NO_x storage and reduction catalyst
NRC	National Research Council
NVH	noise, vibration, and harshness
OBD	on-board diagnostics

OEM	original equipment manufacturer
OHV	overhead valve
PCCI	premixed charge compression ignition
PDA	partial discrete approximation
PFI	port fuel injection
PGM	platinum group metal
PHEV	plug-in hybrid electric vehicle
PM	particulate matter
R&D	research and development
RPE	retail price equivalent
RWD	rear-wheel drive
SAE	Society of Automotive Engineers
SCR	selective catalytic reduction
SGDI	stoichiometric gasoline direct injection
SI	spark ignition
SOC	state of charge
SOHC	single overhead cam
SUV	sport utility vehicle
UDDS	urban dynamometer driving schedule
ULEV	ultralow-emissions vehicle
V6	six cylinder V engine
VEL	valve event and lift
VEM	valve-event modulation
VGT	variable geometry turbochargers
VVL	variable valve lift

E

Comparison of Fuel Consumption and Fuel Economy

Figure E.1 shows the relationship between fuel consumption (FC) and fuel economy (FE), including the slope of the curve that relates them (Johnson, 2009). The slope, which is negative, and the shape of this relationship are important. The slope indicates the change in FC relative to a change in FE—e.g., when the magnitude of the slope is high, such as at 10 mpg, there is large change in FC for a small change in FE. At 50 mpg, however, there is little change in FE since the magnitude of the slope is very low and approaching zero as indicated by the right-hand slope scale on Figure E.1. Fuel consumption decreases slowly after 40 mpg since the slope (lower curve and right-hand scale) of the fuel consumption versus fuel economy (Figure E.1) curve approaches zero. The slope rapidly decreases past 40 mpg since it varies as the inverse of the FE squared, which then results in a small decrease in FC for large FE increases. This fact is very important since fuel consumption is the metric in CAFE. Furthermore, the harmonic average[1] in the CAFE standards is determined as the sales-weighted average of the fuel consumption for the urban and highway schedules converted into fuel economy. Figure 2.2 was derived from Figures 2.1 and E.1 to show how the share of fuel consumption decrease is related to percent increase of fuel economy.[2] The curve in Figure 2.2 is independent of the value of fuel economy and is calculated by the equation in footnote 2. For example, the fuel consumption is 2.5 gal/100 mi at 40 mpg and 1.25 gal/100 mi at 80 mpg, which is a 40 mpg change in fuel economy (100 percent increase in FE) and a change in fuel consumption of only 1.25 gal/100 mi (50 percent decrease in FC), as shown by the lines on the FC vs. FE curve in Figure E.1. In going from 15 to 19 mpg, there is an approximate 1.25 gal/100 mile change in fuel consumption. The nonlinear relationship between fuel consumption and fuel economy gives significantly more weight to lower fuel economy vehicles (15-40 mpg—i.e., 6.5-2.5 gal/100 mi) than to those greater than 40 mpg. Going beyond 40 mpg there is a perception that fuel efficiency is improving faster than the actual change in fuel consumption. For a fleet that contains a large number of vehicles in the 15-35 mpg range, the vehicles with a fuel economy greater than 40 mpg contribute only a small amount to the weighted average CAFE fuel economy, assuming that there are fewer 40-mpg vehicles than 15- to 35-mpg vehicles.

Fuel consumption difference is also the metric that determines the yearly fuel savings in going from a given fuel economy vehicle to a higher fuel economy vehicle:

$$\text{Yearly fuel savings} = \text{yearly miles driven} \times (FC_1 - FC_2)/100 \qquad (E.1)$$

where FC_1 = fuel consumption of existing vehicle, gal/100 mi, and FC_2 = fuel consumption of new vehicle, gal/100 mi. The amount of fuel saved in going from 14 to 16 mpg for 12,000 miles per year is 107 gal. This savings is the same as a change in fuel economy for another vehicle in going from 35 to 50.8 mpg. Equation E.1 and this example again show how important fuel consumption is to judging yearly fuel savings.

REFERENCE

Johnson, J. 2009. Fuel consumption and fuel economy. Presentation to the National Research Council Committee to Assess Fuel Economy Technologies for Medium- and Heavy-Duty Vehicles, April 7, Dearborn, Mich.

---

[1] Harmonic average weighted CAFE = $\dfrac{\sum_1^n N_n}{\sum_1^n N_1 \dfrac{1}{FE_1} + \quad + N_n \dfrac{1}{FE_n}}$

where N_n = number of vehicles in class n; FE_n = fuel economy of class n vehicles; and n = number of separate classes of vehicles.

[2] If $FE_f = (FE_2 - FE_1)/FE_1$ and $FC_f = (FC_1 - FC_2)/FC_2$ where FE_1 and FC_1 = FE and FC for vehicle baseline and FE_2 and FC_2 = FE and FC for vehicle with advanced technology, then, $FC_f = FE_f/(FE_f + 1)$ where FE_f = fractional change in fuel economy and FC_f = fractional change in fuel consumption. This equation can be used for any change in FE or FC to calculate the values shown in Figure 2.2. Also, $FE_f = FC_f/(1 - FC_f)$ and $\%FC = 100\ FC_f$, $\%FE = 100\ FE_f$.

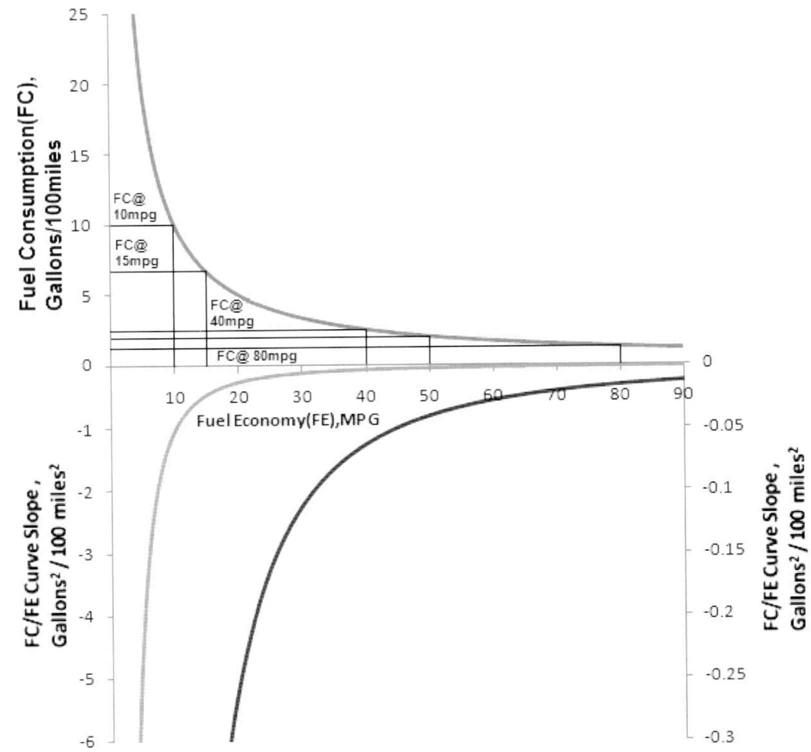

FIGURE E.1 Fuel consumption (FC) versus fuel economy (FE) and slope of FC/FE curve (two curves and two different scales). SOURCE: Johnson (2009). Reprinted with permission.

F

Review of Estimate of Retail Price Equivalent Markup Factors

Vyas et al. (2000) of Argonne National Laboratory (ANL) compared their own markup factors to estimates developed by Energy and Environmental Analysis, Inc. (EEA) and Borroni-Bird. Two different markup factors were compared: (1) the markup over direct manufacturing (variable) costs for components produced in house and (2) the markup for components purchased fully manufactured from outside suppliers. In the ANL analysis, costs of manufacture include materials, assembly labor, and other manufacturing costs but not depreciation, amortization, warranty, or R&D and engineering (Table F.1). Other costs borne by the original equipment manufacturer (OEM) are corporate overhead, benefits (retirement and health care), and distribution, marketing, and dealer costs, including dealer profits.

Because the cost categories used by Borroni-Bird and EEA differed from those used by the ANL study, an exact comparison is not possible (Table F.2). While Vyas et al. (2000) concluded that the three sets of estimates were quite close, the different definitions cloud the issue. For example, Vyas et al. (2000) assumed that half of the costs—shown by Borroni-Bird as transportation/warranty; amortization and depreciation; engineering R&D, pension and health, advertising, and overhead—would be borne by the outside supplier. In their own estimates they allocate all warranty, R&D/engineering, and depreciation and amortization costs to the supplier. Clearly, even components purchased fully manufactured from a Tier 1 supplier will incur costs just for their engineering into the vehicle system and are likely to lead to some warranty costs beyond those covered by the supplier. Still, the bottom-line markup over variable manufacturing costs is very similar: 2.05 for the Borroni-Bird analysis versus 2.00 for the ANL analysis.

The Vyas et al. (2000) memorandum also summarized the cost methodology used by EEA, Inc., in a study for the Office of Technology Assessment (OTA, 1995), although it should be noted that the auto industry has undergone dramatic changes since that time, and the continued applicability of the methodology is debatable. Again, the cost categories differ, but the bottom-line markup over variable manufacturing costs is similar although a bit higher: 2.14 (Table F.3). To get an idea of the markup over outsourced component costs, the ANL analysts again assumed that the supplier would bear the costs of warranty, R&D engineering, and depreciation and amortization. Since EEA methods do not separate warranty costs from manufacturing overhead, Vyas et al. (2000) assumed that warranty costs made up half of the overhead costs. With those assumptions they obtained a markup factor of $100/(33.6 + 6.5 + 6.5 + 10.3/2 + 12.1) = 1.56$. This leaves only a bit more than 5 percent of the total retail price equivalent (RPE) for the costs of integrating components into the overall vehicle design, assembly, and other OEM assembly costs.

The ANL memorandum concludes that all three sources would result in very similar markup factors (Table F.4). However, for markups over Tier 1 supplier costs, the ANL decision on how to allocate the costs has a lot to do with the similarities. A less generous allocation of warranty, assembly, and manufacturing overhead costs to suppliers would result in higher markup factors for outsourced components. Despite these ambiguities, the ANL comparison reasons that the markup for in-house-made components would be about twofold rather than the 1.5-fold markup for components purchased from Tier 1 suppliers.

A markup factor of 1.5 was used by NHTSA (DOT/NHTSA, 2009, p. 173) in its final fuel economy rule for 2011. A somewhat lower RPE markup factor of 1.4 was used by NRC (2002) and by S. Albu, assistant chief, Mobile Source Division, California Air Resources Board, in his presentation to the committee (Albu et al., 2008), while the EPA has used a markup of approximately 1.3 (EPA, 2008).

A markup of approximately 2 over the direct manufacturing cost of parts manufactured in house by an OEM was also supported by Bussmann in a presentation, "Study of industry-average markup factors used to estimate retail price equivalents (RPE)," to the committee on January 24, 2008. In that briefing, Bussman cited a 2003 study of the global automotive industry by McKinsey Global Institute, which came up with a markup factor of 2.08, and his own analysis

TABLE F.1 Components of Manufacturer's Suggested Retail Price (MSRP) Equivalent RPE: ANL Method

Cost Category	Cost Contributor	Relative to Cost of Vehicle Manufacture	Share of MSRP (%)
Vehicle manufacture	Cost of manufacture	1.00	50.0
Production overhead	Warranty	0.10	5.0
	R&D engineering	0.13	6.5
	Depreciation and amortization	0.11	5.5
Corporate overhead	Corporate overhead, retirement, health	0.14	7.0
Selling	Distribution, marketing, dealers	0.47	23.5
Sum of costs		1.95	97.5
Profit	Profit	0.05	2.5
Total contribution to MSRP		2.00	100.0

SOURCE: Vyas et al. (2000).

TABLE F.2 Components of MSRP: Estimated by Borroni-Bird

Cost Category	Cost Contributor	Relative to Cost of Vehicle Manufacture	Share of MSRP (%)
Vehicle manufacture	Materials	0.87	42.4
	Labor, other manufacturing costs	0.13	6.3
Fixed cost	Transportation and warranty	0.09	4.4
Fixed cost	Amortization and depreciation, engineering R&D, pension and health care, advertising, and overhead	0.44	21.5
Selling	Price discounts	0.10	4.9
	Dealer markup	0.36	17.6
Sum of costs		1.99	97.1
Profit		0.06	2.9
MSRP		2.05	100.0

SOURCE: As reported by Vyas et al. (2000).

TABLE F.3 Components of Retail Price Equivalent: EEA, Inc., Method

Cost Category	Cost Contributor	Relative to Cost of Vehicle Manufacture	Share of MSRP (%)
Vehicle manufacture	Division costs	0.72	33.6
	Division overhead	0.14	6.5
	Assembly labor and overhead	0.14	6.5
Overhead	Manufacturing overhead	0.22	10.3
	Amortized engineering, tooling, and facilities	0.26	12.1
Selling	Dealer margin	0.49	22.9
Sum of costs		1.97	92.1
Profit		0.17	7.9
Total		2.14	100.0

SOURCE: EEA, Inc. (1995), as reported by Vyas et al. (2000).

TABLE F.4 Comparison of Markup Factors

Markup Factor for	ANL	Borroni-Bird	EEA
In-house components	2.00	2.05	2.14
Outsourced components	1.50	1.56	1.56

SOURCE: Vyas et al. (2000).

of Chrysler data for 2003-2004, which produced factors of 1.96-1.97. Since these markup factors apply to direct manufacturing costs, they are consistent with the estimates shown in Table F.4. Lyons (2008) used a markup factor of approximately 2.0 but was not specific about the cost components included in the estimate to which this factor was applied.

Information supplied to the committee in the presentation by Duleep on January 25, 2008, implies higher markup factors (Duleep, 2008). Assuming a reference cost of 1.00 for the variable factors used to produce a component (material, labor, energy, factory overhead), EEA calculates the Tier 1 supplier cost by applying multiplicative markups for supplier overhead and profit and an additive factor of 0.1 to 0.2 for tooling, facilities, and engineering (Table F.5). The range is intended to reflect the complexity of the component and the engineering effort required of the supplier to ensure its integration into the full vehicle system. Representing the variable costs by *X*, the total supplier price markup is given by equation 1:

$$\begin{aligned} SupplierRPE_{Low} &= X(1 + 0.20 + 0.05) + 0.10 = \\ &1.00(1.25) + 0.10 = 1.35 \\ SupplierRPE_{High} &= X(1 + 0.20 + 0.05) + 0.20 = \\ &1.00(1.25) + 0.20 = 1.45 \end{aligned} \tag{1}$$

In the EEA method, OEM costs include amortization of tooling, facilities and engineering, and overhead, profit and selling costs, which include marketing, distribution, and dealer costs. EEA assumes an average manufacturer profit of 8 percent, somewhat higher than the 5 percent assumed by ANL and the 6 percent assumed by Borroni-Bird. Amortized costs vary from 5 percent to 15 percent, again depending on the complexity of the part and the costs of integrating it into the vehicle system. Marketing, distribution, and dealer costs are multiplicative and add 25 percent to the OEM costs (Figure F.1).

$$\begin{aligned} RFE_{Low} &= SupplierCost_{Low}\,(1 + 0.20 + 0.08) + 0.05 = \\ &1.35(1.28) + 0.05 = 1.78 \\ RFE_{High} &= SupplierCost_{High}\,(1 + 0.20 + 0.08) + 0.15 = \\ &1.45(1.28) + 0.15 = 2.10 \end{aligned} \tag{2}$$

The resulting markup ranges are 2.22 to 2.51 for the markup over variable costs (corresponding to the ANL "vehicle manufacturing" costs) and 1.65 to 1.73 for the markup over Tier 1 supplier costs (corresponding to the ANL cost of outsourced components). The full breakdown of EEA markup estimates is shown in Table F.5. The markups are comparable to those proposed by Vyas et al. (2000) but higher by a meaningful amount, as shown in Figure F.2. In a note, EEA-ICF, Inc., argues that higher supplier amortized costs are generally associated with lower OEM amortized costs for any give™n part. However, this assertion was not applied here to develop the range of markup factors based on EEA data.

Average RPE factors can be inferred by costing out all the components of a vehicle, summing them to estimate OEM Tier 1 costs or fully burdened in-house manufacturing costs, and then dividing the sum into the selling price of the vehicle. The committee contracted with IBIS Associates (2008) to conduct such an analysis for two popular vehicles: (1) the Honda Accord sedan and (2) the Ford F-150 pickup truck. Current model year (2009) designs and base model trim levels (no nonstandard options) were chosen. Base models

TABLE F.5 Fuel Economy Technology Cost Markup Factors

Item	Cost Low	Cost High	Share Low %	Share High %
Supplier costs				
Factors (materials, labor, energy, factory overhead)	1.00	1.00	45	40
Supplier overhead	0.20	0.20	9	8
Supplier profit	0.05	0.05	2	2
Amortization of tooling + facilities + engineering	0.10	0.20	4	8
Supplier subtotal	1.35	1.45	61	58
Supplier markup	1.35	1.45		
OEM costs				
OEM overhead	0.20	0.20	12	12
OEM profit	0.08	0.08	5	5
Tooling + facilities + engineering amortization	0.05	0.15	2	6
OEM subtotal	1.78	2.01	80	80
OEM markup	1.32	1.38		
Marketing, transport, dealer markup	0.25	0.25	20	20
Total	2.22	2.51	100	100
RPE markup (over factors)	2.22	2.51		
RPE markup (over supplier price)	1.65	1.73		

SOURCE: EEA-ICF, Inc., as reported by Duleep in his presentation to the committee on January 25, 2008.

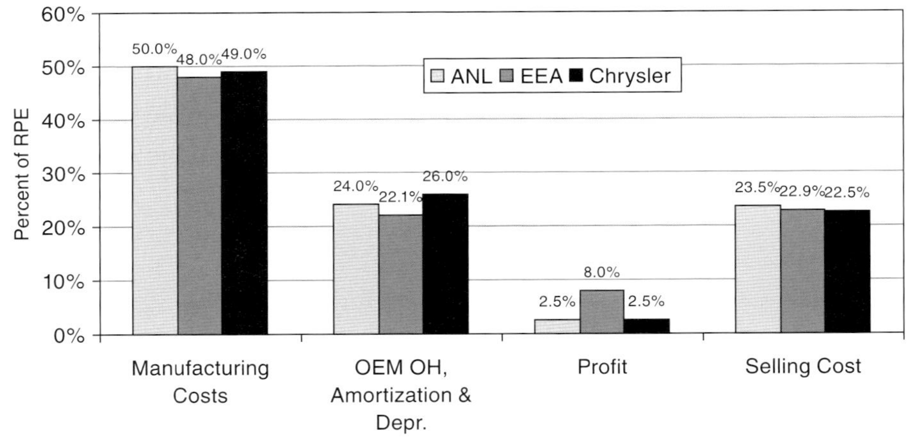

FIGURE F.1 Components of retail price equivalent (RPE) markup. SOURCE: Duleep (2008).

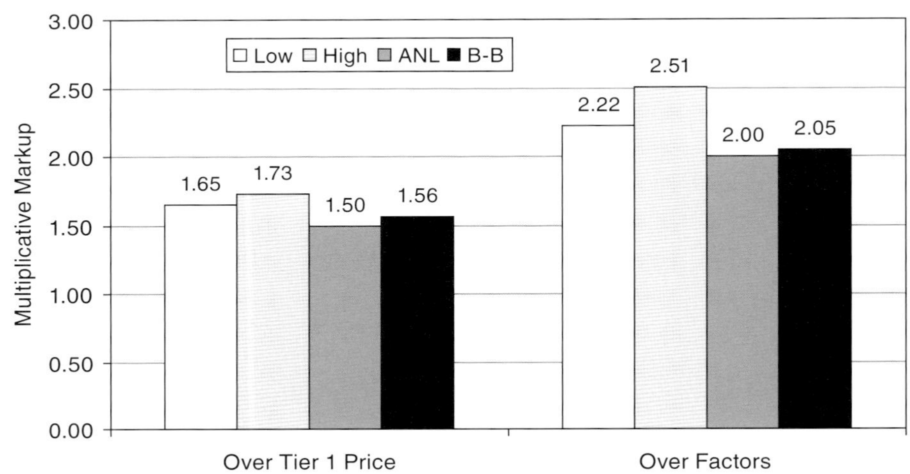

FIGURE F.2 Comparison of Duleep (2008) high/low, Argonne National Laboratory (ANL), and Borroni-Bird (B-B) cost markup factors.

were chosen to reduce the influence of market pricing decisions not driven by manufacturing costs.

Cost estimates were developed for subcomponents in terms of costs paid by OEMs for automotive components and subsystems in five broad systems. Although many of the components are manufactured in house, the costs of these components were estimated using the fixed or indirect manufacturing costs normally borne by a Tier 1 supplier. Results for the base Honda Accord are shown in Table F.6. The base vehicles are the four-door LX sedans produced in Marysville, Ohio, and Lincoln, Alabama. The curb weight of this vehicle is 3,230 lb, with a V6, 3.0-L, dual overhead cam engine, a five-speed manual transmission, and a stamped steel unibody with a lightweight aluminum subframe. Dealer invoice cost for the Accord is $18,830, MSRP is $20,755, and the average market transaction price is $19,370. The cost of all components plus assembly costs is estimated to be $14,564. This results in multipliers of 1.39 to market transaction price and 1.49 to MSRP. The multiplier to dealer invoice cost is 1.35, which means that dealer costs, including profit, amount to about 4 percent of manufacturing costs, not considering any dealer incentives offered by OEMs.

The base 2009 Ford F-150s are two-door XL Regular Cab Styleside short-bed, rear-wheel-drive pickups produced in Dearborn, Michigan, and Kansas City, Missouri. The curb weight of the vehicle is 4,743 lb, with a standard V8, 4.6-L, single overhead valve engine and a four-speed automatic transmission. The truck has a stamped steel body on frame construction. Dealer invoice cost for the F-150 is $20,055, MSRP is $21,565, and the average market transaction price is $21,344. The cost of all components plus assembly is $14,940, as shown in Table F.7. This means an RPE multiplier of 1.52 for market price and 1.54 for MSRP. The markup factor for the dealer invoice is 1.43, so that dealer costs and profit amount to about 9 percent of total manufacturing costs, not including any possible OEM incentives to dealers.

TABLE F.6 Cost Breakdown of Base 2009 Honda Accord LX

	Accord LX Base 2009		
	Mass (kg)	Cost ($)	Detail
Power train	609	6,677	
Engine	206	2,782	I4 2.4 DOHC AL/AL
Battery	20	58	Lead-acid, standard
Fuel storage and delivery	86	388	Gasoline, 18.5 gal
Transmission	70	621	Manual, 5-speed
Thermal management	23	150	
Driveshaft/axle	84	1,189	
Differential	26	203	
Cradle	25	161	Aluminum
Exhaust system	34	300	
Oil and grease	15	25	
Power train electronics	10	400	
Emission control electronics	10	400	
Body	451	2,234	
Body-in-white	307	1,006	Midsize steel unibody
Panels	60	197	Stamped steel midsize
Front/rear bumpers	10	30	Sheet steel
Glass	40	250	Conventional, 4 mm
Paint	12	450	Solvent-borne, average color
Exterior trim	10	50	
Hardware	10	226	
Seals and NVH control	2	24	
Chassis	181	1,643	
Corner suspension	30	217	Lightweight
Braking system	46	404	ABS
Wheels and tires	80	472	Alloy 16″
Steering system	26	549	
Interior	151	2,156	
Instrument panel	24	110	
Trim and insulation	22	429	
Door modules	25	220	
Seating and restraints	60	1,122	
HVAC	20	275	
Electrical	33	1,250	
Interior electrical	11	500	
Chassis electrical	11	500	
Exterior electrical	11	250	
Total components	1,426	13,959	
Final assembly	40	605	
Interior to body	5	140	
Chassis to body	10	90	
Power train to body	10	90	
Electronics to body	5	80	
Other systems to body	10	205	
Total manufacturing	1,466	14,564	

NOTE: DOHC, double overhead cam shaft; HVAC, heating, air conditioning, cooling; NVH, noise, vibration and, harshness; and ABS, automatic braking system.
SOURCE: IBIS (2009).

BIBLIOGRAPHY

Albu, S., California Air Resources Board. 2008. ARB perspective on vehicle technology costs for reducing greenhouse gas emissions. Presentation to the National Research Council Committee on Technologies for Improving Light-Duty Vehicle Fuel Economy on January 24.

Bussmann, W.V., and M.J. Whinihan. 2009. The Estimation of Impacts on Retail Prices of Regulations: A Critique of Automobile Industry Retail Price Equivalent and Indirect Cost Multipliers. Prepared for the Alliance of Automobile Manufacturers, Southfield, Mich. May 6.

DOT/NHTSA (U.S. Department of Transportation/National Highway Traffic Safety Administration). 2009. Average Fuel Economy Standards, Passenger Cars and Light Trucks, Model Year 2001: Final Rule. 49 CFR Parts 523, 531, 533, 534, 536, and 537, Docket No. NHTSA-2009-0062, RIN 2127-AK29. DOT/NHTSA, Washington, D.C. March 23.

Duleep, K.G. 2008. Analysis of technology cost and retail price. Presentation to the National Research Council Committee on Technologies for Improving Light-Duty Vehicle Fuel Economy, January 24.

TABLE F.7 Cost Breakdown of Base 2009 F-150

	F-150 Pickup XL Base		
	Mass (kg)	Cost ($)	Detail
Power train	922	7,666	
Engine	308	3,971	V8 4.6 L SOHC CI/AL
Battery	29	84	Lead-acid, standard
Fuel storage and delivery	102	440	Gasoline, 25 gal
Transmission	118	1,068	Auto 4 pickup truck
Thermal management	45	150	
Driveshaft/axle	150	608	Pickup truck 2WD steel
Differential	37	116	Light truck
Cradle	25	103	Hydroformed steel
Exhaust system	68	300	
Oil and grease	15	25	
Power train electronics	10	400	
Emission control electronics	16	40	
Body	672	2,258	
Body-in-white	500	1,020	Pickup truck body on frame
Panels	55	177	Stamped steel pickup truck
Front/rear bumpers	20	60	Medium truck
Glass	51	250	
Paint	12	450	Solvent-borne, average color
Exterior trim	12	50	
Hardware	13	226	
Seals and NVH control	10	24	
Chassis	348	1,719	
Corner suspension	119	413	Pickup truck 2WD
Braking system	79	520	Light truck ABS 4-wheel
Wheels and tires	105	334	Steel 17"
Steering system	44	453	Pickup truck
Interior	128	1,570	
Instrument panel	24	100	
Trim and insulation	28	350	
Door modules	22	156	
Seating and restraints	40	820	
HVAC	15	144	
Electrical	27	832	
Interior electrical	7	232	
Chassis electrical	10	400	
Exterior electrical	10	200	
Total components	2,098	14,045	
Final assembly	52	905	
Interior to body	10	200	
Chassis to body	10	150	
Power train to body	10	150	
Electronics to body	10	100	
Other systems to body	10	305	
Total manufacturing	2,150	14,950	

NOTE: SOHC, single overhead camshaft.
SOURCE: IBIS (2008).

EPA (U.S. Environmental Protection Agency). 2008. EPA Staff Technical Report: Cost and Effectiveness Estimates of Technologies Used to Reduce Light-Duty Vehicle Carbon Dioxide Emissions. EPA420-R-08-008. Ann Arbor, Mich.

IBIS Associates, Inc., Waltham, Mass. 2008. Data Collection and Analysis: Vehicle Systems Costs. Report to the National Research Council Committee on Assessment of Technologies for Improving Light-Duty Vehicle Fuel Economy. December.

Lyons, J.M., Sierra Research, Inc. 2008. Technology and retail price implications of HR6 CAFE standards based on vehicle simulation modeling (preliminary results). Presentation to the National Research Council Committee on Assessment of Technologies for Improving Light-Duty Vehicle Fuel Economy, January 24.

NRC (National Research Council). 2002. Effectiveness and Impact of Corporate Average Fuel Economy (CAFE) Standards. National Academy Press, Washington, D.C.

OTA (Office of Technology Assessment). 1995. Advanced Automotive Technology: Visions of a Super-Efficient Family Car. OTA-ETI-638. Washington, D.C.

Vyas, A., D. Santini, and R. Cuenca. 2000. Comparison of Indirect Cost Multipliers for Vehicle Manufacturing. Center for Transportation Research, Argonne National Laboratory. Argonne, Ill. April.

G

Compression-Ignition Engine Replacement for Full-Size Pickup/SUV

The analysis and discussion for the main part of Chapter 5 were based on two vehicle classes—namely, a midsize sedan such as the Accord, Camry, Fusion, or Malibu and a midsize SUV such as the Durango, Explorer, or Trailblazer. To enable projections for the entire range of vehicle classes discussed in Chapter 9, it was necessary to create an additional engine specification to provide a CI replacement for the 5.3- to 6.2-L V8 SI engines which would be found in full-size body-on-frame pickup trucks such as the F150, the Silverado, and the Ram 1500 and SUVs such as the Expedition and Tahoe. Table 5.5 in Chapter 5 described a V6 CI engine with displacement between 2.8 and 3.5 L appropriate for midsize SUVs and midsize pickup trucks. For cost reasons, there is a range of displacements for which OEMs would tend to design and build V6 rather than V8 engines since V6s require fewer parts. For CI engines, this V6 range would be from about 2.9 L to perhaps 4.5 L. It was therefore assumed in this additional analysis that the V8 SI engines typically used in full-size pickups would be replaced by a V6 CI engine as long as the torque and power required for equal performance could be achieved. With a base-level specification at a specific torque of 160 N-m/L, the displacement required for a CI V6 to replace an SI V8 of the displacement range 5.3-6.2 L would be 4.4-5.2 L, which is really too large for the V6 configuration. However, from a cost point of view, the V6 configuration would be preferable to a V8 if a V6 concept could be identified that meets the requirements. If no base-level configuration were considered, an advanced-level V6 of 3.5 L could easily provide sufficient torque to replace a 6.2-L SI V8 and could be manufactured with the same set of tooling as the V6 engine whose cost increments are described in Tables 5.5 and 5.8. Therefore, for the full-size pickup class of vehicles, it was assumed in this analysis that the CI replacement for SI V8 engines would be a V6 of displacement up to 3.5 L with advanced-level technology. Cost estimates for such an engine are shown in Tables G.1 to G.3.

TABLE G.1 Incremental CI-Diesel Engine Cost Estimations to Replace SI MPFI OHV Two-Valve 5.3- to 6.2-L V8 Engine in a Full-Size Body-on-Frame Pickup (e.g., Silverado and Ram) or SUV with a 3.5-L V6 DOHC CI

50-State Saleable ULEV II 3.5-L V6 DOHC CI-Diesel Engine, Baseline: SI Gasoline OHV 4-V 5.3- to 6.2-L V8	Estimated Cost Versus Baseline ($)
Common-rail 1,800 bar piezo-actuated fuel system with six injectors (@$75), high-pressure pump ($270), fuel rail, regulator and fuel storage upgrades plus high-energy driver upgrades to the engine control module. Credit for MPFI content deleted ($48).	911
Series sequential turbocharging: One VGT with electronic controls and one fixed-geometry turbocharger with active and passive bypass valves necessary to match high EGR rates at low load conditions ($750). Water-air charge air cooler, circulation pump, thermostat/valve, and plumbing. Engine downsizing credit from V8 ($200).[a]	830
Upgrades to electrical system: starter motor, alternator, battery, and 1.5-kW supplemental electrical cabin heater as is standard in Europe ($99).	167
Cam, crank, connecting rod, bearing, and piston upgrades, oil lines ($62) plus NVH countermeasures to engine ($47) and vehicle ($85).	194
High- and low-pressure EGR system to suppress NO_x at light and heavy loads. Includes hot-side and cold-side electronic rotary diesel EGR valves plus EGR cooler and all plumbing.	226
Add remaining components required for advanced-level technology (details in Table G.3).	308
Emissions control system including the following functionality: DOC, CDPF, selective catalytic reduction (SCR), urea dosing system ($363). Stoichiometric MPFI emissions and evaporative systems credit ($343).	1,040
On-board diagnostics (OBD) and sensing, including four temperature sensors (@$13), wide-range air/fuel ratio sensor ($30), NO_x sensor ($85), two-pressure sensing glow plugs (@$17), six glow plugs (@$3), and Delta-P sensor for DPF ($25). Credit for four switching O_2 sensors (@$9).	227
Total variable cost with credits for SI parts removed excludes any necessary transmission, chassis, or driveline upgrades.	3,903

NOTE: Aftertreatment system cost estimates reflect April 2009 PGM prices. Estimates derived from Martec (2008). CDPF, catalyzed diesel particulate filter; CI, compression ignition; DOC, diesel oxidation catalyst; DOHC, dual over head cam; DPF, diesel particulate filter; DPF, diesel particulate filter; EGR, exhaust gas recirculation; MPFI, multipoint fuel injection; NVH, noise, vibration, harshness; OBD, on-board diagnostics; OHV, over head valve; PGM, platinum group metals; SCR, selective catalytic reduction; SI, spark ignition; ULEV II, ultra-low-emissions vehicle; VGT, variable geometry turbocharger.

[a] Credit for downsizing from V8 to V6 referred to DOHC 4-V V8 downsized to DOHC 4V V6. In this case, credit used by Martec was reduced from $270 to $200 since the parts removed from an OHV 2-V V8 would cost less than those removed from a DOHC 4-V V8.

TABLE G.2 Cost Estimates of Exhaust Emissions Aftertreatment Technologies Capable of Enabling Tier 2, Bin 5 Compliance

Item	Midsize Car (e.g., Malibu), Catalytic Device Sizing Based on 2.0-L (April 2009 PGM prices) ($)	Midsize SUV (e.g., Explorer), Catalytic Device Sizing Based on 3.5- L (April 2009 PGM prices) ($)	Full-Size Pickup (e.g., Explorer), Catalytic Device Sizing Based on 4.4-L (April 2009 PGM prices) ($)
DOC 1			
Monolith and can	52	52	52
PGM loading	139	200	252
DOC 2			
Monolith and can	Not used	52	52
PGM loading	Not used	70	87
EGR catalyst			
Monolith and can	7	Not used	Not used
PGM loading	13	Not used	Not used
Coated DPF			
Advanced cordierite brick and can	124	270	270
PGM loading	131	26	33
NSC system			
Catalyst brick and can	114	Not used	Not used
PGM loading	314	Not used	Not used
SCR-urea system			
SCR brick and can	39	274	274
Urea dosing system	Passive SCR	363	363
Stoichiometric gasoline emissions and evaporative system credit	−245	−343	−343
Emissions system total	688	964	1,040

NOTE: This table complements Table 5.5. Compared to Table 5.5, the columns reflecting November 2007 PGM prices (Columns 2 and 4) have been removed and a new column, Column 4, was added. This column reflects the aftertreatment system cost estimate for the exhaust flow rates of a larger base-level V6 CI engine (i.e., 4.4 L) suitable for replacing 5.5- to 6.2-L two-valve OHV V8 SI engines with 3.5-L advanced-level technology CI engines. Note that, as discussed in Chapter 5, it was assumed that the aftertreatment component sizes for the 3.5-L advanced-level V6 are equal to those of a base-level 4.4-L V6 because the power levels for these two engines would be the same, thus requiring the same exhaust flow rates. All cost estimates are based on April 2009 PGM commodity prices. Column 4 provides the estimate used for the aftertreatment costs in Table G.1.

TABLE G.3 Estimates of Incremental Costs to Implement Developments Whose Estimated Fuel Consumption Reduction Gains Are Summarized in Table 5.2

Item	Midsize Car (e.g., Malibu) 1.6-L L4	Midsize SUV (e.g., Explorer) 2.8-L V6	Full-size Pickup (e.g., Ram 1500) 3.5-L V6	
Downsize engines 2-L L4 to 1.6 L, 3.5-L V6 to 2.8 L, 4.4-L V6 to 3.5 L	50	75	75	Higher load capacity rod bearings and head gasket for higher cylinder pressures (~$12.50/ cylinder)
Two-stage turbocharger system	375	545	0[a]	Additional air flow control valves, piping, cost of additional turbo, water-to-air intercooler with control valve, separate pump
Dual-pressure oil pump	5	6	6	Switchable pressure relief valve for high or low oil pressure
Nonrecirculating LP fuel pump	10	12	12	Variable output LP pump controlled by HP pump output
Low-pressure EGR	—	95	95	Additional piping (~$20) and valves (e.g., integrated back pressure and LP EGR rate ~$75), much more difficult to package for V6 engine with underfloor DPF, cost for L-4 already included in Table 5.4
Direct-acting HP (maximum injection pressures > 2,000 bar) piezo injectors	80	120	120	$20/injector, benefits derived from combination of higher rail pressure and more injector controllability
Total	520	853	308	

NOTE: These developments are CI-diesel downsizing from base level to advanced level, thermodynamic improvements, friction reduction, and engine accessory improvements. Total for full-size body-on-frame pickup ($308 at bottom of Column 4) used in Table G.1. FC, fuel consumption.

[a]Two-stage turbo system already comprehended in Table G.1.

H

Other NRC Assessments of Benefits, Costs, and Readiness of Fuel Economy Technologies

The National Research Council (NRC) has conducted other studies to estimate benefits, costs, and readiness of fuel economy technologies for light-duty vehicles. Indeed, this committee's task is to update the estimates provided in one of the earlier studies, *Effectiveness and Impact of Corporate Average Fuel Economy (CAFE) Standards,* which was issued in 2001. The committee discusses several other studies here. The *Review of the Research Program of the Partnership for a New Generation of Vehicles: Seventh Report* (NRC, 2001) assessed the fuel economy technologies and costs associated with three prototype vehicles built in connection with the Partnership for a New Generation of Vehicles (PNGV) research program to achieve up to three times the fuel economy of a 1994 family sedan. More recent NRC studies that have looked at different aspects of fuel economy technologies include *Transitions to Alternative Transportation Technologies—A Focus on Hydrogen* (NRC, 2008a), *Review of the Research Program of the FreedomCAR and Fuel Partnership: Second Report* (NRC, 2008b), and the report from the America's Energy Future (AEF) Panel on Energy Efficiency, *Real Prospects for Energy Efficiency in the United States* (NAS-NAE-NRC, 2010). Even though the recent report *Transitions to Alternative Transportation Technologies—Plug-In Hybrid Electric Vehicles* (NRC, 2009) was not strictly a report on fuel economy technology, it did address the costs and benefits of plug-in electric vehicles.

While the tasks required under each study are different, some of their analyses of costs, efficiencies, and prospects for the various technologies overlap and are reviewed here. However, the committee does not attempt to review the findings of any studies other than those of the NRC. It simply comments on them, as appropriate, to the degree that the NRC reports are based on them.

REVIEW OF THE RESEARCH PROGRAM OF THE PARTNERSHIP FOR A NEW GENERATION OF VEHICLES, SEVENTH REPORT

The task of the NRC Standing Committee to Review the Research Program of the PNGV (NRC PNGV committee) was to examine the research program, communicate the program's progress to government and industry participants, and identify barriers to the program's success. The PNGV program was a cooperative research and development program between the government and the United States Council for Automotive Research, whose members include the three original equipment manufacturers (OEMs) in the United States: DaimlerChrysler Corporation, Ford Motor Company, and General Motors Corporation. The PNGV was envisioned to allow the parties to cooperate on precompetitive research activities that would ultimately result in the deployment of technologies to reduce our country's fuel consumption and emissions of carbon dioxide. The PNGV aimed to improve the competitiveness of the U.S. manufacturing base for future generations of vehicles and to introduce innovative technologies into conventional vehicles in order to improve fuel consumption or reduce emissions. The final goal of the PNGV program was to develop prototype vehicles that achieve up to three times the average fuel economy of a 1994 family sedan. It was recognized that these new vehicles would have to be sold in high volume in order to have an impact. For this reason, the strategy for the prototype vehicle was to develop an affordable family sedan with a fuel economy of up to 80 mpg that maintained the performance, size, and safety standards of the vehicles of that time. After 2002, the program transitioned to the FreedomCAR and Fuel Research (FreedomCAR) Program, discussed in the following section.

Each of the three automobile companies involved in the PNGV program built its own prototype concept vehicles since this could not be done in the context of precompetitive research. By the time of the seventh NRC report, all three companies had built prototypes that met the then-extant performance, comfort, cargo space, utility, and safety requirements. These prototype vehicles could not, however, meet the price target while simultaneously improving fuel economy to near 80 mpg. The DaimlerChrysler prototype foresaw a price premium of $7,500, while the other two did not announce any price premium associated with their vehicles. All three concept vehicles used hybrid electric power trains with small, turbocharged, compression-ignition direct-injection engines using diesel fuel. All three were start-stop hybrids that shut the engine off when idling. The report from the NRC PNGV committee estimated that dual-mode batteries would probably cost $1,000 to $1,500 per battery unit (1.5 kWh), or $670 to $1,000 per kWh (NRC, 2001). Each company took a different route to reduce the vehicle mass and aerodynamic drag and to supply power for auxiliary loads. The high cost of the lightweight materials and electronic control systems made the price target unattainable. In addition, the cost of the compression-ignition direct-injection engine was greatly increased by the exhaust-gas after-treatment systems to control emissions. In the middle of the PNGV program, the Tier 2 emission standard was promulgated, and the NRC PNGV committee believed that the ability of the diesel engine to meet emissions targets was not clear.

The NRC PNGV committee reported that the PNGV program had made significant progress in implementing desirable technologies as fast as possible. Each of the three automobile manufacturers in the PNGV demonstrated a hybrid electric vehicle before the end of the Partnership in 2004. They had developed the concept vehicles by 2000, but the goal of the development of a preproduction prototype by 2004 was not met because of the termination of the PNGV program. Indeed, the manufacturing and engineering innovations that came out of the PNGV program were implemented before 2000. In the end, the three OEMs demonstrated that a production medium-size passenger car could be produced that achieved 80 mpg, and one OEM (DaimlerChrysler) demonstrated that such a vehicle could be produced at a cost penalty of less than $8,000.

THE FREEDOMCAR AND FUEL RESEARCH PROGRAM REPORT

The task of the NRC Committee on Review of the FreedomCAR and Fuel Research Program (NRC FreedomCAR committee) is to assess the FreedomCAR and Fuel Partnership's management and the research and development activities overseen by the Partnership. The Partnership, started in 2002, built on the earlier PNGV program. FreedomCAR, like PNGV, is a collaboration between the government and industry to support a wide range of pre-competitive research in automotive transportation. The Partnership's goal is to study technologies that will help the United States transition to an automotive fleet free from petroleum use and harmful emissions (NRC, 2005). The vision of the Partnership is to enable a transition pathway that starts with improving the efficiency of today's internal combustion (IC) engines, increasing the use of hybrid electric vehicles, and supporting research in fuel-cell-powered vehicles so that a decision can be reached in 2015 on the economic and technological viability of hydrogen-powered vehicles. In 2009, a greater emphasis began to be placed on plug-in hybrid electric vehicles (PHEVs). The NRC has thus far reviewed the FreedomCAR and Fuel Partnership twice, with reports published in 2005 and 2008. In the second of these reports, one of the NRC FreedomCAR committee's tasks was to comment on the balance and adequacy of the efforts and on the progress achieved since the 2005 report. The conclusions and recommendations of the second report focus on the Partnership's management and oversight but also provide the FreedomCAR committee's opinion on the readiness of new fuel economy technologies.

The NRC FreedomCAR committee report recognizes that more efficient IC engines will contribute the most to reducing fuel consumption and emissions in the near term. The Partnership focuses research on lean-burn, direct-injection engines for both diesel- and gasoline-fueled vehicles, specifically on low-temperature combustion engines and aftertreatment of the exhaust. The report recognizes that, after completing the research necessary to prove a technology's viability, there are typically several years of prototyping and developing manufacturing processes before the technology can be introduced into the vehicle fleet. Because of the urgent need to reduce vehicle fuel consumption, the development phase of these technologies has been accelerated while researchers are still studying the controlling thermochemistry of low-temperature combustion. The result is close coordination between those looking to expand the fundamental knowledge base and those investigating applications. The report from the NRC FreedomCAR committee recommends that the Partnership investigate the impact on emissions of combustion mode switching and transient operation with low-temperature combustion, and it questions how much exhaust energy can actually be recovered. Furthermore, the NRC FreedomCAR committee suggests the Partnership closely analyze the cost-effectiveness of the exhaust gas heat recovery research and the potential fuel efficiency benefits before deciding whether to pursue this research further.

Another goal of the FreedomCAR and Fuel Partnership is to develop, by 2015, battery storage for hybrid electric vehicles that has a 15-year life and a pulse power of 25 kilowatts (kW), with 1 kW of pulse power costing $20. This effort focuses on lithium (Li) ion batteries, which are simultaneously in both the research phase, as the knowledge base for specific electrochemical systems is expanded, and the development phase, as the batteries are built and tested. Significant progress had been made since the first FreedomCAR report (NRC, 2005, 2008b). The Partnership has demonstrated batteries that exceed the requirement for a 300,000-cycle lifetime, that have longer calendar lives, and that operate over a wider temperature range than earlier batteries. The NRC FreedomCAR committee recognized that cost is the primary barrier for introduction of the Li-ion battery to the market and commends the Partnership for researching lower cost materials for the cathode and the microporous separator. The report from the NRC

FreedomCAR committee recommended that the Partnership do a thorough cost analysis of the Li-ion batteries under development to account for recent process and materials costs and for increased production rate costs.

A 50 percent reduction in total vehicle weight at no additional cost is another key goal of the Partnership; it would rely on the widespread application of advanced high-strength steels, aluminum alloys, cast magnesium, and carbon-fiber-reinforced plastics. The NRC FreedomCAR committee concluded that the goal of price parity for the lightweight materials is insurmountable within the time frame of the Partnership (NRC, 2008b). However, the 50 percent weight reduction goal is critical for the Partnership's overall vision of a hydrogen-fueled car. The NRC FreedomCAR committee went beyond that, saying the weight reduction would be mandatory even with the associated cost penalty, because the alternative adjustments to the engine and batteries would cost more. The NRC report recommends maintaining the 50 percent weight reduction goal and analyzing cost-effectiveness to confirm that the added cost of weight reduction can be offset by modifying the fuel cell and battery goals.

THE HYDROGEN REPORT

The tasks of the Committee on Assessment of Resource Needs for Fuel Cell and Hydrogen Technologies (the NRC hydrogen committee) was to establish the maximum practicable number of vehicles that could be fueled by hydrogen by 2020 and to discuss the public and private funding needed to reach that number. The NRC hydrogen committee assumed that (1) the technical goals for fuel cell vehicles, which were less aggressive than those of the FreedomCAR Partnership, are met; (2) that consumers would readily accept such vehicles; (3) that government policies would drive the introduction of fuel cell vehicles and hydrogen production and infrastructure at least to the point where fuel cell vehicles are competitive on the basis of lifecycle cost; and (4) that oil prices are at least \$100 per barrel by 2020 (NRC, 2008a). Thus, the scenarios developed in the hydrogen report are not projections but a maximum possible future market if all assumptions are met. The NRC hydrogen committee concluded that although durable fuel cell systems at significantly lower costs are likely to be increasingly available for light-duty vehicles over the next 5 to 10 years, the FreedomCAR Partnership goals for 2015 are not likely to be met. The NRC hydrogen committee also concluded that commercialization and growth of these hydrogen fuel cell vehicles could get under way by 2015 if supported by strong government policies. Those conclusions are more optimistic than the conclusions on fuel cells contained in this report, whose committee (though it did not consider the potential impact of policies on fuel cell market potential) does not expect progress on fuel cell costs and technology to be as rapid as expected by the NRC hydrogen committee. Further, one OEM that is aggressively pursuing fuel cell vehicles will probably not be in a position to begin significant commercialization until at least 2020, 5 years later than the target date assumed in the hydrogen study.

The task also called for the NRC hydrogen committee to consider whether other technologies could achieve significant CO_2 and oil reductions by 2020. The NRC hydrogen committee considered improvements to spark-ignition (SI) engines, compression-ignition (CI) engines, vehicle transmissions, and hybrid vehicle technologies as well as reductions in weight and other vehicle load reductions. Improvements also could come in the form of reductions in weight and similar improvements. The technical improvements that can be applied to SI engines include variable valve timing and lift, camless valve actuation, cylinder deactivation, the use of gasoline direct injection with turbocharging, and intelligent start-stop, which involves engine shutoff when the vehicle idles. Improvements in vehicle transmissions include the use of conventional 6/7/8-speed automatic transmissions and automated manual transmissions. This report repeats an estimate from Duleep (2007) that combining the projections for improvements in the engine, transmission, weight, parasitic loss (including friction losses, rolling resistance, and air drag), accessories, and idle-stop components could reduce fuel consumption in 2015 by 21 to 29 percent relative to today's vehicles and in 2025 by 31 to 37 percent. Table H.1 shows the improvements estimated for SI engines attributable to these approaches. The NRC hydrogen report also quotes studies by Heywood and colleagues at Massachusetts Institute of Technology (MIT) on the fuel efficiency of light-duty vehicles (Weiss et al., 2000; Heywood, 2007; Kasseris and Heywood, 2007; Kromer and Heywood, 2007). The fuel economy improvements noted in the MIT work result from changes to the engines and transmissions and appropriate reductions in vehicle weight. The MIT work assumes that the improvements are aimed entirely at reducing fuel consumption. Table H.2 shows the improvements in fuel economy compared to a 2005 SI engine vehicle that MIT estimates could be achieved by 2030, although the NRC hydrogen committee assumed that these levels of fuel economy would not be available as quickly.

TABLE H.1 Potential Reductions in Fuel Consumption (gallons per mile) for Spark-Ignition Vehicles Expected from Advances in Conventional Vehicle Technology by Category, Projected to 2025

	2006-2015 (%)	2016-2025 (%)
Engine and transmission	12-16	18-22
Weight, drag, and tire loss reduction	6-9	10-13
Accessories	2-3	3-4
Intelligent start-stop	3-4	3-4

NOTE: Values for 2016-2025 include those of 2006-2015.
SOURCE: Duleep (2007).

TABLE H.2 Comparison of Projected Improvements in Vehicle Fuel Consumption from Advances in Conventional Vehicle Technology

	Fuel Consumption (L/100 km)	Relative to 2005 Gasoline[a]	Relative to 2030 Gasoline[a]	Relative to 2005 Gasoline[b]	Relative to 2030 Gasoline[b]
2005 Gasoline	8.8	1.00			
2005 Diesel	7.4	0.84			
2005 Turbo	7.9	0.9			
2005 Hybrid	5.7	0.65			
2030 Gasoline	5.5	0.63	1.00		
2030 Diesel	4.7	0.53	0.85	0.61	1.00
2030 Turbo	4.9	0.56	0.89	0.45	0.77
2030 Hybrid	3.1	0.35	0.56	0.54	0.88
2030 Plug-in	1.9	0.21	0.34	0.38	0.615

[a]From Kromer and Heywood (2007).
[b]From Weiss et al. (2000).

Although the NRC hydrogen committee acknowledges the potential for hybrids outlined in Kromer and Heywood, it concluded that advances in hybrid technology are more likely to lower the cost of battery packs than to increase fuel economy significantly. This would increase their appeal to consumers relative to conventional vehicles and, thus, their market share (Kromer and Heywood, 2007). To simplify the analysis in the hydrogen report, the NRC hydrogen committee assumed that hybrids reduce fuel consumption a constant 29 percent annually relative to conventional vehicles, which also improve each year. This value is within the range of the potential for power split hybrids in the present report.

Thus, the NRC hydrogen committee judged that hybrid electric vehicles could, if focused on vehicle efficiency, consistently reduce fuel consumption 29 percent relative to comparable evolutionary internal combustion engine vehicles (ICEVs). Although this judgment is conservative compared to that of Kromer and Heywood, it still leads to a 60-mpg average for new spark-ignition hybrids by 2050. This means that hybrid technologies will have reached their greatest fuel consumption reductions by 2009 and that future improvements in hybrid vehicle fuel economy would be primarily attributable to the same technologies that reduce fuel consumption in conventional vehicles. Thus, hybrid vehicles reduce fuel consumption by 2.6 percent per year from 2010 through 2025, 1.7 percent per year in 2025-2035, and 0.5 percent per year between 2035 and 2050, the same as do evolutionary ICEVs.

PLUG-IN HYBRID ELECTRIC REPORT

After the publication of the NRC report *Transitions to Alternative Transportation Technologies—A Focus on Hydrogen* (NRC, 2008), the U.S. Department of Energy asked the Committee on Assessment of Resource Needs for Fuel Cell and Hydrogen Technologies to expand its analysis to include plug-in hybrid electric vehicles. The committee reconvened to examine the issues associated with PHEVs and wrote *Transitions to Alternative Transportation Technologies—Plug-in Hybrid Electric Vehicles* (referred to here as the PHEV report) to that additional task (NRC, 2009).

In accordance with the committee's statement of task, the PHEV report does the following:

- Reviews the current and projected status of PHEV technologies.
- Considers the factors that will affect how rapidly PHEVs would enter the marketplace, including the interface with the electric transmission and distribution system.
- Determines a maximum practical penetration rate for PHEVs consistent with the time frame of the 2008 Hydrogen Report and other factors considered in that report.
- Incorporates PHEVs into the models used in the 2008 Hydrogen Report to estimate the costs and impacts on petroleum consumption and carbon dioxide (CO_2) emissions.

As in this report, the PHEV report considered two types of PHEVs, a PHEV10 with an all-electric range of 10 miles and a PHEV40 with an all-electric range of 40 miles. Both reports use the same architectures as this committee, which include a spark-ignited internal combustion engine, two electrical machines, power electronics, and a Li-ion battery. Only the first task relates to our report, and comparing the two, it is necessary to separate the current technology status and the projections. The assessment of current technologies in the PHEV report is in close agreement with the assessment of this committee. Both discuss the different battery chemistries and the advantages and problems of each and point out how PHEVs differ from batteries for

HEVs, because the critical parameter is the energy available as opposed to the power needs. The discussion of power electronics and motors and generators within the PHEV report again generally parallels what is in this report. There are some differences in terms of the technological needs. For example, the PHEV report assumes that liquid cooling is assumed to be required for the PHEV40 battery packs whereas this report assumes air cooling will be sufficient.

The PHEV report was required to project and analyze the technology costs to 2050, while this report stopped at 2025. The methodology used is similar, and in both cases the costs were built up by adding the costs of the new components needed compared to an internal combustion engine vehicle. Costs were deducted for components such as engine simplification and the elimination of the transmission. The information was obtained from OEMs and suppliers in a similar way. For the PHEV10 the cost estimates in this report are within 5 percent of those in the PHEV report and within 3 percent for the 2020 to 2030 time frame. For the PHEV 40 the committee's costs are significantly lower: by 45 percent for current costs and 42 percent for the 2020 to 2030 time frame. In view of the uncertainties of actual costs and how these would translate as retail price equivalents, the difference can be attributed to a difference in professional judgment.

A more difficult question is the rate at which the cost of the battery will come down, and what makes projections even harder is the injection of a substantial amount of capital by the administration and the enthusiasm of investors. Basically there are two ways of looking at future cost declines:

- People making these very large investments in both vehicles and lithium ion batteries must expect the market to take off. Since the success of vehicle electrification depends on reductions in the price of battery by factors of two or three, investors and the administration must be optimistic that large cost reductions will occur.
- A more pessimistic perspective is that lithium ion is a well-developed technology with billions of individual cells being produced.

How much improvement can one realistically expect in the 10-year horizon of the report? Both reports take a fairly conservative viewpoint in terms of the cost reductions of batteries over time and, taking into account developments in the last year, both reports may turn out to be overly conservative.

AEF ENERGY EFFICIENCY PANEL REPORT

The America's Energy Future Energy Efficiency Panel examined the technical potential for reducing energy demand by improving efficiency in transportation, lighting, heating, cooling, and industrial processes using existing technologies, technologies developed but not yet widely utilized, and prospective technologies. In its report, *Real Prospects for Energy Efficiency in the United States* (NAS-NAE-NRC, 2010), the panel estimated the current contributions and future potential of existing technologies. In addition, the energy efficiency panel estimated the potential for new technologies that could begin to be commercially deployed in the next decade, the associated impacts of these technologies, and the projected costs per unit of reduction in energy demand. The panel's work on light-duty vehicles is summarized in the following sections.

Gasoline SI Engine

Gasoline SI engine efficiency improvements contemplated by the NRC energy efficiency panel included engine friction reduction, smart cooling systems, variable valve timing (VVT), two- and three-step variable valve lift (VVL), cylinder deactivation, direct injection (DI), and turbocharging with engine downsizing. Most of these are already in low-volume production, and all could be deployed in large volumes in the next decade. In 15 to 20 years, technologies such as camless valve actuation, continuous variable valve lift (CVVL), and homogeneous-charge compression ignition (HCCI) could be deployed. The conclusions hoped for in connection with the deployment of camless valve actuation and HCCI are more optimistic than those anticipated for fuel cells in this report. The NRC energy efficiency panel survey shows the above technologies have the potential to reduce vehicle fuel consumption by 10 to 15 percent by 2020 and by an additional 15 to 20 percent by 2030 (EEA, 2007; Kasseris and Heywood, 2007; Ricardo, Inc., 2008; and NRC, 2008a).

Diesel CI Engine

Owing to high compression ratios and reduced pumping losses, turbocharged diesel engines offer a 20 to 25 percent efficiency advantage over gasoline SI engines when adjusted for the higher energy density of diesel fuel. The primary efficiency improvements in CI engines are likely to come from increased power density, improved engine system management, more sophisticated fuel injection systems, and improved combustion processes. New exhaust after-treatment technologies are emerging that reduce emissions of particulate matter and oxides of nitrogen to levels comparable to those of SI engines. One challenge for diesel engines noted by the NRC energy efficiency panel is the added costs and fuel economy penalties associated with the aftertreatment systems for reducing these emissions (Bandivadekar et al., 2008; Johnson, 2008; Ricardo, Inc., 2008).

Gasoline Hybrid Electric Vehicle

The primary efficiency benefits of a gasoline hybrid electric vehicle (HEV) noted by the NRC energy efficiency panel are realized by eliminating idling, including regenerative braking, downsizing the engine, and operating at more efficient engine conditions than current SI engines.

The NRC energy efficiency panel classifies hybrids on how well their electric motor and generator function. Belt-driven starter-generator systems eliminate engine idle to reduce fuel consumption by 4 to 6 percent. Integrated starter-generator systems that recover energy from regenerative braking, along with the start-stop function, can achieve a fuel consumption reduction of 10 to 12 percent. A parallel full hybrid with power assist, such as Honda's integrated motor assist system, can reduce fuel consumption by more than 20 to 25 percent, whereas more complex systems using two motors such as Toyota's hybrid synergy drive can reduce fuel consumption more than 30 percent. Some diesel HEV prototypes are now being developed. Diesel HEVs could be 10 percent more efficient than an equivalent gasoline hybrid, which translates to a 20 percent lower diesel fuel consumption when greater fuel density is factored in. A diesel HEV would be significantly more expensive than a gasoline HEV.

Vehicle Technologies and Transmission Improvements

The NRC energy efficiency panel notes that reducing the vehicle weight by 10 percent is commonly thought to reduce fuel consumption by 5 to 7 percent when accompanied by appropriate engine downsizing to maintain constant performance. Preliminary vehicle simulation results suggest that the relative benefits of weight reduction may be smaller for some types of hybrid vehicles (An and Santini, 2004; Wohlecker et al., 2007). In a conventional vehicle the energy used to accelerate the mass is mostly dissipated in the brakes, while in a hybrid a significant fraction of this braking energy is recovered, sent back to the battery, and reused. Thus weight reduction in hybrid vehicles has a much smaller effect on reducing fuel consumption than such reduction in non-hybrid vehicles. Additional weight reduction can be achieved by vehicle redesign and downsizing as well as by substituting lighter-weight materials in vehicle construction. For example, downsizing a passenger car by one EPA size-class can reduce vehicle weight by approximately 10 percent (Cheah et al., 2007). Additional sources of fuel consumption benefits noted by the NRC energy efficiency panel are from improvements in tires. A recent NRC report on tires and passenger vehicle fuel economy (NRC, 2006) agrees with estimates in the literature (Schuring and Futamura, 1990) that the vehicle fuel consumption will be reduced by 1 or 2 percent for a reduction of 0.001 in the coefficient of rolling resistance of passenger tires—equivalent to a 10 percent reduction in overall rolling resistance. The NRC energy efficiency panel also discussed transmission efficiency improvements likely in the next 10 to 20 years through an increase in the number of gears and through improvements in bearings, gears, sealing elements, and the hydraulic system. Table H.3 lists the efficiency improvements considered by the NRC energy efficiency panel that can be expected from different transmission systems in this time frame. Note that while a continuously variable transmission (CVT) allows the engine to operate near its maximum efficiency, the current estimates of CVT efficiency are lower than the corresponding efficiencies of 6- or 7-speed automatic transmissions. CVTs have been in low-volume production for well over a decade.

TABLE H.3 Expected Transmission System Efficiency Improvements

Transmission	Efficiency (%)
Current automatic transmission (4- and 5-speed)	84-89
Automatic transmission (6- or 7-speed)	93-95
Dual-clutch transmission (wet clutch)	86-94
Dual-clutch transmission (dry clutch)	90-95
Continuously variable transmission	87-90

SOURCE: NAS-NAE-NRC (2010), quoting Ricardo, Inc. (2008) and EEA (2007).

Summary and Costs of Potential Light-Duty Vehicle Efficiency Improvements

Table H.4 shows plausible levels of petroleum reduction potential through vehicle technology improvements estimated by the NRC energy efficiency panel. The NRC energy efficiency panel developed its estimates from a number of sources (An and Santini, 2004; Wohlecker et al., 2007; Cheah et al., 2007; NPC, 2007; and NRC, 2004). The estimates shown in Table H.4 assume that vehicle size and performance, such as the power-to-weight ratio and acceleration, are kept constant at today's levels. The evolutionary improvements briefly outlined above and discussed in more detail in the NRC energy efficiency panel report can reduce the fuel consumption of a gasoline ICE vehicle by up to 35 percent in the next 25 years. The diesel engine currently offers a 20 percent reduction in fuel consumption over a gasoline engine and, while the diesel engine will continue to evolve, the gap between gasoline and diesel vehicle fuel consumption is likely to narrow to a 15 percent improvement. Hybrid vehicles (including PHEVs) have a greater potential for improvement and can deliver deeper reductions in vehicle fuel consumption, although they continue to depend on petroleum (or alternative liquid fuels, such as biofuels). Battery electric vehicles (BEVs) and fuel cell vehicles (FCVs) are two longer-term technologies.

The cost estimates developed by the NRC energy efficiency panel shown in Table H.4 represent the approximate incremental retail price of future vehicle systems, including emissions control costs, compared to a 2005 baseline gasoline ICE vehicle (NHTSA, 2007; EEA, 2007; Bandivadekar et al., 2008). The first column shown is for a midsize car; the second column is for a typical pickup truck or SUV. These retail prices are based on the costs associated with producing a vehicle at the manufacturing plant gate. To account for distribution costs and manufacturer and dealer profit margins, production costs were multiplied by a factor of

TABLE H.4 Plausible Reductions in Petroleum Use from Vehicle Efficiency Improvements over the Next 25 Years and Estimated Incremental Cost of Advanced Vehicles Relative to a Baseline 2005 Standard Gasoline Vehicle

	Petroleum Consumption (gasoline equivalent)		Incremental Retail Price (2007 dollars)	
Propulsion System	Relative to Current Gasoline ICE	Relative to 2035 Gasoline ICE	Car	Light Truck
Current gasoline	1	—	0	0
Current diesel	0.8	—	1,700	2,100
Current HEV	0.75	—	4,900	6,300
2035 gasoline	0.65	1	2,000	2,400
2035 diesel	0.55	0.85	3,600	4,500
2035 HEV	0.4	0.6	4,500	5,500
2035 PHEV	0.2	0.3	7,800	10,500
2035 BEV	None	—	16,000	24,000
2035 hydrogen FCV	None	—	7,300	10,000

NOTE: BEV, battery electric vehicle; FCV, fuel cell vehicle; HEV, hybrid electric vehicle; ICE, internal combustion engine.
SOURCE: Report from the NRC Panel on Energy Efficiency (NAS-NAE-NRC, 2010) quoting Bandivadekar et al. (2008).

1.4 to provide representative retail price estimates (Evans, 2008). The timescales indicated for these future technology vehicles are not precise. The rate of price reduction will depend on the deployment rate (Bandivadekar et al., 2008; Evans, 2008).

The results in Table H.4 show that alternative powertrains such as improved gasoline and diesel engines and hybrids entering the fleet today cost from 10 percent to 30 percent more than a current gasoline vehicle. This price difference is estimated to drop to 5 percent to 15 percent in the mid-term future. Longer-term options such as plug-in hybrid and FCVs are estimated to cost between 25 and 30 percent more than a future gasoline vehicle. Battery electric vehicles with standard vehicle performance and size remain costly, approaching double the cost of a future gasoline vehicle. A more plausible market opportunity for BEVs is small city cars with reduced range. However, these also will need significantly improved battery performance and battery costs to become competitive.

Based on the estimates in Table H.4, the NRC energy efficiency panel concludes that evolutionary improvements in gasoline ICE vehicles are likely to prove the most cost-effective way to reduce petroleum consumption. Since these vehicles will be sold in large quantities in the near term, it is critical that their efficiency improvements are directed toward reducing fuel consumption. While the current hybrids appear less competitive than a comparable diesel vehicle, they are likely to become more cost competitive over time. PHEVs, BEVs, and FCVs appear to be more costly alternatives for reducing petroleum consumption and greenhouse gas emissions. Among these three technologies, PHEVs are likely to become available in the near to midterm, whereas BEVs and FCVs are mid- to long-term alternatives.

BIBLIOGRAPHY

An, F., and D. Santini. 2004. Mass impacts on fuel economies of conventional vs. hybrid electric vehicles. SAE Technical Paper 2004-01-0572. SAE International, Warrendale, Pa.

An, F., J.M. DeCicco, and M.H. Ross. 2001. Assessing the Fuel Economy Potential of Light-Duty Vehicles. SAE International, Warrendale, Pa.

Bandivadekar, A.P. 2008. Evaluating the Impact of Advanced Vehicle and Fuel Technologies in the U.S. Light-Duty Vehicle Fleet. Ph.D. thesis. MIT Engineering Systems Division, Cambridge, Mass.

Bandivadekar, A., K. Bodek, L. Cheah, C. Evans, T. Groode, J. Heywood, E. Kasseris, K. Kromer, and M. Weiss. 2008. On the Road in 2035: Reducing Transportation's Petroleum Consumption and GHG Emissions. Massachusetts Institute of Technology (MIT) Laboratory for Energy and the Environment, Cambridge, Mass. July.

Cheah L., C. Evans, A. Bandivadekar, and J. Heywood. 2007. Factor of Two: Halving the Fuel Consumption of New U.S. Automobiles by 2035. LFEE Report 2007-04 RP. MIT Laboratory for Energy and the Environment. Cambridge, Mass: Massachusetts Institute of Technology. Available at http://web.mit.edu/sloan-auto-lab/research/beforeh2/files/cheah_factorTwo.pdf.

Duleep, K. 2007. The Hydrogen Transition and Competing Automotive Technology. Presentation to National Research Council Panel on Fuel Cell Vehicles and Hydrogen on April 18, Washington, D.C.

EEA (Energy and Environmental Analysis, Inc.). 2007. Update for Advanced Technologies to Improve Fuel Economy of Light Duty Vehicles. Draft Final Report. Submitted to U.S. Department of Energy. EEA, Washington, D.C. August.

Evans, C. 2008. Putting Policy in Drive: Coordinating Measures to Reduce Fuel Use and Greenhouse Gas Emissions from U.S. Light-Duty Vehicles. M.S. Thesis, Technology and Policy Program. June. Massachusetts Institute of Technology, Cambridge.

Heywood, J.B. 2007. Strategies to Reduce Transportation Petroleum Consumption and Greenhouse Gas Emissions. Presentation at FreedomCAR and Fuels Program Meeting.

Johnson, T. 2008. Diesel Emission Control Technology in Review. SAE Technical Paper 2008-01-0069. Society of Automotive Engineers, Warrendale, Pa. April.

Kasseris, E.P., and J.B. Heywood. 2007. Comparative Analysis of Automotive Powertrain Choices for the Next 25 Years. SAE International, Warrendale, Pa.

Kromer, M.A., and J.B. Heywood. 2007. Electric Powertrains: Opportunities and Challenges in the U.S. Light-Duty Vehicle Fleet. MIT Laboratory for Energy and the Environment, Cambridge, Mass.

Lutsey, N. 2008. Prioritizing Climate Change Mitigation Alternatives: Comparing Transportation Technologies to Options in Other Sectors. Ph.D. dissertation. University of California, Davis. May.

Lutsey, N., and D. Sperling. 2005. Energy efficiency, fuel economy, and policy implications. Transportation Research Record 1941:8-25.

NAS-NAE-NRC (National Academy of Sciences-National Academy of Engineering-National Research Council). 2010. Real Prospects for Energy Efficiency in the United States. The National Academies Press, Washington, D.C.

NPC (National Petroleum Council). 2007. Technologies for Transportation Efficiency. Topic paper of the Transportation Efficiency Subgroup. Council Committee on Global Oil and Gas. NPC, Washington, D.C. July.

NRC (National Research Council). 2001. Review of the Research Program of the Partnership for a New Generation of Vehicles, Seventh Report. The National Academies Press, Washington, D.C.

NRC. 2004. The Hydrogen Economy: Opportunities, Costs, Barriers, and R&D Needs. The National Academies Press, Washington, D.C.

NRC. 2005. Review of the Research Program of the FreedomCAR and Fuel Partnership, First Report. The National Academies Press, Washington, D.C.

NRC. 2006. Tires and Passenger Vehicle Fuel Economy: Informing Consumers, Improving Performance—Special Report 286. The National Academies Press, Washington, D.C.

NRC. 2008a. Transitions to Alternative Transportation Technologies—A Focus on Hydrogen. The National Academies Press, Washington, D.C.

NRC. 2008b. Review of the Research Program of the FreedomCAR and Fuel Partnership: Second Report. The National Academies Press, Washington, D.C.

NRC. 2009. Transitions to Alternative Transportation Technologies—Plug-in Hybrid Electric Vehicles. The National Academies Press, Washington, D.C.

Ricardo, Inc. 2008. A Study of Potential Effectiveness of Carbon Dioxide Reducing Vehicle Technologies. Prepared for EPA Office of Transportation and Air Quality. January. Available at http://www.epa.gov/OMS/technology/420r08004a.pdf. Accessed July 10, 2009.

Schuring, D., and S. Futamura. 1990. Rolling loss of pneumatic highway tires in the eighties. Rubber Chemistry and Technology 62(3):315-367.

Weiss, M.A., et al. 2000. On the Road in 2020: A Life-Cycle Analysis of New Automobile Technologies. Massachusetts Institute of Technology, Cambridge, Mass.

Wohlecker, R., M. Johannaber, and M. Espig. 2007. Determination of weight elasticity of fuel economy for ICE, hybrid and fuel cell vehicles. SAE Technical Paper 2007-01-0343. SAE International, Warrendale, Pa.

I

Results of Other Major Studies

Tables I.1 through I.8, which indicate the costs and fuel consumption benefits from other major studies, are included here to facilitate the comparison to other sources of technology cost and effectiveness. However, the reader is encouraged to look at the original source material to gain a better understanding of the different assumptions made in each study. For example, some sources consider incremental benefits, while others do not. Certain items, such as improved accessories, may include different technologies, which makes an apples-to-apples comparison difficult. Retail price equivalent factors also vary from source to source, reinforcing the need to review the original materials as well as the tables.

REFERENCES

DOT/NHTSA (U.S. Department of Transportation/National Highway Traffic Safety Administration). 2009. Average Fuel Economy Standards, Passenger Cars and Light Trucks, Model Year 2001: Final Rule. 49 CFR Parts 523, 531, 533, 534, 536, and 537, Docket No. NHTSA-2009-0062, RIN 2127-AK29, March 23. Washington D.C.

EEA (Energy and Environmental Analysis, Inc.). 2007. Technologies to Improve Light-Duty Vehicle Fuel Economy. Draft report to the National Research Council Committee on Fuel Economy of Light-Duty Vehicles. EEA, Arlington, Va. September.

EPA (U.S. Environmental Protection Agency). 2008. EPA Staff Technical Report: Cost and Effectiveness Estimates of Technologies Used to Reduce Light-Duty Vehicle Carbon Dioxide Emissions. EPA420-R-08-008. EPA, Washington, D.C.

Martec Group, Inc. 2008. Variable Costs of Fuel Economy Technologies. Prepared for Alliance of Automobile Manufacturers. June 1; amended September 26 and December 10.

NESCCAF (Northeast States Center for a Clean Air Future). 2004. Reducing Greenhouse Gas Emissions from Light-Duty Motor Vehicles. March.

NRC (National Research Council). 2002. Effectiveness and Impact of Corporate Average Fuel Economy (CAFE) Standards. National Academy Press, Washington, D.C.

Ricardo, Inc. 2008. A Study of Potential Effectiveness of Carbon Dioxide Reducing Vehicle Technologies. Prepared for the U.S. Environmental Protection Agency, EPA420-R-08-004, Contract No. EP-C-06-003, Work Assignment No. 1-14, Ricardo, Inc., Ann Arbor, Mich.

Sierra Research. 2008. Basic Analysis of the Cost and Long-Term Impact of the Energy Independence and Security Act Fuel Economy Standards. Sierra Research, Sacramento, Calif. April 24.

TABLE I.1 Technology Effectiveness, Incremental (Percent) Fuel Consumption Benefit from DOT/NHTSA (2009)

NHTSA - 2011 Rule

Technologies		Perf. SubcompactCar I4				Perf. Compact Car V6				Perf. Midsize Car V6				Perf. Large Car V8			
		Incremental Value		Net Value		Incremental Value		Net Value		Incremental Value		Net Value		Incremental Value		Net Value	
Spark Ignition Techs	Abbreviation	Low	High	Low	High	Low	High	Low	High	Low	High	Low	High	Low	High	Low	High
Low Friction Lubricants	LUB	0.5		-	-	0.5		-	-	0.5		-	-	0.5		-	-
Engine Friction Reduction	EFR	1.0	2.0	-	-	1.0	2.0	-	-	1.0	2.0	-	-	1.0	2.0	-	-
VVT - Coupled Cam Phasing (CCP), SOHC	CCP	1.0	3.0	-	-	1.0	3.0	-	-	1.0	3.0	-	-	1.0	3.0	-	-
Discrete Variable Valve Lift (DVVL), SOHC	DVVL	1.0	3.0	-	-	1.0	3.0	-	-	1.0	3.0	-	-	1.0	3.0	-	-
Cylinder Deactivation, SOHC	DEAC	-	-	-	-	2.5	3.0	-	-	2.5	3.0	-	-	2.5	3.0	-	-
VVT - In take Cam Phasing (ICP)	ICP	1.0	2.0	-	-	1.0	2.0	-	-	1.0	2.0	-	-	1.0	2.0	-	-
VVT - Dual Cam Phasing (DCP)	DCP	2.0	3.0	-	-	2.0	3.0	-	-	2.0	3.0	-	-	2.0	3.0	-	-
Discrete Variable Valve Lift (DVVL), DOHC	DVVL	1.0	3.0	-	-	1.0	3.0	-	-	1.0	3.0	-	-	1.0	3.0	-	-
Continuously Variable Valve Lift (CVVL)	CVVL	1.5	3.5	-	-	1.5	3.5	-	-	1.5	3.5	-	-	1.5	3.5	-	-
Cylinder Deactivation, OHV	DEAC	-	-	-	-	3.9	5.5	-	-	3.9	5.5	-	-	3.9	5.5	-	-
VVT - Coupled Cam Phasing (CCP), OHV	CCP	1.0	1.5	-	-	1.0	1.5	-	-	1.0	1.5	-	-	1.0	1.5	-	-
Discrete Variable Valve Lift (DVVL), OHV	DVVL	0.5	2.6	-	-	0.5	2.6	-	-	0.5	2.6	-	-	0.5	2.6	-	-
Conversion to DOHC with DCP	CDOHC	1.0	2.6	-	-	1.0	2.6	-	-	1.0	2.6	-	-	1.0	2.6	-	-
Stoichiometric Gasoline Direct Injection (GDI)	SGDI	1.9	2.9	5.0	13.0	1.9	2.9	7.0	14.0	1.9	2.9	7.0	14.0	1.9	2.9	7.0	14.0
Turbocharging and Downsizing	TRBDS	4.5	5.2	11.0	17.0	2.1	2.2	11.0	17.0	2.1	2.2	11.0	17.0	2.1	2.2	11.0	17.0
Diesel Techs																	
Conversion to Diesel	DSL	15.0	15.3	21.2	25.9	12.3	13.1	21.2	25.9	11.1	12.0	20.2	24.9	11.1	12.0	20.2	24.9
Conversion to Diesel following TRBDS	DSL	6.6	7.7	21.2	25.9	6.6	7.7	21.2	25.9	5.3	6.5	20.2	24.9	5.3	6.5	20.2	24.9
Conversion to Advanced Diesel	ADSL	-	-	-	-	-	-	-	-	-	-	-	-	-	-	-	-
Electrification/Accessory Techs																	
Electric Power Steering (EPS)	EPS	1.0	2.0	-	-	1.0	2.0	-	-	1.0	2.0	-	-	1.0	2.0	-	-
Improved Accessories	IACC	1.0	2.0	-	-	1.0	2.0	-	-	1.0	2.0	-	-	1.0	2.0	-	-
12V BAS Micro-Hybrid	MHEV	1.0	2.9	-	-	1.0	2.9	-	-	3.4	4.0	-	-	3.4	4.0	-	-
Higher Voltage/Improved Alternator	HVIA	0.2	0.9	-	-	0.2	0.9	-	-	0.2	0.6	-	-	0.2	0.6	-	-
Integrated Starter Generator	ISG	1.8	2.6	-	-	1.8	2.6	-	-	1.8	1.9	-	-	1.8	2.6	-	-
Transmission Techs																	
Continuously Variable Transmission (CVT)	CVT	0.7	2.0	-	-	0.7	2.0	-	-	0.7	2.0	-	-	-	-	-	-
6/7/8-Speed Auto. Trans. with Improved Internals	NAUTO	1.4	3.4	-	-	1.4	3.4	-	-	1.4	3.4	-	-	1.4	3.4	-	-
Dual Clutch Transmission (DCT)	DCT	2.7	4.1	5.5	9.7	2.7	4.1	5.5	9.7	2.7	4.1	5.5	9.7	2.7	4.1	5.5	9.7
Hybrid Techs																	
Power Split Hybrid	PSHEV	14.6	15.2	21.0	26.5	14.6	15.0	21.0	26.5	13.1	14.6	21.0	26.5	13.7	15.7	21.0	26.5
2-Mode Hybrid	2MHEV	-	-	-	-	-	-	-	-	-	-	-	-	-	-	-	-
Plug-in hybrid	PHEV	62.0	65.0	65.0	69.5	62.0	65.0	65.0	69.5	61.0	65.0	65.0	69.5	-	-	-	-
Vehicle Techs																	
Mass Reduction - 1%	MR1	-	-	-	-	-	-	-	-	-	-	-	-	-	-	-	-
Mass Reduction - 2%	MR2	-	-	-	-	-	-	-	-	-	-	-	-	-	-	-	-
Mass Reduction - 5%	MR5	-	-	-	-	-	-	-	-	-	-	-	-	-	-	-	-
Mass Reduction - 10%	MR10	-	-	-	-	-	-	-	-	-	-	-	-	-	-	-	-
Mass Reduction - 20%	MR20	-	-	-	-	-	-	-	-	-	-	-	-	-	-	-	-
Low Rolling Resistance Tires	ROLL	-	-	-	-	-	-	-	-	-	-	-	-	-	-	-	-
Low Drag Brakes	LDB	-	-	-	-	-	-	-	-	-	-	-	-	-	-	-	-
Secondary Axle Disconnect	SAX	1.0	1.5	-	-	1.0	1.5	-	-	1.0	1.5	-	-	1.0	1.5	-	-
Aero Drag Reduction 10%	AERO	2.0	3.0	-	-	2.0	3.0	-	-	2.0	3.0	-	-	2.0	3.0	-	-

continued

TABLE I.1 Continued

NHTSA - 2011 Rule

Technologies		Subcompact Car				Compact Car				Midsize Car				Large Car			
		I4				I4				I4				V6			
		Incremental Value		Net Value		Incremental Value		Net Value		Incremental Value		Net Value		Incremental Value		Net Value	
Spark Ignition Techs	Abbreviation	Low	High	Low	High	Low	High	Low	High	Low	High	Low	High	Low	High	Low	High
Low Friction Lubricants	LUB	0.5		-	-	0.5		-	-	0.5		-	-	0.5		-	-
Engine Friction Reduction	EFR	1.0	2.0	-	-	1.0	2.0	-	-	1.0	2.0	-	-	1.0	2.0	-	-
VVT- Coupled Cam Phasing (CCP), SOHC	CCP	1.0	3.0	-	-	1.0	3.0	-	-	1.0	3.0	-	-	1.0	3.0	-	-
Discrete Variable Valve Lift (DVVL), SOHC	DVVL	1.0	3.0	-	-	1.0	3.0	-	-	1.0	3.0	-	-	1.0	3.0	-	-
Cylinder Deactivation, SOHC	DEAC	-	-	-	-	-	-	-	-			-	-	2.5	3.0	-	-
VVT - In take Cam Phasing (ICP	ICP	1.0	2.0	-	-	1.0	2.0	-	-	1.0	2.0	-	-	1.0	2.0	-	-
VVT - Dual Cam Phasing (DCP)	DCP	2.0	3.0	-	-	2.0	3.0	-	-	2.0	3.0	-	-	2.0	3.0	-	-
Discrete Variable Valve Lift (DVVL), DOHC	DVVL	1.0	3.0	-	-	1.0	3.0	-	-	1.0	3.0	-	-	1.0	3.0	-	-
Continuously Variable Valve Lift (CVVL	CVVL	1.5	3.5	-	-	1.5	3.5	-	-	1.5	3.5	-	-	1.5	3.5	-	-
Cylinder Deactivation, OHV	DEAC	-	-	-	-	-	-	-	-	-	-	-	-	3.9	5.5	-	-
VVT - Coupled Cam Phasing (CCP), OHV	CCP	1.0	1.5	-	-	1.0	1.5	-	-	1.0	1.5	-	-	1.0	1.5	-	-
Discrete Variable Valve Lift (DVVL), OHV	DVVL	0.5	2.6	-	-	0.5	2.6	-	-	0.5	2.6	-	-	0.5	2.6	-	-
Conversion to DOHC with DCP	CDOHC	1.0	2.6	-	-	1.0	2.6	-	-	1.0	2.6	-	-	1.0	2.6	-	-
Stoichiometric Gasoline Direct Injection (GDI	SGDI	1.9	2.9	5.0	13.0	1.9	2.9	5.0	13.0	1.9	2.9	5.0	13.0	1.9	2.9	5.0	13.0
Turbocharging and Downsizing	TRBDS	4.5	5.2	11.0	17.5	4.5	5.2	11.0	17.5	4.5	5.2	11.0	17.5	2.1	2.2	11.0	17.5
Diesel Techs																	
Conversion to Diesel	DSL	15.0	15.3	21.2	25.9	15.0	15.3	21.2	25.9	13.8	14.2	20.2	24.9	11.1	12.0	20.2	24.9
Conversion to Diesel following TRBDS	DSL	6.6	7.7	21.2	25.9	6.6	7.7	21.2	25.9	5.3	6.5	20.2	24.9	5.3	6.5	20.2	24.9
Conversion to Advanced Diesel	ADSL	-	-	-	-	-	-	-	-	-	-	-	-	-	-	-	-
Electrification/Accessory Techs																	
Electric Power Steering (EPS)	EPS	1.0	2.0	-	-	1.0	2.0	-	-	1.0	2.0	-	-	1.0	2.0	-	-
Improved Accessories	IACC	1.0	2.0	-	-	1.0	2.0	-	-	1.0	2.0	-	-	1.0	2.0	-	-
12V BAS Micro-Hybrid	MHEV	1.0	2.9	-	-	1.0	2.9	-	-	3.4	4.0	-	-	3.4	4.0	-	-
Higher Voltage/Improved Alternator	HVIA	0.2	0.9	-	-	0.2	0.9	-	-	0.2	0.6	-	-	0.2	0.6	-	-
Integrated Starter Generator	ISG	5.7	6.5	-	-	5.7	6.5	-	-	5.7	6.5	-	-	5.7	6.5	-	-
Transmission Techs																	
Continuously Variable Transmission (CVT	CVT	0.7	2.0	-	-	0.7	2.0	-	-	0.7	2.0	-	-	0.7	2.0	-	-
6/7/8-Speed Auto. Trans. with Improved Internal	NAUTO	1.4	3.4	-	-	1.4	3.4	-	-	1.4	3.4	-	-	1.4	3.4	-	-
Dual Clutch Transmission (DCT)	DCT	5.5	7.5	8.2	12.9	5.5	7.5	8.2	12.9	2.7	4.1	5.5	9.7	2.7	4.1	5.5	9.7
Hybrid Techs																	
Power Split Hybrid	PSHEV	13.5	13.9	23.0	28.5	13.5	13.9	23.0	28.5	11.8	12.8	23.0	28.5	11.8	12.8	23.0	28.5
2-Mode Hybrid	2MHEV	-	-	-	-	-	-	-	-	-	-	-	-	-	-	-	-
Plug-in hybrid	PHEV	61.0	63.0	65.0	69.5	61.0	63.0	65.0	69.5	60.0	63.0	65.0	69.5	-	-	-	-
Vehicle Techs																	
Mass Reduction - 1%	MR1	-	-	-	-	-	-	-	-	-	-	-	-	-	-	-	-
Mass Reduction - 2%	MR2	-	-	-	-	-	-	-	-	-	-	-	-	-	-	-	-
Mass Reduction - 5%	MR5	-	-	-	-	-	-	-	-	-	-	-	-	-	-	-	-
Mass Reduction - 10%	MR10	-	-	-	-	-	-	-	-	-	-	-	-	-	-	-	-
Mass Reduction - 20%	MR20	-	-	-	-	-	-	-	-	-	-	-	-	-	-	-	-
Low Rolling Resistance Tires	ROLL	1.0	2.0	-	-	1.0	2.0	-	-	1.0	2.0	-	-	1.0	2.0	-	-
Low Drag Brakes	LDB	-	-	-	-	-	-	-	-	-	-	-	-	-	-	-	-
Secondary Axle Disconnec	SAX	1.0	1.5	-	-	1.0	1.5	-	-	1.0	1.5	-	-	1.0	1.5	-	-
Aero Drag Reduction 10%	AERO	2.0	3.0	-	-	2.0	3.0	-	-	2.0	3.0	-	-	2.0	3.0	-	-

continued

TABLE I.1 Continued

NHTSA - 2011 Rule

Technologies		Minivan LT V6 Incremental Value		Minivan LT V6 Net Value		Small LT I4 Incremental Value		Small LT I4 Net Value		Midsize LT V6 Incremental Value		Midsize LT V6 Net Value		Large LT V8 Incremental Value		Large LT V8 Net Value	
Spark Ignition Techs	Abbreviation	Low	High	Low	High	Low	High	Low	High	Low	High	Low	High	Low	High	Low	High
Low Friction Lubricants	LUB	0.5		-	-	0.5		-	-	0.5		-	-	0.5		-	-
Engine Friction Reduction	EFR	1.0	2.0	-	-	1.0	2.0	-	-	1.0	2.0	-	-	1.0	2.0	-	-
VVT- Coupled Cam Phasing (CCP), SOHC	CCP	1.0	3.0	-	-	1.0	3.0	-	-	1.0	3.0	-	-	1.0	3.0	-	-
Discrete Variable Valve Lift (DVVL), SOHC	DVVL	1.0	3.0	-	-	1.0	3.0	-	-	1.0	3.0	-	-	1.0	3.0	-	-
Cylinder Deactivation, SOHC	DEAC	2.5	3.0	-	-	-	-	-	-	2.5	3.0	-	-	2.5	3.0	-	-
VVT - In take Cam Phasing (ICP)	ICP	1.0	2.0	-	-	1.0	2.0	-	-	1.0	2.0	-	-	1.0	2.0	-	-
VVT - Dual Cam Phasing (DCP)	DCP	2.0	3.0	-	-	2.0	3.0	-	-	2.0	3.0	-	-	2.0	3.0	-	-
Discrete Variable Valve Lift (DVVL), DOHC	DVVL	1.0	3.0	-	-	1.0	3.0	-	-	1.0	3.0	-	-	1.0	3.0	-	-
Continuously Variable Valve Lift (CVVL)	CVVL	1.5	3.5	-	-	1.5	3.5	-	-	1.5	3.5	-	-	1.5	3.5	-	-
Cylinder Deactivation, OHV	DEAC	3.9	5.5	-	-	-	-	-	-	3.9	5.5	-	-	3.9	5.5	-	-
VVT - Coupled Cam Phasing (CCP), OHV	CCP	1.0	1.5	-	-	1.0	1.5	-	-	1.0	1.5	-	-	1.0	1.5	-	-
Discrete Variable Valve Lift (DVVL), OHV	DVVL	0.5	2.6	-	-	0.5	2.6	-	-	0.5	2.6	-	-	0.5	2.6	-	-
Conversion to DOHC with DCP	CDOHC	1.0	2.6	-	-	1.0	2.6	-	-	1.0	2.6	-	-	1.0	2.6	-	-
Stoichiometric Gasoline Direct Injection (GDI)	SGDI	1.9	2.9	7.0	14.0	1.9	2.9	4.5	13.0	1.9	2.9	7.0	14.0	1.9	2.9	7.0	14.0
Turbocharging and Downsizing	TRBDS	2.1	2.2	11.0	17.5	4.5	5.2	11.0	17.5	2.1	2.2	11.0	17.5	2.1	2.2	11.0	17.5
Diesel Techs		-	-	-	-	-	-	-	-	-	-	-	-	-	-	-	-
Conversion to Diesel	DSL	11.1	12.0	20.2	24.9	13.8	14.2	20.2	24.9	9.9	12.0	20.2	23.9	10.0	10.9	19.2	23.9
Conversion to Diesel following TRBDS	DSL	5.3	6.5	20.2	24.9	5.3	6.5	20.2	24.9	4.0	6.5	20.2	23.9	4.0	5.3	19.2	23.9
Conversion to Advanced Diesel	ADSL	-	-	-	-	-	-	-	-	-	-	-	-	-	-	-	-
Electrification/Accessory Techs		-	-	-	-	-	-	-	-	-	-	-	-	-	-	-	-
Electric Power Steering (EPS)	EPS	1.0	2.0	-	-	1.0	2.0	-	-	1.0	2.0	-	-	-	-	-	-
Improved Accessories	IACC	1.0	2.0	-	-	1.0	2.0	-	-	-	-	-	-	-	-	-	-
12V BAS Micro-Hybrid	MHEV	3.4	4.0	-	-	1.0	2.9	-	-	3.4	4.0	-	-	-	-	-	-
Higher Voltage/Improved Alternator	HVIA	0.2	0.6	-	-	0.2	0.9	-	-	0.2	0.6	-	-	-	-	-	-
Integrated Starter Generator	ISG	5.7	6.5	-	-	5.7	6.5	-	-	5.7	6.5	-	-	-	-	-	-
Transmission Techs		-	-	-	-	-	-	-	-	-	-	-	-	-	-	-	-
Continuously Variable Transmission (CVT)	CVT	0.7	2.0	-	-	0.7	2.0	-	-	-	-	-	-	-	-	-	-
6/7/8-Speed Auto. Trans. with Improved Internals	NAUTO	1.4	3.4	-	-	1.4	3.4	-	-	1.4	3.4	-	-	1.4	3.4	-	-
Dual Clutch Transmission (DCT)	DCT	2.7	4.1	5.5	9.7	2.7	4.1	5.5	9.7	2.7	4.1	5.5	9.7	2.7	4.1	5.5	9.7
Hybrid Techs		-	-	-	-	-	-	-	-	-	-	-	-	-	-	-	-
Power Split Hybrid	PSHEV	11.8	12.8	23.0	28.5	13.5	13.9	23.0	28.5	13.3	16.2	23.0	28.5	-	-	-	-
2-Mode Hybrid	2MHEV	-	-	-	-	1.5	4.3	17.5	21.0	0.3	2.9	17.5	21.0	7.9	8.7	13.5	17.0
Plug-in hybrid	PHEV	-	-	-	-	61.0	63.0	65.0	69.5	-	-	-	-	-	-	-	-
Vehicle Techs		-	-	-	-	-	-	-	-	-	-	-	-	-	-	-	-
Mass Reduction - 1%	MR1	-	-	-	-	-	-	-	-	0.4		-	-	0.4		-	-
Mass Reduction - 2%	MR2	-	-	-	-	-	-	-	-	0.4		-	-	0.4		-	-
Mass Reduction - 5%	MR5	-	-	-	-	-	-	-	-	1.0		-	-	1.0		-	-
Mass Reduction - 10%	MR10	-	-	-	-	-	-	-	-	-	-	-	-	-	-	-	-
Mass Reduction - 20%	MR20	-	-	-	-	-	-	-	-	-	-	-	-	-	-	-	-
Low Rolling Resistance Tires	ROLL	1.0	2.0	-	-	1.0	2.0	-	-	1.0	2.0	-	-	-	-	-	-
Low Drag Brakes	LDB	-	-	-	-	-	-	-	-	0.5	1.0	-	-	0.5	1.0	-	-
Secondary Axle Disconnect	SAX	1.0	1.5	-	-	1.0	1.5	-	-	1.0	1.5	-	-	1.0	1.5	-	-
Aero Drag Reduction 10%	AERO	2.0	3.0	-	-	2.0	3.0	-	-	2.0	3.0	-	-	2.0	3.0	-	-

TABLE I.2 Technology Effectiveness, Incremental (Percent) Fuel Consumption Benefit from NRC (2002)

NRC - 2002				
Technologies				
Spark Ignition Techs	Abbreviation	Low	High	**AVG**
Low Friction Lubricants	LUB	1.0		**1.0**
Engine Friction Reduction	EFR	1.0	5.0	**3.0**
VVT- Coupled Cam Phasing (CCP), SOHC	CCP	1.0	2.0	**1.5**
Discrete Variable Valve Lift (DVVL), SOHC	DVVL	1.0	2.0	**1.5**
Cylinder Deactivation, SOHC	DEAC	-	-	**-**
VVT - In take Cam Phasing (ICP)	ICP	2.0	3.0	**2.5**
VVT - Dual Cam Phasing (DCP)	DCP	2.0	3.0	**2.5**
Discrete Variable Valve Lift (DVVL), DOHC	DVVL	1.0	2.0	**1.5**
Continuously Variable Valve Lift (CVVL)	CVVL	1.0	2.0	**1.5**
Cylinder Deactivation, OHV	DEAC	3.0	6.0	**4.5**
VVT - Coupled Cam Phasing (CCP), OHV	CCP	2.0	3.0	**2.5**
Discrete Variable Valve Lift (DVVL), OHV	DVVL	1.0	2.0	**1.5**
Conversion to DOHC with DCP	CDOHC	-	-	**-**
Stoichiometric Gasoline Direct Injection (GDI)	SGDI	-	-	**-**
Turbocharging and Downsizing	TRBDS	5.0	7.0	**6.0**
Diesel Techs		**Non-incremental**		
Conversion to Diesel	DSL	-	-	**-**
Conversion to Diesel following TRBDS	DSL	-	-	**-**
Conversion to Advanced Diesel	ADSL	-	-	**-**
Electrification/Accessory Techs		**Non-incremental**		
Electric Power Steering (EPS)	EPS	1.5	2.5	**2.0**
Improved Accessories	IACC	1.0	2.0	**1.5**
12V BAS Micro-Hybrid	MHEV	-	-	**-**
Higher Voltage/Improved Alternator	HVIA	-	-	**-**
Integrated Starter Generator	ISG	4.0	7.0	**5.5**
Transmission Techs		**Non-incremental**		
Continuously Variable Transmission (CVT)	CVT	4.0	8.0	**6.0**
6/7/8-Speed Auto. Trans. with Improved Internals	NAUTO	1.0	2.0	**1.5**
Dual Clutch Transmission (DCT)	DCT	3.0	5.0	**4.0**
Hybrid Techs		**Non-incremental**		
Power Split Hybrid	PSHEV	-	-	**-**
2-Mode Hybrid	2MHEV	-	-	**-**
Plug-in hybrid	PHEV	-	-	**-**
Vehicle Techs		**Non-incremental**		
Mass Reduction - 1%	MR1	-	-	**-**
Mass Reduction - 2%	MR2	-	-	**-**
Mass Reduction - 5%	MR5	-	-	**-**
Mass Reduction - 10%	MR10	-	-	**-**
Mass Reduction - 20%	MR20	-	-	**-**
Low Rolling Resistance Tires	ROLL	1.0	1.5	**1.3**
Low Drag Brakes	LDB	-	-	**-**
Secondary Axle Disconnect	SAX	-	-	**-**
Aero Drag Reduction 10%	AERO	-	-	**-**

TABLE I.3 Technology Effectiveness, Incremental (Percent) Fuel Consumption Benefit from EPA (2008)

EPA 2008

Technologies		Small Car I4			Large Car V6		
Spark Ignition Techs	Abbreviation	Low	High	**AVG**	Low	High	**AVG**
Low Friction Lubricants	LUB	0.5		**0.5**	0.5		**0.5**
Engine Friction Reduction	EFR	1.0	3	**2.0**	1.0	3	**2.0**
VVT- Coupled Cam Phasing (CCP), SOHC	CCP	3.0	-	**3.0**	4.0	-	**4.0**
Discrete Variable Valve Lift (DVVL), SOHC	DVVL	4.0	-	**4.0**	3.0	-	**3.0**
Cylinder Deactivation, SOHC	DEAC	-	-	**-**	6.0	-	**6.0**
VVT - In take Cam Phasing (ICP)	ICP	2.0	-	**2.0**	1.0	-	**1.0**
VVT - Dual Cam Phasing (DCP)	DCP	3.0	-	**3.0**	4.0	-	**4.0**
Discrete Variable Valve Lift (DVVL), DOHC	DVVL	4.0	-	**4.0**	3.0	-	**3.0**
Continuously Variable Valve Lift (CVVL)	CVVL	5.0	-	**5.0**	6.0	-	**6.0**
Cylinder Deactivation, OHV	DEAC	-	-	**-**	6.0	-	**6.0**
VVT - Coupled Cam Phasing (CCP), OHV	CCP	3.0	-	**3.0**	4.0	-	**4.0**
Discrete Variable Valve Lift (DVVL), OHV	DVVL	4.0	-	**4.0**	4.0	-	**4.0**
Conversion to DOHC with DCP	CDOHC	-	-	**-**	-	-	**-**
Stoichiometric Gasoline Direct Injection (GDI)	SGDI	1.0	2.0	**1.5**	1.0	2.0	**1.5**
Turbocharging and Downsizing[1]	TRBDS	5.0	7.0	**6.0**	5.0	7.0	**6.0**
Diesel Techs		**Non-incremental**		**-**	**-**	**-**	**-**
Conversion to Diesel	DSL	25.0	35.0	**30.0**	30.0	40.0	**35.0**
Conversion to Diesel following TRBDS	DSL	-	-	**-**	-	-	**-**
Conversion to Advanced Diesel	ADSL	-	-	**-**	-	-	**-**
Electrification/Accessory Tech		**Non-incremental**		**-**	**-**	**-**	**-**
Electric Power Steering (EPS)	EPS	1.5	-	**1.5**	1.5	2.0	**1.8**
Improved Accessories[2]	IACC	1.0	2.0	**1.5**	1.0	2.0	**1.5**
12V BAS Micro-Hybrid	MHEV	-	-	**-**	-	-	**-**
Higher Voltage/Improved Alternator	HVIA	-	-	**-**	-	-	**-**
Integrated Starter Generator	ISG	30.0	-	**30.0**	25.0	-	**25.0**
Transmission Tech		**Non-incremental**		**-**	**-**	**-**	**-**
Continuously Variable Transmission (CVT)[3]	CVT	6.0	-	**6.0**	6.0	-	**6.0**
6/7/8-Speed Auto. Trans. with Improved Internals[3]	NAUTO	4.5	6.0	**5.3**	4.5	6.0	**5.3**
Dual Clutch Transmission (DCT)[3]	DCT	9.5	14.5	**12.0**	9.5	14.5	**12.0**
Hybrid Techs		**Non-incremental**		**-**	**-**	**-**	**-**
Power Split Hybrid	PSHEV	35.0	-	**35.0**	35.0	-	**35.0**
2-Mode Hybrid	2MHEV	-	-	**-**	40.0	-	**40.0**
Plug-in hybrid	PHEV	58.0	-	**58.0**	58.0	-	**58.0**
Vehicle Techs		**Non-incremental**		**-**	**-**	**-**	**-**
Mass Reduction - 1%	MR1	-	-	**-**	-	-	**-**
Mass Reduction - 2%	MR2	-	-	**-**	-	-	**-**
Mass Reduction - 5%	MR5	-	-	**-**	-	-	**-**
Mass Reduction - 10%	MR10	-	-	**-**	-	-	**-**
Mass Reduction - 20%	MR20	-	-	**-**	-	-	**-**
Low Rolling Resistance Tire	ROLL	1.0	2.0	**1.5**	1.0	2.0	**1.5**
Low Drag Brakes	LDB	-	-	**-**	-	-	**-**
Secondary Axle Disconnect	SAX	1.0	-	**1.0**	1.0	-	**1.0**
Aero Drag Reduction 10%	AERO	-	-	**-**	-	-	**-**

TABLE I.4 Technology Effectiveness, Incremental (percent) Fuel Consumption Benefit from Ricardo, Inc. (2008), NESCCAF (2004), Sierra Research (2008), and EEA (2007)

Ricardo, Inc.

Technologies		Standard Car			Full Size Car		Small MPV			Large MPV		Truck	
		I4, 2.4L-4V, DCP, 5spd AT, 3.39 FDR			V6, 3.5L-4V, 5spd AT, 2.87 FDR		I4, 2.4L-4V, DCP, 4spd AT, 3.91 FDR			V6, 3.8L-2V, OHV, 4spd AT, 3.43 FDR		V8, 5.4L-3V, CCP, 4spd AT, 3.73 FDR	
Spark Ignition Techs	Abbreviation	from EPA report	from report for NAS	Low	from EPA report	from report for NAS	from EPA report	from report for NAS (pack 5)	(pack 15)	from EPA report	from report for NAS	from EPA report	from report for NAS
Low Friction Lubricants	LUB	2.0	-	-	2.0	-	-	-	-	2.0	-	2.0	-
Engine Friction Reduction	EFR		-	-		-		-	-		-		-
VVT- Coupled Cam Phasing (CCP), SOHC	CCP	-	-	-	-	-	-	-	-	-	-	-	-
Discrete Variable Valve Lift (DVVL), SOHC	DVVL	-	-	-	-	-	-	-	-	-	-	-	-
Cylinder Deactivation, SOHC	DEAC	-	-	-	-	-	-	-	-	-	-	-	-
VVT - In take Cam Phasing (ICP)	ICP	-	-	-	-	-	-	-	-	-	-	-	-
VVT - Dual Cam Phasing (DCP)	DCP	-	-	-	5.0	-	-	-	-	5.0	-	-	-
Discrete Variable Valve Lift (DVVL), DOHC	DVVL	2.0	1.5	-	-	-	-	-	-	-	-	-	-
Continuously Variable Valve Lift (CVVL)	CVVL	-	-	-	3.0	-	-	-	-	-	-	-	-
Cylinder Deactivation, OHV	DEAC	-	-	-	-	-	-	-	-	-	4.0	-	-
VVT - Coupled Cam Phasing (CCP), OHV	CCP	-	-	-	-	-	-	-	-	-	2.5	-	-
Discrete Variable Valve Lift (DVVL), OHV	DVVL	-	-	-	-	-	-	-	-	-	-	-	-
Conversion to DOHC with DCP	CDOHC	-	-	-	-	-	-	-	-	-	-	-	-
Stoichiometric Gasoline Direct Injection (GDI)	SGDI	-	-	-	2.0	-	2.0	-	-	3.0	2.8	-	-
Turbocharging and Downsizing	TRBDS	-	-	-	-	-	-	-	4.4	3.0	-	-	-
Diesel Techs													
Conversion to Diesel	DSL	-	-	-	-	-	-	12.8	-	-	-	16.0	11.4
Conversion to Diesel following TRBDS	DSL	-	-	-	-	-	-	-	-	-	-	-	-
Conversion to Advanced Diesel	ADSL	-	-	-	-	-	-	-	-	-	-	-	-
Electrification/Accessory Techs													
Electric Power Steering (EPS)	EPS	-	-	-	-	-	-	-	-	-	-	-	-
Improved Accessories	IACC	-	-	-	-	-	-	1.7	1.4	-	-	-	-
12V BAS Micro-Hybrid	MHEV	3.0	3.0	-	-	-	3.0	-	-	-	2.0	-	-
Higher Voltage/Improved Alternator	HVIA	-	-	-	-	-	-	-	-	-	-	-	-
Integrated Starter Generator	ISG	-	-	-	-	-	-	-	-	-	-	-	-
Transmission Techs													
Continuously Variable Transmission (CVT)	CVT	-	-	-	-	-	-	-	-	-	-	-	-
6/7/8-Speed Auto. Trans. with Improved Internals	NAUTO	-	-	-	-	-	-	-	-	2.0	2.4	-	-
Dual Clutch Transmission (DCT)	DCT	8.0	6.7	-	7.0	-	-	6.8	6.4	-	-	9.0	9.3
Hybrid Techs													
Power Split Hybrid	PSHEV	-	-	-	-	-	-	-	-	-	-	-	-
2-Mode Hybrid	2MHEV	-	-	-	-	-	-	-	-	-	-	-	-
Plug-in hybrid	PHEV	-	-	-	-	-	-	-	-	-	-	-	-
Vehicle Techs													
Mass Reduction - 1%	MR1	-	-	-	-	-	-	-	-	-	-	-	-
Mass Reduction - 2%	MR2	-	-	-	-	-	-	-	-	-	-	-	-
Mass Reduction - 5%	MR5	-	-	-	-	-	-	-	-	-	-	-	-
Mass Reduction - 10%	MR10	-	-	-	-	-	-	-	-	-	-	-	-
Mass Reduction - 20%	MR20	-	-	-	-	-	-	-	-	-	-	-	-
Low Rolling Resistance Tires	ROLL	-	1.3	-	-	1.3	-	1.6	1.9	-	1.4	-	-
Low Drag Brakes	LDB	-	-	-	-	-	-	-	-	-	-	-	-
Secondary Axle Disconnect	SAX	-	-	-	-	-	-	-	-	-	-	-	-
Aero Drag Reduction 10%	AERO	-	-	-	-	-	-	-	-	-	-	2.0	1.4

continued

TABLE I.4 Continued

NESCCAF						
Technologies		Small Car	Large Car	MiniVan	SmallTruck/SUV	Large Truck/SUV
		I4	V6	V6	V6	V8
Spark Ignition Tech	Abbreviation					
Low Friction Lubricants	LUB	0.5	0.5	0.5	0.5	0.5
Engine Friction Reduction	EFR	0.5	0.5	0.5	0.5	0.5
VVT- Coupled Cam Phasing (CCP), SOHC	CCP	3.0	4.0	2.0	2.0	4.0
Discrete Variable Valve Lift (DVVL), SOHC	DVVL	4.0	4.0	3.0	4.0	4.0
Cylinder Deactivation, SOHC	DEAC	-	6.0	5.0	6.0	4.0
VVT - In take Cam Phasing (ICP)	ICP	2.0	1.0	1.0	1.0	2.0
VVT - Dual Cam Phasing (DCP)	DCP	3.0	4.0	2.0	3.0	4.0
Discrete Variable Valve Lift (DVVL), DOHC	DVVL	4.0	4.0	3.0	4.0	4.0
Continuously Variable Valve Lift (CVVL)	CVVL	5.0	6.0	4.0	5.0	5.0
Cylinder Deactivation, OHV	DEAC	-	6.0	5.0	6.0	4.0
VVT - Coupled Cam Phasing (CCP), OHV	CCP	-	-	-	-	-
Discrete Variable Valve Lift (DVVL), OHV	DVVL	4.0	4.0	3.0	4.0	4.0
Conversion to DOHC with DCP	CDOHC	-	-	-	-	-
Stoichiometric Gasoline Direct Injection (GDI)	SGDI	0.0	1.0	-1.0	-1.0	0.0
Turbocharging and Downsizin	TRBDS	6.0	8.0	6.0	6.0	-
Diesel Techs		-	-	-	-	-
Conversion to Diesel	DSL	-	-	-	-	-
Conversion to Diesel following TRBDS	DSL	-	-	-	-	-
Conversion to Advanced Diesel	ADSL	13.0	15.0	18.0	21.0	17.0
Electrification/Accessory Tech		-	-	-	-	-
Electric Power Steering (EPS)	EPS	1.0	-	-	-	1.0
Improved Accessories	IACC	3.0	-	-	-	2.0
12V BAS Micro-Hybrid	MHEV	-	-	-	-	-
Higher Voltage/Improved Alternator	HVIA	1.0	-	-	-	0.0
Integrated Starter Generato	ISG	-	-	-	-	-
Transmission Tech		-	-	-	-	-
Continuously Variable Transmission (CVT)	CVT	4.0	3.0	4.0	-	-
6/7/8-Speed Auto. Trans. with Improved Internals	NAUTO	3.0	3.0	3.0	3.0	2.0
Dual Clutch Transmission (DCT	DCT	8.0	7.0	8.0	8.0	5.0
Hybrid Techs		-	-	-	-	-
Power Split Hybrid	PSHEV	53.0	53.0	53.0	53.0	53.0
2-Mode Hybric	2MHEV	-	-	-	-	-
Plug-in hybric	PHEV	-	-	-	-	-
Vehicle Techs		-	-	-	-	-
Mass Reduction - 1%	MR1	0.5	-	-	-	0.6
Mass Reduction - 2%	MR2	1.0	-	-	-	1.1
Mass Reduction - 5%	MR5	2.6	-	-	-	2.9
Mass Reduction - 10%	MR10	5.3	-	-	-	5.7
Mass Reduction - 20%	MR20	-	-	-	-	-
Low Rolling Resistance Tires	ROLL	1.8	-	-	-	2.0
Low Drag Brakes	LDB	-	-	-	-	-
Secondary Axle Disconnect	SAX	-	-	-	-	-
Aero Drag Reduction 10%	AERO	1.7	-	-	-	1.9

continued

TABLE I.4 Continued

Sierra Research			
Technologies		Midsize	Truck
		- assume engine size adj. for constant acceleration	
Spark Ignition Techs	Abbreviation		
Low Friction Lubricants	LUB	0.5	0.5
Engine Friction Reduction	EFR	-	-
VVT- Coupled Cam Phasing (CCP), SOHC	CCP	-	-
Discrete Variable Valve Lift (DVVL), SOHC	DVVL	-	-
Cylinder Deactivation, SOHC	DEAC	7.5	8.8
VVT - In take Cam Phasing (ICP)	ICP	-	-
VVT - Dual Cam Phasing (DCP)	DCP	-	-
Discrete Variable Valve Lift (DVVL), DOHC	DVVL	6.3	6.8
Continuously Variable Valve Lift (CVVL)	CVVL	11.4	12.4
Cylinder Deactivation, OHV	DEAC	7.5	8.8
VVT - Coupled Cam Phasing (CCP), OHV	CCP	-	-
Discrete Variable Valve Lift (DVVL), OHV	DVVL	-	-
Conversion to DOHC with DCP	CDOHC	-	-
Stoichiometric Gasoline Direct Injection (GDI)	SGDI	5.9	6.2
Turbocharging and Downsizing	TRBDS	-0.3	0.3
Diesel Techs			
Conversion to Diesel	DSL	-	-
Conversion to Diesel following TRBDS	DSL	21.3	18.6
Conversion to Advanced Diesel	ADSL	-	-
Electrification/Accessory Techs			
Electric Power Steering (EPS)	EPS	1.8	1.1
Improved Accessories	IACC	-	-
12V BAS Micro-Hybrid	MHEV	-	-
Higher Voltage/Improved Alternator	HVIA	0.9	0.6
Integrated Starter Generator	ISG	-	-
Transmission Techs			
Continuously Variable Transmission (CVT)	CVT	-	-
6/7/8-Speed Auto. Trans. with Improved Internals	NAUTO	-	-
Dual Clutch Transmission (DCT)	DCT	4.0	4.4
Hybrid Techs			
Power Split Hybrid	PSHEV	28.7	22.1
2-Mode Hybrid	2MHEV	-	-
Plug-in hybrid	PHEV	-	-
Vehicle Techs			
Mass Reduction - 1%	MR1	-	-
Mass Reduction - 2%	MR2	-	-
Mass Reduction - 5%	MR5	-	-
Mass Reduction - 10%	MR10	-	-
Mass Reduction - 20%	MR20	-	-
Low Rolling Resistance Tires	ROLL	-	-
Low Drag Brakes	LDB	-	-
Secondary Axle Disconnect	SAX	-	-
Aero Drag Reduction 10%	AERO	-	-

continued

TABLE I.4 Continued

		EEA		
Technologies		-constant engine size		percent relative to PFI, fixed valve timing
		Values were converted to FC%		
Spark Ignition Techs	Abbreviation	Low	High	AVG
Low Friction Lubricants	LUB	0.9	1.1	**1.0**
Engine Friction Reduction	EFR	1.8	6.0	**3.9**
VVT- Coupled Cam Phasing (CCP), SOHC	CCP	1.3	1.9	**1.6**
Discrete Variable Valve Lift (DVVL), SOHC	DVVL	n/a	n/a	**n/a**
Cylinder Deactivation, SOHC	DEAC	5.3	7.1	**6.2**
VVT - In take Cam Phasing (ICP)	ICP	1.1	1.8	**1.4**
VVT - Dual Cam Phasing (DCP)	DCP	1.8	2.5	**2.2**
Discrete Variable Valve Lift (DVVL), DOHC	DVVL	2.9	3.8	**3.4**
Continuously Variable Valve Lift (CVVL)	CVVL	6.5	8.3	**7.4**
Cylinder Deactivation, OHV	DEAC	5.3	7.1	**6.2**
VVT - Coupled Cam Phasing (CCP), OHV	CCP	1.3	1.9	**1.6**
Discrete Variable Valve Lift (DVVL), OHV	DVVL	n/a	n/a	**n/a**
Conversion to DOHC with DCP	CDOHC	n/a	n/a	**n/a**
Stoichiometric Gasoline Direct Injection (GDI)	SGDI	2.9	3.8	**3.4**
Turbocharging and Downsizing	TRBDS	n/a	n/a	**n/a**
Diesel Techs				
Conversion to Diesel	DSL	24.8	30.1	**n/a**
Conversion to Diesel following TRBDS	DSL	n/a	n/a	**n/a**
Conversion to Advanced Diesel	ADSL	n/a	n/a	**n/a**
Electrification/Accessory Techs				
Electric Power Steering (EPS)	EPS	1.8	2.2	**2.0**
Improved Accessories	IACC	n/a	n/a	**n/a**
12V BAS Micro-Hybrid	MHEV	4.0	4.6	**4.3**
Higher Voltage/Improved Alternator	HVIA	0.3	0.7	**0.5**
Integrated Starter Generator	ISG	2.9	11.5	**7.2**
Transmission Techs				
Continuously Variable Transmission (CVT)	CVT	4.8	7.8	**6.3**
6/7/8-Speed Auto. Trans. with Improved Internals	NAUTO	4.0	5.5	**4.8**
Dual Clutch Transmission (DCT)	DCT	6.1	7.0	**6.5**
Hybrid Techs				
Power Split Hybrid	PSHEV	-	-	**-**
2-Mode Hybrid	2MHEV	-	-	**-**
Plug-in hybrid	PHEV	-	-	**-**
Vehicle Techs				
Mass Reduction - 1%	MR1	n/a	n/a	**n/a**
Mass Reduction - 2%	MR2	n/a	n/a	**n/a**
Mass Reduction - 5%	MR5	3.0	3.2	**3.1**
Mass Reduction - 10%	MR10	5.8	6.2	**6.0**
Mass Reduction - 20%	MR20	1.3	1.5	**-**
Low Rolling Resistance Tires	ROLL	n/a	n/a	n/a
Low Drag Brakes	LDB	n/a	n/a	**n/a**
Secondary Axle Disconnect	SAX	1.8	2.2	**n/a**
Aero Drag Reduction 10%	AERO	3.5	4.2	**3.8**

TABLE I.5 Incremental Costs ($) from DOT/NHTSA (2009)

NHTSA 2011

Technologies		Subcompact Car				Compact Car				Midsize Car				Large Car			
		I4				I4				I4				V6			
Spark Ignition Techs	Abbreviation	Low	High	**AVG**	**Net**	Low	High	**AVG**	**Net**	Low	High	**AVG**	**Net**	Low	High	**AVG**	**Net**
Low Friction Lubricants	LUB	5.0		**5.0**	-	5.0		**5.0**	-	5.0		**5.0**	-	5.0		**5.0**	-
Engine Friction Reduction	EFR	52.0	196.0	**124.0**	-	52.0	196.0	**124.0**	-	52.0	196.0	**124.0**	-	78.0	294.0	**186.0**	-
VVT- Coupled Cam Phasing (CCP), SOHC	CCP	61.0	-	**61.0**	-	61.0	-	**61.0**	-	61.0	-	**61.0**	-	122.0	-	**122.0**	-
Discrete Variable Valve Lift (DVVL), SOHC	DVVL	201.0	-	**201.0**	-	201.0	-	**201.0**	-	201.0	-	**201.0**	-	306.0	-	**306.0**	-
Cylinder Deactivation, SOHC	DEAC	-	-	**-**	-	-	-	**-**	-	-	-	**-**	-	75.0	-	**75.0**	-
VVT - In take Cam Phasing (ICP)	ICP	61.0	-	**61.0**	-	61.0	-	**61.0**	-	61.0	-	**61.0**	-	122.0	-	**122.0**	-
VVT - Dual Cam Phasing (DCP)	DCP	61.0	-	**61.0**	-	61.0	-	**61.0**	-	61.0	-	**61.0**	-	122.0	-	**122.0**	-
Discrete Variable Valve Lift (DVVL), DOHC	DVVL	201.0	-	**201.0**	-	201.0	-	**201.0**	-	201.0	-	**201.0**	-	306.0	-	**306.0**	-
Continuously Variable Valve Lift (CVVL)	CVVL	306.0	-	**306.0**	-	306.0	-	**306.0**	-	306.0	-	**306.0**	-	432.0	-	**432.0**	-
Cylinder Deactivation, OHV	DEAC	-	-	**-**	-	-	-	**-**	-	-	-	**-**	-	306.0	-	**306.0**	-
VVT - Coupled Cam Phasing (CCP), OHV	CCP	61.0	-	**61.0**	-	61.0	-	**61.0**	-	61.0	-	**61.0**	-	122.0	-	**122.0**	-
Discrete Variable Valve Lift (DVVL), OHV	DVVL	201.0	-	**201.0**	-	201.0	-	**201.0**	-	201.0	-	**201.0**	-	76.0	-	**76.0**	-
Conversion to DOHC with DCP	CDOHC	373.0	-	**373.0**	-	373.0	-	**373.0**	-	373.0	-	**373.0**	-	590.0	-	**590.0**	-
Stoichiometric Gasoline Direct Injection (GDI)	SGDI	293.0	440.0	**366.5**	-	293.0	440.0	**366.5**	-	293.0	440.0	**366.5**	-	384.0	558.0	**471.0**	-
Turbocharging and Downsizing	TRBDS	1223.0	-	**1223.0**	-	1223.0	-	**1223.0**	-	1223.0	-	**1223.0**	-	822.0	-	**822.0**	-
Diesel Techs																	
Conversion to Diesel	DSL	2963.0	3254.0	**3108.5**	**4000.0**	2963.0	3254.0	**3108.5**	**4000.0**	2963.0	3254.0	**3108.5**	**4000.0**	4105.0	4490.0	**4297.5**	**5600.0**
Conversion to Diesel following TRBDS	DSL	1567.0	1858.0	**1712.5**	**4000.0**	1567.0	1858.0	**1712.5**	**4000.0**	1567.0	1858.0	**1712.5**	**4000.0**	3110.0	3495.0	**3302.5**	**5600.0**
Conversion to Advanced Diesel	ADSL	-	-	**-**	-	-	-	**-**	-	-	-	**-**	-	-	-	**-**	-
Electrification/Accessory Techs															-		
Electric Power Steering (EPS)	EPS	105.0	120.0	**112.5**	-	105.0	120.0	**112.5**	-	105.0	120.0	**112.5**	-	105.0	120.0	**112.5**	-
Improved Accessories	IACC	173.0	211.0	**192.0**	-	173.0	211.0	**192.0**	-	173.0	211.0	**192.0**	-	173.0	211.0	**192.0**	-
12V BAS Micro-Hybrid	MHEV	372.0	-	**372.0**	-	408.0	-	**408.0**	-	453.0	-	**453.0**	-	490.0	-	**490.0**	-
Higher Voltage/Improved Alternator	HVIA	84.0	-	**84.0**	-	84.0	-	**84.0**	-	84.0	-	**84.0**	-	84.0	-	**84.0**	-
Integrated Starter Generator	ISG	1713.0	-	**1713.0**	-	2019.0	-	**2019.0**	-	2190.0	-	**2190.0**	-	2386.0	-	**2386.0**	-
Transmission Techs																	
Continuously Variable Transmission (CVT)	CVT	300.0	-	**300.0**	-	300.0	-	**300.0**	-	300.0	-	**300.0**	-	300.0	-	**300.0**	-
6/7/8-Speed Auto. Trans. with Improved Internals	NAUTO	323.0	-	**323.0**	-	323.0	-	**323.0**	-	323.0	-	**323.0**	-	323.0	-	**323.0**	-
Dual Clutch Transmission (DCT)	DCT	68.0	-	**68.0**	**500.0**	68.0	-	**68.0**	**500.0**	218.0	-	**218.0**	**600.0**	218.0	-	**218.0**	**600.0**
Hybrid Techs																	
Power Split Hybrid	PSHEV	1409.0	1462.0	**1435.5**	**4300.0**	1742.0	1795.0	**1768.5**	**4900.0**	2175.0	2228.0	**2201.5**	**5600.0**	2534.0	2587.0	**2560.5**	**6200.0**
2-Mode Hybrid	2MHEV	-	-	**-**	-	-	-	**-**	-	-	-	**-**	-	-	-	**-**	-
Plug-in hybrid	PHEV	19648.0	19701.0	**19674.5**	**22500.0**	23670.0	23723.0	**23696.5**	**26700.0**	26702.0	26755.0	**26728.5**	**30000.0**	-	-	**-**	-
Vehicle Techs																	
Mass Reduction - 1%	MR1	-	-	**-**	-	-	-	**-**	-	-	-	**-**	-	-	-	**-**	-
Mass Reduction - 2%	MR2	-	-	**-**	-	-	-	**-**	-	-	-	**-**	-	-	-	**-**	-
Mass Reduction - 5%	MR5	-	-	**-**	-	-	-	**-**	-	-	-	**-**	-	-	-	**-**	-
Mass Reduction - 10%	MR10	-	-	**-**	-	-	-	**-**	-	-	-	**-**	-	-	-	**-**	-
Mass Reduction - 20%	MR20	-	-	**-**	-	-	-	**-**	-	-	-	**-**	-	-	-	**-**	-
Low Rolling Resistance Tires	ROLL	6.0	9.0	**7.5**	-	6.0	9.0	**7.5**	-	6.0	9.0	**7.5**	-	6.0	9.0	**7.5**	-
Low Drag Brakes	LDB	-	-	**-**	-	-	-	**-**	-	-	-	**-**	-	-	-	**-**	-
Secondary Axle Disconnect	SAX	117.0	-	**117.0**	-	117.0	-	**117.0**	-	117.0	-	**117.0**	-	117.0	-	**117.0**	-
Aero Drag Reduction 10%	AERO	60.0	116.0	**88.0**	-	60.0	116.0	**88.0**	-	60.0	116.0	**88.0**	-	60.0	116.0	**88.0**	-

continued

TABLE I.5 Continued

NHTSA 2011																	
Technologies		Performance Subcompact Car				Performance Compact Car				Performance Midsize Car				Performance Large Car			
		I4				V6				V6				V8			
Spark Ignition Techs	Abbreviation	Low	High	AVG	Net	Low	High	AVG	Net	Low	High	AVG	Net	Low	High	AVG	Net
Low Friction Lubricants	LUB	5.0		**5.0**	-	5.0		**5.0**	-	5.0		**5.0**	-	5.0		**5.0**	-
Engine Friction Reduction	EFR	52.0	196.0	**124.0**	-	78.0	294.0	**186.0**	-	78.0	294.0	**186.0**	-	104.0	392.0	**248.0**	-
VVT- Coupled Cam Phasing (CCP), SOHC	CCP	61.0	-	**61.0**	-	122.0	-	**122.0**	-	122.0	-	**122.0**	-	122.0	-	**122.0**	-
Discrete Variable Valve Lift (DVVL), SOHC	DVVL	201.0	-	**201.0**	-	306.0	-	**306.0**	-	306.0	-	**306.0**	-	396.0	-	**396.0**	-
Cylinder Deactivation, SOHC	DEAC	-	-	**-**	-	75.0	-	**75.0**	-	75.0	-	**75.0**	-	75.0	-	**75.0**	-
VVT - In take Cam Phasing (ICP)	ICP	61.0	-	**61.0**	-	122.0	-	**122.0**	-	122.0	-	**122.0**	-	122.0	-	**122.0**	-
VVT - Dual Cam Phasing (DCP)	DCP	61.0	-	**61.0**	-	122.0	-	**122.0**	-	122.0	-	**122.0**	-	122.0	-	**122.0**	-
Discrete Variable Valve Lift (DVVL), DOHC	DVVL	201.0	-	**201.0**	-	306.0	-	**306.0**	-	306.0	-	**306.0**	-	396.0	-	**396.0**	-
Continuously Variable Valve Lift (CVVL)	CVVL	306.0	-	**306.0**	-	432.0	-	**432.0**	-	432.0	-	**432.0**	-	582.0	-	**582.0**	-
Cylinder Deactivation, OHV	DEAC	-	-	**-**	-	306.0	-	**306.0**	-	306.0	-	**306.0**	-	400.0	-	**400.0**	-
VVT - Coupled Cam Phasing (CCP), OHV	CCP	61.0	-	**61.0**	-	122.0	-	**122.0**	-	122.0	-	**122.0**	-	122.0	-	**122.0**	-
Discrete Variable Valve Lift (DVVL), OHV	DVVL	201.0	-	**201.0**	-	76.0	-	**76.0**	-	76.0	-	**76.0**	-	76.0	-	**76.0**	-
Conversion to DOHC with DCP	CDOHC	373.0	-	**373.0**	-	590.0	-	**590.0**	-	590.0	-	**590.0**	-	746.0	-	**746.0**	-
Stoichiometric Gasoline Direct Injection (GDI)	SGDI	293.0	440.0	**366.5**	-	384.0	558.0	**471.0**	-	384.0	558.0	**471.0**	-	512.0	744.0	**628.0**	-
Turbocharging and Downsizing	TRBDS	1223.0	-	**1223.0**	-	822.0	-	**822.0**	-	822.0	-	**822.0**	-	1229.0	-	**1229.0**	-
Diesel Techs																	
Conversion to Diesel	DSL	2963.0	3254.0	**3108.5**	**4000.0**	4105.0	4490.0	**4297.5**	**5600.0**	4105.0	4490.0	**4297.5**	**5600.0**	5125.0	5617.0	**5371.0**	**7000.0**
Conversion to Diesel following TRBDS	DSL	1567.0	1858.0	**1712.5**	**4000.0**	3110.0	3495.0	**3302.5**	**5600.0**	3110.0	3495.0	**3302.5**	**5600.0**	3723.0	4215.0	**3969.0**	**7000.0**
Conversion to Advanced Diesel	ADSL	-	-	**-**	-	-	-	**-**	-	-	-	**-**	-	-	-	**-**	-
Electrification/Accessory Techs															-		
Electric Power Steering (EPS)	EPS	105.0	120.0	**112.5**	-	105.0	120.0	**112.5**	-	105.0	120.0	**112.5**	-	105.0	120.0	**112.5**	-
Improved Accessories	IACC	173.0	211.0	**192.0**	-	173.0	211.0	**192.0**	-	173.0	211.0	**192.0**	-	173.0	211.0	**192.0**	-
12V BAS Micro-Hybrid	MHEV	406.0	-	**406.0**	-	443.0	-	**443.0**	-	494.0	-	**494.0**	-	549.0	-	**549.0**	-
Higher Voltage/Improved Alternator	HVIA	84.0	-	**84.0**	-	84.0	-	**84.0**	-	84.0	-	**84.0**	-	84.0	-	**84.0**	-
Integrated Starter Generator	ISG	1789.0	1864.0	**1826.5**	-	2054.0	-	**2054.0**	-	2183.0	-	**2183.0**	-	2351.0	-	**2351.0**	-
Transmission Techs																	
Continuously Variable Transmission (CVT)	CVT	300.0	-	**300.0**	-	300.0	-	**300.0**	-	300.0	-	**300.0**	-	-	-	**-**	-
6/7/8-Speed Auto. Trans. with Improved Internals	NAUTO	323.0	638.0	**480.5**	-	323.0	638.0	**480.5**	-	323.0	638.0	**480.5**	-	323.0	638.0	**480.5**	-
Dual Clutch Transmission (DCT)	DCT	97.0	218.0	**157.5**	**600.0**	97.0	218.0	**157.5**	**600.0**	97.0	218.0	**157.5**	**600.0**	97.0	218.0	**157.5**	**600.0**
Hybrid Techs																	
Power Split Hybrid	PSHEV	2838.0	2966.0	**2902.0**	**5900.0**	3144.0	3197.0	**3170.5**	**6400.0**	4093.0	4146.0	**4119.5**	**7500.0**	5467.0	5520.0	**5493.5**	**8800.0**
2-Mode Hybrid	2MHEV	-	-	**-**	-	-	-	**-**	-	-	-	**-**	-	-	-	**-**	-
Plug-in hybrid	PHEV	23736.0	23864.0	**23800.0**	**26800.0**	26790.0	26843.0	**26816.5**	**30100.0**	30110.0	30163.0	**30136.5**	**33600.0**	-	-	**-**	-
Vehicle Techs																	
Mass Reduction - 1%	MR1	-	-	**-**	-	-	-	**-**	-	-	-	**-**	-	-	-	**-**	-
Mass Reduction - 2%	MR2	-	-	**-**	-	-	-	**-**	-	-	-	**-**	-	-	-	**-**	-
Mass Reduction - 5%	MR5	-	-	**-**	-	-	-	**-**	-	-	-	**-**	-	-	-	**-**	-
Mass Reduction - 10%	MR10	-	-	**-**	-	-	-	**-**	-	-	-	**-**	-	-	-	**-**	-
Mass Reduction - 20%	MR20	-	-	**-**	-	-	-	**-**	-	-	-	**-**	-	-	-	**-**	-
Low Rolling Resistance Tires	ROLL	-	-	**-**	-	-	-	**-**	-	-	-	**-**	-	-	-	**-**	-
Low Drag Brakes	LDB	-	-	**-**	-	-	-	**-**	-	-	-	**-**	-	-	-	**-**	-
Secondary Axle Disconnect	SAX	117.0	-	**117.0**	-	117.0	-	**117.0**	-	117.0	-	**117.0**	-	117.0	-	**117.0**	-
Aero Drag Reduction 10%	AERO	60.0	116.0	**88.0**	-	60.0	116.0	**88.0**	-	60.0	116.0	**88.0**	-	60.0	116.0	**88.0**	-

continued

TABLE I.5 Continued

NHTSA 2011

Technologies		Minivan LT V6				Small LT I4				Midsize LT V6				Large LT V8			
Spark Ignition Techs	Abbreviation	Low	High	AVG	Net	Low	High	AVG	Net	Low	High	AVG	Net	Low	High	AVG	Net
Low Friction Lubricants	LUB	5.0		**5.0**	-	5.0		**5.0**	-	5.0		**5.0**	-	5.0		**5.0**	-
Engine Friction Reduction	EFR	78.0	294.0	**186.0**	-	52.0	196.0	**124.0**	-	78.0	294.0	**186.0**	-	104.0	392.0	**248.0**	-
VVT- Coupled Cam Phasing (CCP), SOHC	CCP	122.0	-	**122.0**	-	61.0	-	**61.0**	-	122.0	-	**122.0**	-	122.0	-	**122.0**	-
Discrete Variable Valve Lift (DVVL), SOHC	DVVL	306.0	-	**306.0**	-	201.0	-	**201.0**	-	306.0	-	**306.0**	-	396.0	-	**396.0**	-
Cylinder Deactivation, SOHC	DEAC	75.0	-	**75.0**	-	-	-	-	-	75.0	-	**75.0**	-	75.0	-	**75.0**	-
VVT - In take Cam Phasing (ICP)	ICP	122.0	-	**122.0**	-	61.0	-	**61.0**	-	122.0	-	**122.0**	-	122.0	-	**122.0**	-
VVT - Dual Cam Phasing (DCP)	DCP	122.0	-	**122.0**	-	61.0	-	**61.0**	-	122.0	-	**122.0**	-	122.0	-	**122.0**	-
Discrete Variable Valve Lift (DVVL), DOHC	DVVL	306.0	-	**306.0**	-	201.0	-	**201.0**	-	306.0	-	**306.0**	-	396.0	-	**396.0**	-
Continuously Variable Valve Lift (CVVL)	CVVL	432.0	-	**432.0**	-	306.0	-	**306.0**	-	432.0	-	**432.0**	-	582.0	-	**582.0**	-
Cylinder Deactivation, OHV	DEAC	306.0	-	**306.0**	-	-	-	-	-	306.0	-	**306.0**	-	400.0	-	**400.0**	-
VVT - Coupled Cam Phasing (CCP), OHV	CCP	122.0	-	**122.0**	-	61.0	-	**61.0**	-	122.0	-	**122.0**	-	122.0	-	**122.0**	-
Discrete Variable Valve Lift (DVVL), OHV	DVVL	76.0	-	**76.0**	-	201.0	-	**201.0**	-	76.0	-	**76.0**	-	76.0	-	**76.0**	-
Conversion to DOHC with DCP	CDOHC	590.0	-	**590.0**	-	373.0	-	**373.0**	-	590.0	-	**590.0**	-	746.0	-	**746.0**	-
Stoichiometric Gasoline Direct Injection (GDI)	SGDI	384.0	558.0	**471.0**	-	293.0	440.0	**366.5**	-	384.0	558.0	**471.0**	-	512.0	744.0	**628.0**	-
Turbocharging and Downsizing	TRBDS	822.0	-	**822.0**	-	1223.0	-	**1223.0**	-	822.0	-	**822.0**	-	1229.0	-	**1229.0**	-
Diesel Techs																	
Conversion to Diesel	DSL	4105.0	4490.0	**4297.5**	**5600.0**	2963.0	3254.0	**3108.5**	**4000.0**	4105.0	4490.0	**4297.5**	**5600.0**	5125.0	5617.0	**5371.0**	**7000.0**
Conversion to Diesel following TRBDS	DSL	3110.0	3495.0	**3302.5**	**5600.0**	1567.0	1858.0	**1712.5**	**4000.0**	3110.0	3495.0	**3302.5**	**5600.0**	3723.0	4215.0	**3969.0**	**7000.0**
Conversion to Advanced Diesel	ADSL	-	-	-	-	-	-	-	-	-	-	-	-	-	-	-	-
Electrification/Accessory Techs																	
Electric Power Steering (EPS)	EPS	105.0	120.0	**112.5**	-	105.0	120.0	**112.5**	-	105.0	120.0	**112.5**	-	-	-	-	-
Improved Accessories	IACC	173.0	211.0	**192.0**	-	173.0	211.0	**192.0**	-	-	-	-	-	-	-	-	-
12V BAS Micro-Hybrid	MHEV	490.0	-	**490.0**	-	427.0	-	**427.0**	-	502.0	-	**502.0**	-	-	-	-	-
Higher Voltage/Improved Alternator	HVIA	84.0	-	**84.0**	-	84.0	-	**84.0**	-	84.0	-	**84.0**	-	-	-	-	-
Integrated Starter Generator	ISG	2386.0	-	**2386.0**	-	2029.0	-	**2029.0**	-	2457.0	-	**2457.0**	-	-	-	-	-
Transmission Techs																	
Continuously Variable Transmission (CVT)	CVT	300.0	-	**300.0**	-	300.0	-	**300.0**	-	-	-	-	-	-	-	-	-
6/7/8-Speed Auto. Trans. with Improved Internals	NAUTO	323.0	-	**323.0**	-	323.0	638.0	**480.5**	-	323.0	638.0	**480.5**	-	323.0	638.0	**480.5**	-
Dual Clutch Transmission (DCT)	DCT	218.0	-	**218.0**	**600.0**	97.0	218.0	**157.5**	**600.0**	97.0	218.0	**157.5**	**600.0**	97.0	218.0	**157.5**	**600.0**
Hybrid Techs																	
Power Split Hybrid	PSHEV	2534.0	2587.0	**2560.5**	**6200.0**	1932.0	1985.0	**1958.5**	**5200.0**	3173.0	3188.0	**3180.5**	**6400.0**	-	-	-	-
2-Mode Hybrid	2MHEV	-	-	-	-	6376.0	6429.0	**6402.5**	**9800.0**	8313.0	8328.0	**8320.5**	**12100.0**	14106.0	15171.0	**14638.5**	**15300.0**
Plug-in hybrid	PHEV	-	-	-	-	24376.0	24329.0	**24352.5**	**27500.0**	-	-	-	-	-	-	-	-
Vehicle Techs																	
Mass Reduction - 1%	MR1	-	-	-	-	-	-	-	-	1.0	2.0	**1.5**	-	1.0	2.0	**1.5**	-
Mass Reduction - 2%	MR2	-	-	-	-	-	-	-	-	1.0	2.0	**1.5**	-	1.0	2.0	**1.5**	-
Mass Reduction - 5%	MR5	-	-	-	-	-	-	-	-	2.0	4.0	**3.0**	-	2.0	4	**3.0**	-
Mass Reduction - 10%	MR10	-	-	-	-	-	-	-	-	-	-	-	-	-	-	-	-
Mass Reduction - 20%	MR20	-	-	-	-	-	-	-	-	-	-	-	-	-	-	-	-
Low Rolling Resistance Tires	ROLL	6.0	9.0	**7.5**	-	6.0	9.0	**7.5**	-	6.0	9.0	**7.5**	-	-	-	-	-
Low Drag Brakes	LDB	-	-	-	-	-	-	-	-	89.0	-	**89.0**	-	-	-	-	-
Secondary Axle Disconnect	SAX	117.0	-	**117.0**	-	117.0	-	**117.0**	-	117.0	-	**117.0**	-	89.0	-	**89.0**	-
Aero Drag Reduction 10%	AERO	60.0	116.0	**88.0**	-	60.0	116.0	**88.0**	-	60.0	116.0	**88.0**	-	117.0	-	**117.0**	-

TABLE I.6 Incremental Costs ($) from NRC (2002)

NRC 2002				
Technologies				
Spark Ignition Techs	Abbreviation	Low	High	AVG
Low Friction Lubricants	LUB	-		**-**
Engine Friction Reduction	EFR	35.0	140.0	**87.5**
VVT- Coupled Cam Phasing (CCP), SOHC	CCP	35.0	140.0	**87.5**
Discrete Variable Valve Lift (DVVL), SOHC	DVVL	70.0	120.0	**95.0**
Cylinder Deactivation, SOHC	DEAC	112.0	252.0	**182.0**
VVT - In take Cam Phasing (ICP)	ICP	35.0	140.0	**87.5**
VVT - Dual Cam Phasing (DCP)	DCP	35.0	140.0	**87.5**
Discrete Variable Valve Lift (DVVL), DOHC	DVVL	70.0	120.0	**95.0**
Continuously Variable Valve Lift (CVVL)	CVVL	-	-	**-**
Cylinder Deactivation, OHV	DEAC	112.0	252.0	**182.0**
VVT - Coupled Cam Phasing (CCP), OHV	CCP	35.0	140.0	**87.5**
Discrete Variable Valve Lift (DVVL), OHV	DVVL	70.0	120.0	**95.0**
Conversion to DOHC with DCP	CDOHC	-	-	**-**
Stoichiometric Gasoline Direct Injection (GDI)	SGDI	-	-	**-**
Turbocharging and Downsizinç	TRBDS	350.0	560.0	**455.0**
Diesel Techs				
Conversion to Diesel	DSL	-	-	**-**
Conversion to Diesel following TRBDS	DSL	-	-	**-**
Conversion to Advanced Diesel	ADSL	-	-	**-**
Electrification/Accessory Techs				
Electric Power Steering (EPS)	EPS	105.0	150.0	**127.5**
Improved Accessories	IACC	84.0	112.0	**98.0**
12V BAS Micro-Hybrid	MHEV	-	-	**-**
Higher Voltage/Improved Alternator	HVIA	-	-	**-**
Integrated Starter Generator	ISG	210.0	350.0	**280.0**
Transmission Techs				
Continuously Variable Transmission (CVT)	CVT	140.0	350.0	**245.0**
6/7/8-Speed Auto. Trans. with Improved Internals	NAUTO	140.0	280.0	**210.0**
Dual Clutch Transmission (DCT)	DCT	70.0	280.0	**175.0**
Hybrid Techs				
Power Split Hybrid	PSHEV	-	-	**-**
2-Mode Hybrid	2MHEV	-	-	**-**
Plug-in hybrid	PHEV	-	-	**-**
Vehicle Techs				
Mass Reduction - 1%	MR1	-	-	**-**
Mass Reduction - 2%	MR2	-	-	**-**
Mass Reduction - 5%	MR5	210.0	350.0	**280.0**
Mass Reduction - 10%	MR10	-	-	**-**
Mass Reduction - 20%	MR20	-	-	**-**
Low Rolling Resistance Tires	ROLL	14.0	56.0	**35.0**
Low Drag Brakes	LDB	-	-	**-**
Secondary Axle Disconnec	SAX	-	-	**-**
Aero Drag Reduction 10%	AERO	0.0	140.0	**70.0**

TABLE I.7 Incremental Costs ($) from EPA (2008)

EPA

Technologies		Small Car			Large Car			Minivan			Small Truck			Large Truck		
		I4			V6			V6			V6			V8		
Spark Ignition Techs	Abbreviation	Low	High	**AVG**	Low	High	**AVG**	Low	High	**AVG**	Low	High	**AVG**	Low	High	**AVG**
Low Friction Lubricants	LUB	3	-	**3**	3	-	**3**	3	-	**3**	3	-	**3**	3	-	**3**
Engine Friction Reduction	EFR	0	84	**42**	0	126	**63**	0	126	**63**	0	126	**63**	0	168	**84**
VVT- Coupled Cam Phasing (CCP), SOHC	CCP	59	-	**59**	119	-	**119**	119	-	**119**	119	-	**119**	119	-	**119**
Discrete Variable Valve Lift (DVVL), SOHC	DVVL	169	-	**169**	246	-	**246**	246	-	**246**	246	-	**246**	322	-	**322**
Cylinder Deactivation, SOHC	DEAC	-	-	**-**	203	-	**203**	203	-	**203**	203	-	**203**	229	-	**229**
VVT - In take Cam Phasing (ICP)	ICP	59	-	**59**	119	-	**119**	119	-	**119**	119	-	**119**	119	-	**119**
VVT - Dual Cam Phasing (DCP)	DCP	89	-	**89**	209	-	**209**	209	-	**209**	209	-	**209**	209	-	**209**
Discrete Variable Valve Lift (DVVL), DOHC	DVVL	169	-	**169**	246	-	**246**	246	-	**246**	246	-	**246**	322	-	**322**
Continuously Variable Valve Lift (CVVL)	CVVL	254	-	**254**	466	-	**466**	466	-	**466**	466	-	**466**	508	-	**508**
Cylinder Deactivation, OHV	DEAC	-	-	**-**	203	-	**203**	203	-	**203**	203	-	**203**	229	-	**229**
VVT - Coupled Cam Phasing (CCP), OHV	CCP	59	-	**59**	59	-	**59**	59	-	**59**	59	-	**59**	59	-	**59**
Discrete Variable Valve Lift (DVVL), OHV	DVVL	169	-	**169**	246	-	**246**	246	-	**246**	246	-	**246**	322	-	**322**
Conversion to DOHC with DCP	CDOHC	-	-	**-**	-	-	**-**	-	-	**-**	-	-	**-**	-	-	**-**
Stoichiometric Gasoline Direct Injection (GDI)	SGDI	122	420	**271**	204	525	**364.5**	204	525	**364.5**	204	525	**364.5**	228	525	**376.5**
Turbocharging and Downsizing	TRBDS	690	-	**690**	120	-	**120**	120	-	**120**	120	-	**120**	810	-	**810**
Diesel Techs																
Conversion to Diesel	DSL	2790	-	**2790**	3045	-	**3045**	3120	-	**3120**	3405	-	**3405**	4065	-	**4065**
Conversion to Diesel following TRBDS	DSL	-	-	**-**	-	-	**-**	-	-	**-**	-	-	**-**	-	-	**-**
Conversion to Advanced Diesel	ADSL	-	-	**-**	-	-	**-**	-	-	**-**	-	-	**-**	-	-	**-**
Electrification/Accessory Techs																
Electric Power Steering (EPS)	EPS	118	197	**157.5**	118	197	**157.5**	118	197	**157.5**	118	197	**157.5**	118	197	**157.5**
Improved Accessories	IACC	-	-	**-**	-	-	**-**	-	-	**-**	-	-	**-**	-	-	**-**
12V BAS Micro-Hybrid	MHEV	-	-	**-**	-	-	**-**	-	-	**-**	-	-	**-**	-	-	**-**
Higher Voltage/Improved Alternator	HVIA	-	-	**-**	-	-	**-**	-	-	**-**	-	-	**-**	-	-	**-**
Integrated Starter Generator	ISG	2477	-	**2477**	3153	-	**3153**	-	-	**-**	-	-	**-**	-	-	**-**
Transmission Techs																
Continuously Variable Transmission (CVT)	CVT	231	-	**231**	270	-	**270**	270	-	**270**	-	-	**-**	-	-	**-**
6/7/8-Speed Auto. Trans. with Improved Internals	NAUTO	76	167	**121.5**	76	167	**121.5**	76	167	**121.5**	76	167	**121.5**	76	167	**121.5**
Dual Clutch Transmission (DCT)	DCT	141	-	**141**	141	-	**141**	141	-	**141**	141	-	**141**	141	-	**141**
Hybrid Techs																
Power Split Hybrid	PSHEV	3754	-	**3754**	-	-	**-**	-	-	**-**	-	-	**-**	-	-	**-**
2-Mode Hybrid	2MHEV	-	-	**-**	4655	-	**4655**	4655	-	**4655**	4655	-	**4655**	6006	-	**6006**
Plug-in hybrid	PHEV	4500	-	**4500**	6750	-	**6750**	6750	-	**6750**	6750	-	**6750**	10200	-	**10200**
Vehicle Techs																
Mass Reduction - 1%	MR1	-	-	**-**	-	-	**-**	-	-	**-**	-	-	**-**	-	-	**-**
Mass Reduction - 2%	MR2	-	-	**-**	-	-	**-**	-	-	**-**	-	-	**-**	-	-	**-**
Mass Reduction - 5%	MR5	-	-	**-**	-	-	**-**	-	-	**-**	-	-	**-**	-	-	**-**
Mass Reduction - 10%	MR10	-	-	**-**	-	-	**-**	-	-	**-**	-	-	**-**	-	-	**-**
Mass Reduction - 20%	MR20	-	-	**-**	-	-	**-**	-	-	**-**	-	-	**-**	-	-	**-**
Low Rolling Resistance Tires	ROLL	6	-	**6**	6	-	**6**	6	-	**6**	6	-	**6**	6	-	**6**
Low Drag Brakes	LDB	-	-	**-**	-	-	**-**	-	-	**-**	87	-	**87**	87	-	**87**
Secondary Axle Disconnect	SAX	676	-	**676**	676	-	**676**	676	-	**676**	676	-	**676**	-	-	**-**
Aero Drag Reduction 10%	AERO	-	-	**-**	-	-	**-**	0	75	**37.5**	0	75	**37.5**	0	75	**37.5**

TABLE I.8 Technology Effectiveness, Incremental (Percent) Fuel Consumption Benefit from EEA (2007), Sierra Research (2008), Martec (2008), and NESCCAF (2004)

EEA													
Technologies		I4			I-6			V6			V8		
Spark Ignition Techs	Abbreviation	Low	High	**AVG**	Low	High	**AVG**	Low	High	**AVG**	Low	High	**AVG**
Low Friction Lubricants	LUB	14	18	**16**	17	23	**20**	17	23	**20**	20	28	**24**
Engine Friction Reduction	EFR	18	60	**39**	23	85	**54**	23	85	**54**	27	88	**58**
VVT - Coupled Cam Phasing (CCP), SOHC	CCP	50	54	**52**	50	54	**52**	100	108	**104**	100	108	**104**
Discrete Variable Valve Lift (DVVL), SOHC	DVVL	-	-	-	-	-	-	-	-	-	-	-	-
Cylinder Deactivation, SOHC	DEAC	-	-	-	302	318	**310**	302	318	**310**	205	225	**215**
VVT - In take Cam Phasing (ICP)	ICP	50	54	**52**	50	54	**52**	100	108	**104**	100	108	**104**
VVT - Dual Cam Phasing (DCP)	DCP	76	84	**80**	76	84	**80**	178	190	**184**	178	190	**184**
Discrete Variable Valve Lift (DVVL), DOHC	DVVL	142	158	**150**	188	212	**200**	198	222	**210**	255	285	**270**
Continuously Variable Valve Lift (CVVL)	CVVL	314	346	**330**	380	420	**400**	440	480	**460**	575	625	**600**
Cylinder Deactivation, OHV	DEAC	-	-	-	302	318	**310**	302	318	**310**	205	225	**215**
VVT - Coupled Cam Phasing (CCP), OHV	CCP	-	-	-	-	-	-	-	-	-	-	-	-
Discrete Variable Valve Lift (DVVL), OHV	DVVL	82	94	**88**	120	140	**130**	120	140	**130**	144	170	**157**
Conversion to DOHC with DCP	CDOHC	-	-	-	-	-	-	-	-	-	-	-	-
Stoichiometric Gasoline Direct Injection (GDI)	SGDI	145	155	**150**	193	207	**200**	193	207	**200**	240	260	**250**
Turbocharging and Downsizing	TRBDS	480	520	**500**	540	580	**560**	550	610	**580**	630	690	**660**
Diesel Techs													
Conversion to Diesel	DSL	-	-	**2200.0**	-	-	-	-	-	**3200.0**	-	-	-
Conversion to Diesel following TRBDS	DSL	-	-	-	-	-	-	-	-	-	-	-	-
Conversion to Advanced Diesel	ADSL	-	-	-	-	-	-	-	-	-	-	-	-
Electrification/Accessory Techs												-	
Electric Power Steering (EPS)	EPS	75	85	80	75	85	80	75	85	80	75	85	80
Improved Accessories	IACC	-	-	-	-	-	-	-	-	-	-	-	-
12V BAS Micro-Hybrid	MHEV	320	380	350	320	380	350	320	380	350	320	380	350
Higher Voltage/Improved Alternator	HVIA	16	18	17	16	18	17	16	18	17	16	18	17
Integrated Starter Generator	ISG	-	-	-	-	-	-	-	-	-	-	-	-
Transmission Techs													
Continuously Variable Transmission (CVT)	CVT	225	255	240	360	400	380	360	400	380	360	400	380
6/7/8-Speed Auto. Trans. with Improved Internals	NAUTO	190	325	258	190	325	258	190	325	258	190	325	258
Dual Clutch Transmission (DCT)	DCT	195	225	210	195	225	210	195	225	210	195	225	210
Hybrid Techs													
Power Split Hybrid	PSHEV	-	-	-	-	-	-	-	-	-	-	-	-
2-Mode Hybrid	2MHEV	-	-	-	-	-	-	-	-	-	-	-	-
Plug-in hybrid	PHEV	-	-	-	-	-	-	-	-	-	-	-	-
Vehicle Techs													
Mass Reduction - 1%	MR1	-	-	-	-	-	-	-	-	-	-	-	-
Mass Reduction - 2%	MR2	-	-	-	-	-	-	-	-	-	-	-	-
Mass Reduction - 5%	MR5	100	140	120	100	140	120	100	140	120	100	140	120
Mass Reduction - 10%	MR10	380	440	410	380	440	410	380	440	410	380	440	410
Mass Reduction - 20%	MR20	-	-	-	-	-	-	-	-	-	-	-	-
Low Rolling Resistance Tires	ROLL	18	22	20	18	22	20	18	22	20	18	22	20
Low Drag Brakes	LDB	-	-	-	-	-	-	-	-	-	-	-	-
Secondary Axle Disconnect	SAX	-	-	-	-	-	-	-	-	-	-	-	-
Aero Drag Reduction 10%	AERO	23	33	28	23	33	28	23	33	28	23	33	28

continued

TABLE I.8 Continued

Sierra Research			
Technologies		Midsize	Truck
Spark Ignition Techs	Abbreviation	Low	Low
Low Friction Lubricants	LUB	13	16
Engine Friction Reduction	EFR	-	-
VVT- Coupled Cam Phasing (CCP), SOHC	CCP	-	-
Discrete Variable Valve Lift (DVVL), SOHC	DVVL	-	-
Cylinder Deactivation, SOHC	DEAC	335	410
VVT - In take Cam Phasing (ICP)	ICP	-	-
VVT - Dual Cam Phasing (DCP)	DCP	-	-
Discrete Variable Valve Lift (DVVL), DOHC	DVVL	-	-
Continuously Variable Valve Lift (CVVL)	CVVL	-	-
Cylinder Deactivation, OHV	DEAC	335	410
VVT - Coupled Cam Phasing (CCP), OHV	CCP	-	-
Discrete Variable Valve Lift (DVVL), OHV	DVVL	-	-
Conversion to DOHC with DCP	CDOHC	-	-
Stoichiometric Gasoline Direct Injection (GDI)	SGDI	515	630
Turbocharging and Downsizing	TRBDS	814	996
Diesel Techs			
Conversion to Diesel	DSL	-	-
Conversion to Diesel following TRBDS	DSL	5775	7063
Conversion to Advanced Diesel	ADSL	-	-
Electrification/Accessory Techs			
Electric Power Steering (EPS)	EPS	76	140
Improved Accessories	IACC	-	-
12V BAS Micro-Hybrid	MHEV	-	-
Higher Voltage/Improved Alternator	HVIA	68	83
Integrated Starter Generator	ISG	-	-
Transmission Techs			
Continuously Variable Transmission (CVT)	CVT	-	-
6/7/8-Speed Auto. Trans. with Improved Internals	NAUTO	-	-
Dual Clutch Transmission (DCT)	DCT	450	551
Hybrid Techs			
Power Split Hybrid	PSHEV	-	-
2-Mode Hybrid	2MHEV	-	-
Plug-in hybrid	PHEV	-	-
Vehicle Techs			
Mass Reduction - 1%	MR1	-	-
Mass Reduction - 2%	MR2	-	-
Mass Reduction - 5%	MR5	-	-
Mass Reduction - 10%	MR10	-	-
Mass Reduction - 20%	MR20	-	-
Low Rolling Resistance Tires	ROLL	-	-
Low Drag Brakes	LDB	-	-
Secondary Axle Disconnec	SAX	-	-
Aero Drag Reduction 10%	AERO	-	-

continued

TABLE I.8 Continued

Martec Research

Technologies		MPFI, DOHC, 4V L4	MPFI, DOHC, 4V V6	MPFI, DOHC, 4V V8
Spark Ignition Techs	Abbreviation			
Low Friction Lubricants	LUB	-	-	-
Engine Friction Reduction	EFR	-	-	-
VVT- Coupled Cam Phasing (CCP), SOHC	CCP	-	-	-
Discrete Variable Valve Lift (DVVL), SOHC	DVVL	-	-	-
Cylinder Deactivation, SOHC	DEAC	-	-	-
VVT - In take Cam Phasing (ICP)	ICP	-	-	-
VVT - Dual Cam Phasing (DCP)	DCP	-	-	-
Discrete Variable Valve Lift (DVVL), DOHC	DVVL	-	480	-
Continuously Variable Valve Lift (CVVL)	CVVL	428	675	825
Cylinder Deactivation, OHV	DEAC	-	-	-
VVT - Coupled Cam Phasing (CCP), OHV	CCP	-	-	-
Discrete Variable Valve Lift (DVVL), OHV	DVVL	-	-	-
Conversion to DOHC with DCP	CDOHC	-	-	-
Stoichiometric Gasoline Direct Injection (GDI)	SGDI	440	558	746
Turbocharging and Downsizing	TRBDS	-	855	1289
Diesel Techs				
Conversion to Diesel	DSL	-	-	-
Conversion to Diesel following TRBDS	DSL	-	3542	5198
Conversion to Advanced Diesel	ADSL	-	-	-
Electrification/Accessory Techs				
Electric Power Steering (EPS)	EPS	-	-	-
Improved Accessories	IACC	-	-	-
12V BAS Micro-Hybrid	MHEV	627	-	-
Higher Voltage/Improved Alternator	HVIA	-	-	-
Integrated Starter Generator	ISG	617	-	-
Transmission Techs				
Continuously Variable Transmission (CVT)	CVT	-	-	-
6/7/8-Speed Auto. Trans. with Improved Internals	NAUTO	638	638	638
Dual Clutch Transmission (DCT)	DCT	450	450	450
Hybrid Techs				
Power Split Hybrid	PSHEV	5246	7871	9681
2-Mode Hybrid	2MHEV	-	-	-
Plug-in hybrid	PHEV	-	-	-
Vehicle Techs				
Mass Reduction - 1%	MR1	-	-	-
Mass Reduction - 2%	MR2	-	-	-
Mass Reduction - 5%	MR5	-	-	-
Mass Reduction - 10%	MR10	-	-	-
Mass Reduction - 20%	MR20	-	-	-
Low Rolling Resistance Tires	ROLL	-	-	-
Low Drag Brakes	LDB	-	-	-
Secondary Axle Disconnec	SAX	-	-	-
Aero Drag Reduction 10%	AERO	-	-	-

continued

TABLE I.8 Continued

NESCCAF		
Technologies		Large Car
		V6
Spark Ignition Techs	Abbreviation	
Low Friction Lubricants	LUB	16
Engine Friction Reduction	EFR	16
VVT- Coupled Cam Phasing (CCP), SOHC	CCP	173
Discrete Variable Valve Lift (DVVL), SOHC	DVVL	278
Cylinder Deactivation, SOHC	DEAC	173
VVT - In take Cam Phasing (ICP)	ICP	105
VVT - Dual Cam Phasing (DCP)	DCP	210
Discrete Variable Valve Lift (DVVL), DOHC	DVVL	383
Continuously Variable Valve Lift (CVVL)	CVVL	623
Cylinder Deactivation, OHV	DEAC	173
VVT - Coupled Cam Phasing (CCP), OHV	CCP	-
Discrete Variable Valve Lift (DVVL), OHV	DVVL	-
Conversion to DOHC with DCP	CDOHC	-
Stoichiometric Gasoline Direct Injection (GDI)	SGDI	278
Turbocharging and Downsizing	TRBDS	-420
Diesel Techs		
Conversion to Diesel	DSL	-
Conversion to Diesel following TRBDS	DSL	-
Conversion to Advanced Diesel	ADSL	1125
Electrification/Accessory Techs		
Electric Power Steering (EPS)	EPS	60
Improved Accessories	IACC	75
12V BAS Micro-Hybrid	MHEV	-
Higher Voltage/Improved Alternator	HVIA	60
Integrated Starter Generator	ISG	-
Transmission Techs		
Continuously Variable Transmission (CVT)	CVT	263
6/7/8-Speed Auto. Trans. with Improved Internals	NAUTO	-
Dual Clutch Transmission (DCT)	DCT	-
Hybrid Techs		
Power Split Hybrid	PSHEV	5246
2-Mode Hybrid	2MHEV	-
Plug-in hybrid	PHEV	-
Vehicle Techs		
Mass Reduction - 1%	MR1	-
Mass Reduction - 2%	MR2	-
Mass Reduction - 5%	MR5	321
Mass Reduction - 10%	MR10	-
Mass Reduction - 20%	MR20	-
Low Rolling Resistance Tires	ROLL	96
Low Drag Brakes	LDB	-
Secondary Axle Disconnect	SAX	-
Aero Drag Reduction 10%	AERO	134

J

Probabilities in Estimation of Fuel Consumption Benefits and Costs

The committee estimated cumulative fuel consumption by successively multiplying the base fuel consumption by one less the estimated fractional reductions associated with specific technologies. The estimates of cumulative cost impacts are obtained by successively adding individual retail price equivalent change estimates. The committee has provided rough confidence intervals for the individual fractional reductions. The confidence intervals are based on the committee's judgment and have not been derived in a rigorous, reproducible method. The committee's intent in providing the confidence intervals is to convey its opinion that all such estimates are subject to uncertainty. The committee believes it is important to communicate the degree of uncertainty in estimates of fuel consumption potential and cost even though it cannot make these estimates with precision or scientific rigor. Given the judgmental nature of our fuel consumption and cost estimates, the committee has attempted to aggregate them with an appropriate degree of mathematical rigor. The following describes the method used by the committee to aggregate its estimates of uncertainty for individual technologies to estimate the confidence intervals for the full technology pathways shown in Chapter 9.

Assuming the individual estimates of cost impacts are independent, the variance of the sum of n cost estimates is equal to the sum of the variances. Thus the standard deviation of the sum is the square root of the sum of the squared standard deviations. Let $\pm 1.64\omega$ be the committee's estimated confidence interval for the retail price impact of technology i. The confidence interval for the sum of i price impact estimates would be $\pm$ 1.64ω, where ω_n is defined as follows.

$$\omega_n = \sqrt{\sum_{i-1}^{n} \omega_i^2} \qquad \text{Equation 1}$$

Let f_i be the impact of technology i on fuel consumption, where $f_i = 1 - \Delta_I$ and Δ_I is the expected fractional reduction expected from technology I, and let p_i be the expected increase in retail price equivalent. Let $\pm$ 1.64σ_i be the committee's estimated confidence interval for technology i and assume that σ_i^2 is a reasonable estimate of the variance of the estimate, whose distribution is assumed to be symmetric. Furthermore, it is assumed that the individual technology estimates are independent. The exact formula for the variance of the product of n independent random variables was derived by Goodman (1962), who also pointed out that if the square of the coefficients of variation (σ_i^2/f^2) of the variables is small, then an approximation to the exact variance should be reasonably accurate. The committee's estimates of fuel consumption reduction are on the order of f = 1 – 0.05, in general, while its estimates of the confidence intervals 1.64σ are on the order of 0.02. Thus the square of the coefficients of variation are on the order of 0.00015/0.9025 = 0.00016. However, Goodman also notes that his approximate formula tends to underestimate the variance, in general. As a consequence, we use his exact formula, shown below in Equation 2.

$$Var\left(\prod_{i=1}^{n} f_i\right) = \prod_{i=1}^{n} f_i^2 \left[\prod_{i=1}^{n}\left(\frac{\sigma_i^2}{f_i^2}+1\right)-1\right] \qquad \text{Equation 2}$$

$$1.64 \times StdDev\left(f_n\right) = 1.64 \times \sqrt{Var\left(\prod_{i=1}^{n} f_i\right)}$$

Equation 1 can be used to calculate a confidence interval for either the cumulative fuel consumption or cumulative cost impacts by calculating the square root of the variance and multiplying by 1.64. The committee believes that its 1.64σ_i bounds represent, very approximately, a 90 percent confidence interval. Assuming that the cost and fuel consumption estimates are also independent, the probability that fuel consumption is within its 90 percent confidence bounds and cost is within its confidence bounds at the same time implies that the joint confidence interval is an 81 percent confidence interval.

$$\text{Prob}\left(f_i - 1.64\sigma_i < f_i < f_i + 1.64\sigma_i\right) = 0.9$$
$$\text{Prob}\left(p_i - 1.64\sigma_i < p_i < p_i + 1.64\sigma_i\right) = 0.9$$
$$\text{Prob}\left(f_i - 1.64\sigma_i < f_i < f_i + 1.64\sigma_i\right)$$
$$\text{Prob}\left(p_i - 1.64\sigma_i < p_i < p_i + 1.64\sigma_i\right) = 0.9 \times 0.9 = 0.81$$

The committee did not address what specific probability distribution the uncertainty about fuel consumption and cost impacts might take. However, if one assumes they follow a normal distribution, then the ratio of a 90 percent confidence interval to an 81 percent confidence interval would be approximately 1.64/1.31 = 1.25. Thus, an appropriately rough adjustment factor to convert the individual confidence intervals to a joint confidence interval of 90 percent would widen them by about 25 percent.

REFERENCE

Goodman, L.A. 1962. The variance of a product of K random variables. Journal of the American Statistical Association 57(297):54-60.

K

Model Description and Results for the EEA-ICF Model

METHODOLOGY OVERVIEW

The lumped parameter approach to fuel consumption modeling uses the same basic principles as all simulation models, but instead of calculating fuel consumption second by second, as is sometimes done, it uses an average cycle. Such an approach has been used widely by industry and regulatory agencies, most recently by the U.S. Environmental Protection Agency (EPA) to help assess the 2012-2016 proposed fuel economy standards (EPA, 2008). The method can be generally described as a first-principles-based energy balance, which accounts for all the different categories of energy loss, including the following:

- Losses based on the second law of thermodynamics,
- Heat loss from the combusted gases to the exhaust and coolant,
- Pumping loss,
- Mechanical friction loss,
- Transmission losses,
- Accessory loads,
- Vehicle road load tire and aerodynamic drag losses, and
- Vehicle inertial energy lost to the brakes.

Conceptually, each technology improvement is characterized by the percent change to each of the loss categories. If multiple technologies are employed to reduce the same category of loss, each successive technology has a smaller impact as the category of loss becomes closer to zero.

EEA-ICF Inc.[1] has developed a lumped parameter model that is broadly similar in scope and content to the EPA model (Duleep, 2007). In this model, all of the baseline vehicle energy losses are determined computationally, and many of the technology effects on each source of loss have been determined from data presented at technical conferences. However, the EPA does not document how the various losses were determined for the baseline vehicle: It says only that the vehicle has a fixed percentage of fuel lost to each category. The EPA also does not document how the technology-specific improvements in each category of loss were characterized. It appears that the losses for both the baseline vehicle and the effects of technology improvements were based not on computed values but on expert opinion.

MODEL COMPUTATIONS

Here the committee summarizes the EEA-ICF model. GM researchers Sovran and Bohn (1981) used numerical integration over the Federal Test Procedure city and highway driving cycles to determine the energy required at the wheel to move a vehicle over the driving cycle as a function of its weight, frontal area, drag coefficient, and tire rolling resistance coefficient. This procedure is used to compute the energy requirement at the wheel for the given baseline vehicle and translated to energy at the engine output shaft by using transmission and driveline efficiency factors (which differ by transmission type and number of gears) derived from the open literature. Accessory energy requirements are added as a fixed energy amount that is a function of engine size. This determines total engine output energy; average cycle power is then computed by distributing the energy over the cycle time when positive engine output is required—that is, the time spent at closed throttle braking and idle are accounted for separately. Average cycle RPM excluding idle was obtained for specific vehicles from simulation models on specific vehicles, and these data are scaled by the ratio of the N/V for the data vehicle and the baseline vehicle. The data are used to determine average brake mean effective pressure (BMEP) for the positive power portion of the cycle.

[1] Energy and Environmental Analysis, Inc. (EEA) was acquired by ICF International during the course of this study. In this appendix, reference is made to EEA-ICF, although in the report as a whole reference is made simply to EEA.

Fuel consumption is determined by the following relationship:

$$IMEP = BMEP + FMEP + PMEP$$

where I is for indicated, F is for friction, P is for pumping, and MEP is the mean effective pressure in each category. The fuel consumption model is derived from a methodology to estimate an engine map using a semiempirical model developed by researchers at Ford and the University of Nottingham (Shayler et al., 1999). In this formulation, fuel consumption is proportional to IMEP divided by indicated thermal efficiency (sometimes called the Willans line), friction is determined empirically from engine layout and is a function of RPM only, and PMEP is simply intake manifold pressure (atmospheric pressure). Intake manifold pressure is solved for any given BMEP, since IMEP is also proportional to intake pressure. This model explicitly derives thermal efficiency, friction loss, and pumping loss for the baseline vehicle. Fuel consumption at idle and closed throttle braking are modeled as functions of engine displacement only. The baseline engine is always modeled with fixed valve lift and timing, and the pumping loss is adjusted for the presence of variable valve timing if applicable. The model can be construed as a two-point approximation of a complete engine map and is a very reasonable representation of fuel consumption at light and moderate loads where there is no fuel enrichment.

The technologies are characterized by their effect on each of the losses explicitly accounted for in the model, and the representation is similar in concept to the representation in the EPA model. In the EEA-ICF analysis, the committee collected information on the effect of each engine technology on peak engine efficiency, pumping loss, and friction loss as a cycle average from technical papers that describe measured changes in these attributes from prototype or production systems. When these losses are not explicitly measured, they are computed from other published values such as the change in compression ratio, the change in torque, or the measured change in fuel consumption.

Comparison of Results to Detailed Simulation Model Outputs

Both EEA-ICF and EPA have compared the lumped parameter results with new full-scale simulation modeling results on several vehicle classes with different combinations of planned technological improvements. The simulations were done by the consulting firm Ricardo, Inc., and documented in a separate report (Ricardo, 2008). The Ricardo work modeled five baseline vehicles (standard car, large car, small MPV, large MPV, and large truck) and 26 technology combinations, covering gasoline and diesel power trains used in the EPA model, but there was no simulation of hybrids.

In a majority of the comparisons done by EPA, the lumped parameter model estimates were close to the Ricardo estimates, and the EPA concluded the results of their model were plausible, although a few technology packages required additional investigation. The EPA has indicated that it will continue to use the lumped parameter approach as an analytical tool, perhaps adjusting it to improve its fidelity as more simulation results become available.

EEA-ICF also performed analysis for the NRC Committee on Assessment of Technologies for Improving Light-Duty Vehicle Fuel Economy (Duleep, 2008a, 2008b). Based on the committee's experience, when a number of engine, transmission, and other technology improvements are simultaneously added to a baseline vehicle, the net fuel economy benefit can be approximated by taking 90 percent of the additive sum of the individual technology benefits, as developed by EEA-ICF. The committee used this technique to develop a quick approximation of the level of agreement likely between the Ricardo simulations and the EEA-ICF lumped parameter model. It was able to perform a quick analysis of only 23 of 26 packages developed by Ricardo, since there were no data on HCCI engines, which were used in three of the Ricardo technology packages.

Ricardo included one technology for which the committee had no specific data. It called this "fast warm-up" technology because it involved the control of coolant flow to the engine immediately after cold start. Based on the data presented by Ricardo, the benefit of the technology was estimated at 1 percent, including the benefit of the electric water pump. All other technology benefits were based on the data from ICF-EEA previous reports to DOE on fuel economy technology. These benefit estimates were adjusted for the presence or absence of technologies on the baseline vehicle, since all benefits in the DOE reports have been typically defined relative to an engine with fixed valve timing and a four-speed automatic transmission. The results are illustrated in Figure K.1, and the plot shows the difference between the Ricardo results and the quick approximation method.

In 16 of the 23 cases, the Ricardo estimate is within +5 percent of the quick estimate. In two cases, the Ricardo estimates were more than 10 percent lower than the quick estimates, as shown in Figure K.1. In five cases, the Ricardo estimates were 10 percent (or more) higher than the quick estimate. The difference implies that the benefits are larger than the simple sum of individual technology benefits and that technology synergies are positive. The committee also examined the technology packages in the two "low" and five "high" outliers. Both low outliers had technology packages with a continuously variable transmission (CVT) as one of the technologies. The five high outliers had no major technology improvement in common.

More detailed analysis was also done with the EEA-ICF lumped parameter model. Constraints on resources and time allowed the committee to analyze only 9 of the 23 cases with the lumped parameter model, but the 9 cases included both high and low outliers from the previous analysis. Three technology packages were analyzed for a standard car, which used a Toyota Camry baseline; three for a compact

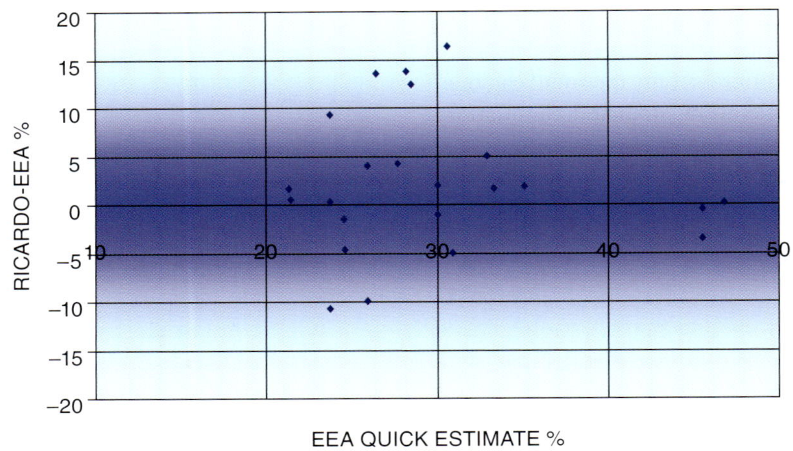

FIGURE K.1 Comparison of the difference between the Ricardo, Inc., results and the quick approximation method.

van, which used a Chrysler Voyager baseline; and three for a standard pickup, which used a Ford F-150 baseline. Table K.1 shows the results and compares them with those of the quick method. The more detailed modeling reduced the average difference between the Ricardo estimates and the committee estimates for the Toyota Camry and the Chrysler compact van but increased the difference for the Ford F-150 truck. The largest observed difference is for Package 10 on the Ford, where the baseline 5.4-L V8 is replaced by a 3.6-L V6 turbo GDI engine and the downsizing is consistent with the 33 percent reduction that was used.

Comparison of Model Results to NRC Estimates

The NRC study has developed a series of technology paths whose combined effect on fuel consumption was estimated from expert inputs on the marginal benefits of each successive technology given technologies already adopted. Paths were specified for five different vehicles: small cars, intermediate/large cars, high-performance sedans, body-on-frame small trucks, and large trucks. There were no substantial differences in the paths or the resulting fuel consumption estimates across the five vehicles: All estimated decreases in fuel consumption were between 27 and 29 percent for

TABLE K.1 Comparison of Fuel Economy Improvements (in Percent) from Ricardo, Inc., Modeling, EEA-ICF Quick Analysis, and the EEA-ICF Model

Vehicle	Technology Package	Ricardo Estimate	EEA Quick Result	EEA Model Result
Toyota Camry	Z	33.0	23.7	32.6
	1	13.0	23.7	23.1
	2	22.0	22.4	21.9
	RMS difference		8.15	5.85
Chrysler Voyager	4	26.0	30.9	29.9
	6b	35.5	33.3	35.5
	16	41.0	28.5	36.6
	RMS difference		7.85	3.39
Ford F-150	9	32.0	30.0	28.3
	10	42.0	28.2	26.4
	16	23.0	21.3	23.4
	RMS difference		8.12	9.25

NOTE: RMS, root mean square difference between the EEA-ICF estimate and the Ricardo estimate. The differences seem to be in the same range as the differences between the EPA estimates with their lumped parameter model and the Ricardo estimates. It is also important to note that the EPA model results are more consistent with the results of the EEA-ICF model. The "low" Ricardo result for Package 1 on the Camry is also significantly lower than the EPA estimate of 20.5 percent fuel economy benefit, which is closer to the EEA-ICF estimate of 23 percent than to the Ricardo 13 percent estimate. Similarly, the high Ricardo estimate for Package 10 on the Ford F-150 is also substantially higher than the EPA estimate of 30.5 percent fuel efficiency gain, which is, in turn, higher than the committee estimate of 26.4 percent but much lower than the Ricardo estimate of 42 percent.

spark-ignition engines and 36 and 40 percent for diesel engines. Since the "performance sedan" and intermediate sedan specifications were not very different, only the small car, one intermediate car, and two trucks were simulated. Simulation was done for the spark ignition engine and the diesel engine paths, but not for the hybrid path.

Table K.2 lists the model results versus the committee estimates for the eight cases (four for spark ignition and four for diesel). In general, the model forecasts are very close to but typically slightly lower than the forecasts of experts, although well within the range of uncertainty included in the committee estimate. Only one vehicle, the full-size truck, shows a larger difference on the diesel path. Historically, the committee's method of forecasting the marginal benefit of technology along a specified path has been criticized as potentially leading to an overestimation of benefits for spark ignition engines since it could lead to infeasible solutions if total pumping loss reduction estimated exceeded the actual pumping loss. The simulation model output's explicit tracking of the losses addresses this issue directly to ensure that no basic scientific relationships are violated.

Fuel consumption is decreased by reducing the tractive energy required to move the vehicle (by reducing weight, aerodynamic drag, or rolling resistance), reducing losses to the transmission and drive line, reducing accessory energy consumption, or reducing engine fuel consumption during idle and closed throttle braking. Fuel consumption can also be reduced by increasing engine efficiency over the cycle, which is accomplished by increasing peak efficiency or by reducing mechanical friction and pumping loss. Figures K.2 through K.5 show the technology path steps and track the reductions from both approaches separately, with the reduction in energy required to drive through the test cycle shown on top and the engine efficiency shown below. Peak engine efficiency actually decreases slightly due to turbocharging and downsizing, but the cycle efficiency increases from about 24 to 29 percent owing to reduction in pumping and friction loss (blue part of the bar). The general trends are very similar across all four vehicle types, but the key feature is that pumping and friction loss are not reduced to physically impossible levels for the solution.

TABLE K.2 Comparison of Fuel Consumption Reductions (in Percent) for NRC Estimates and the EEA-ICF Model

Spark Ignition Path	NAS	EEA-ICF
Small car	27	26.7
Intermediate/large car	29	27.3
BOF small truck	27	27.3
BOF large truck	29	26.2
Diesel path		
Small car	37	35.7
Intermediate/large car	37	36.2
BOF small truck	37	36.6
BOF large truck	40	36.5

NOTE: BOF, body on frame.

REFERENCES

Duleep, K.G. 2007. Overview of lumped parameter model. Presentation to the National Research Council Committee for the Assessment of Technologies for Improving Light-Duty Vehicle Fuel Economy on October 26, Washington, D.C.

Duleep, K.G. 2008a. EEA-ICF Analysis of Ricardo simulation outputs. Presentation to the National Research Council Committee for the Assessment of Technologies for Improving Light-Duty Vehicle Fuel Economy on February 26, Washington, D.C.

Duleep, K.G. 2008b. EEA-ICF analysis update. Presentation to the National Research Council Committee for the Assessment of Technologies for Improving Light-Duty Vehicle Fuel Economy on April 1, Washington, D.C.

EPA (U.S. Environmental Protection Agency). 2008a. EPA Staff Technical Report: Cost and Effectiveness Estimates of Technologies Used to Reduce Light-Duty Vehicle Carbon Dioxide Emissions. EPA420-R-08-008. Ann Arbor, Mich.

Ricardo, Inc. 2008. A Study of the Potential Effectiveness of Carbon Dioxide Reducing Vehicle Technologies. Report to the Environmental Protection Agency. June 26.

Sovran, G., and M. Bohn, 1981. Formulae for the tractive energy requirements of the vehicles driving the EPA schedules. SAE Paper 810184. SAE International, Warrendale, Pa.

Shayler, P., J. Chick, and D. Eade. 1999. A method of predicting brake specific fuel consumption maps. SAE Paper 1999-01-0556. SAE International, Warrendale, Pa.

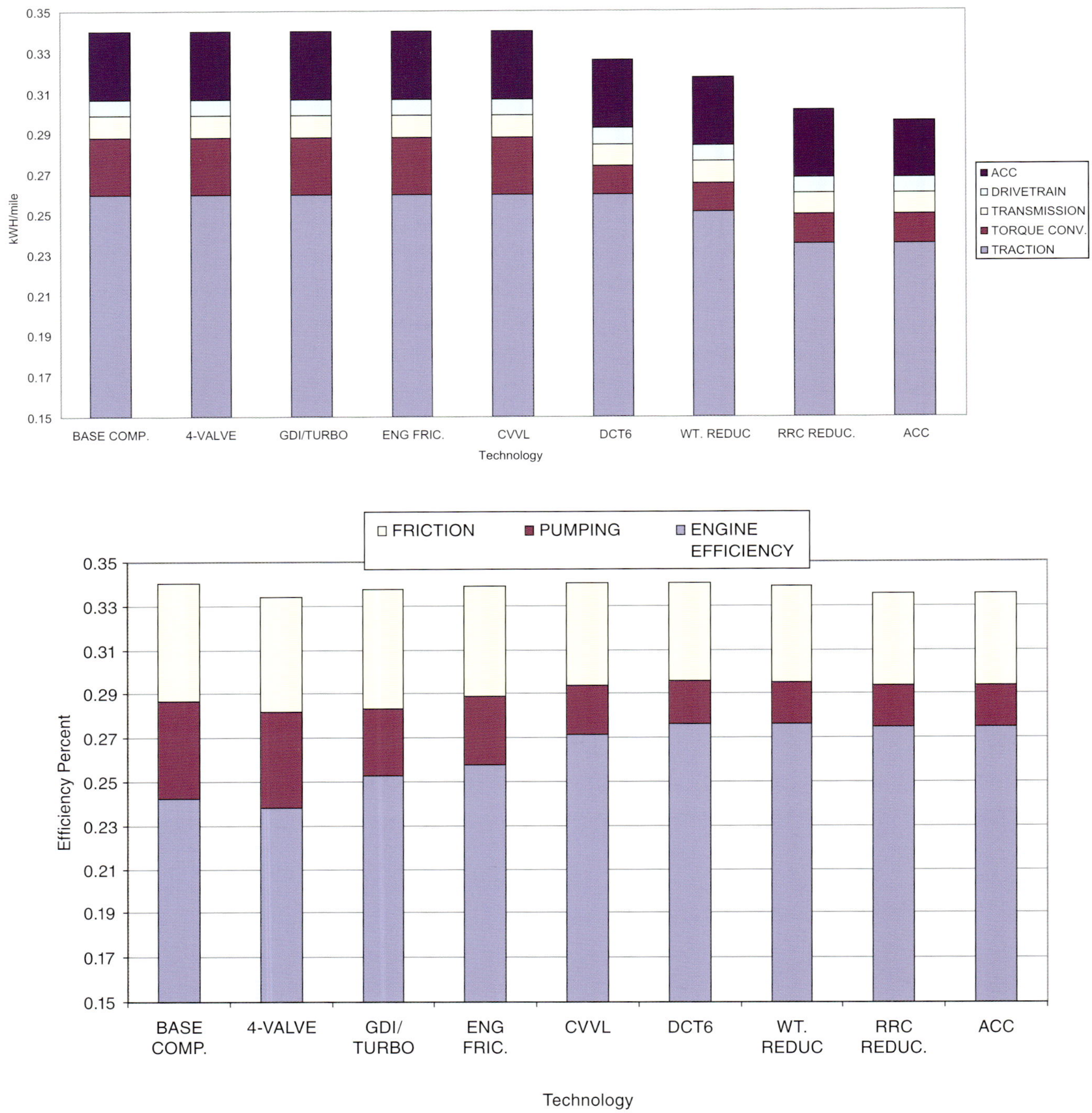

FIGURE K.2 Technology path steps and reduction in energy required to drive through the test cycle (top) and the engine efficiency (bottom), body-on-frame small truck.

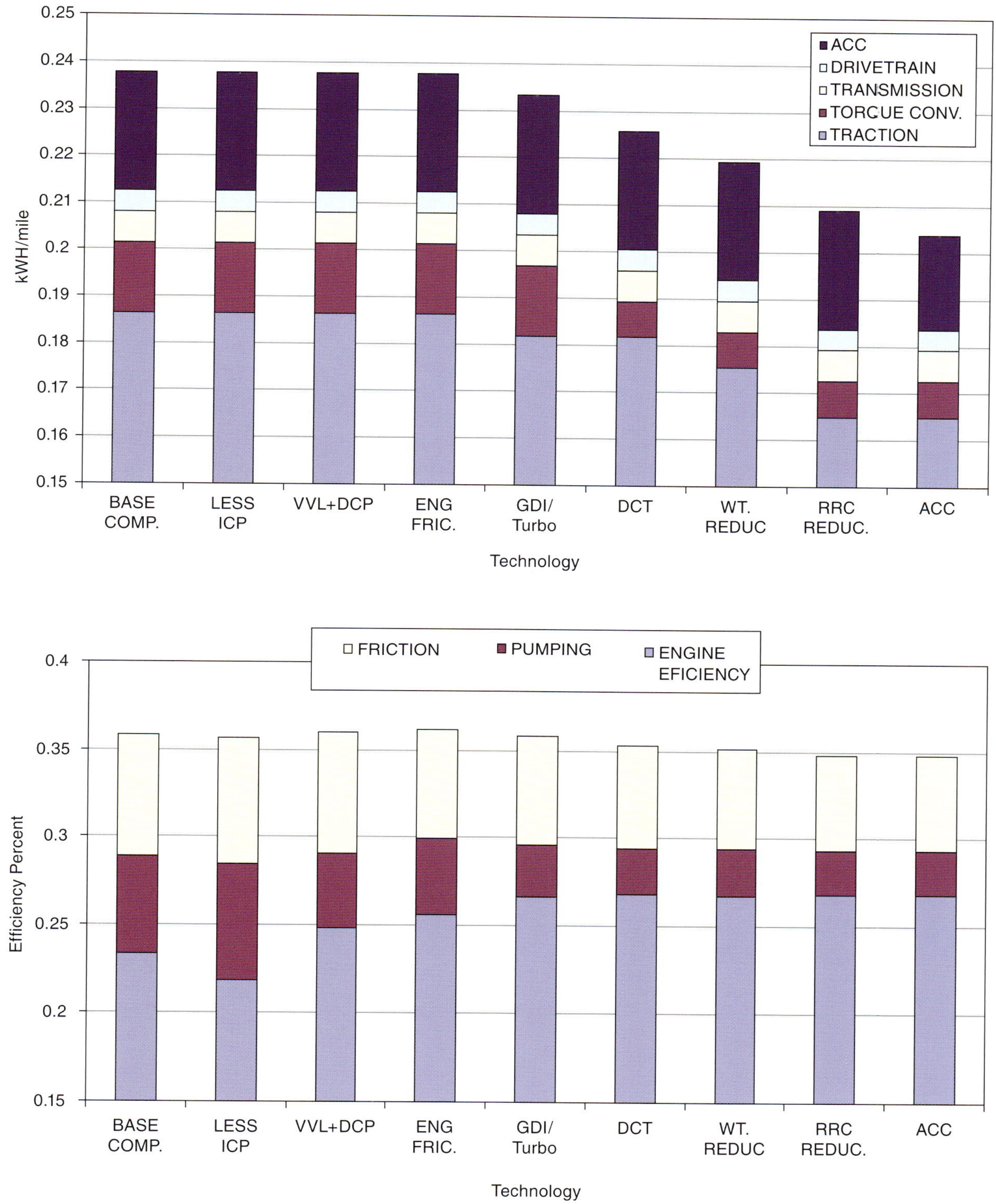

FIGURE K.3 Technology path steps and reduction in energy required to drive through the test cycle (top) and the engine efficiency (bottom), midsize sedan.

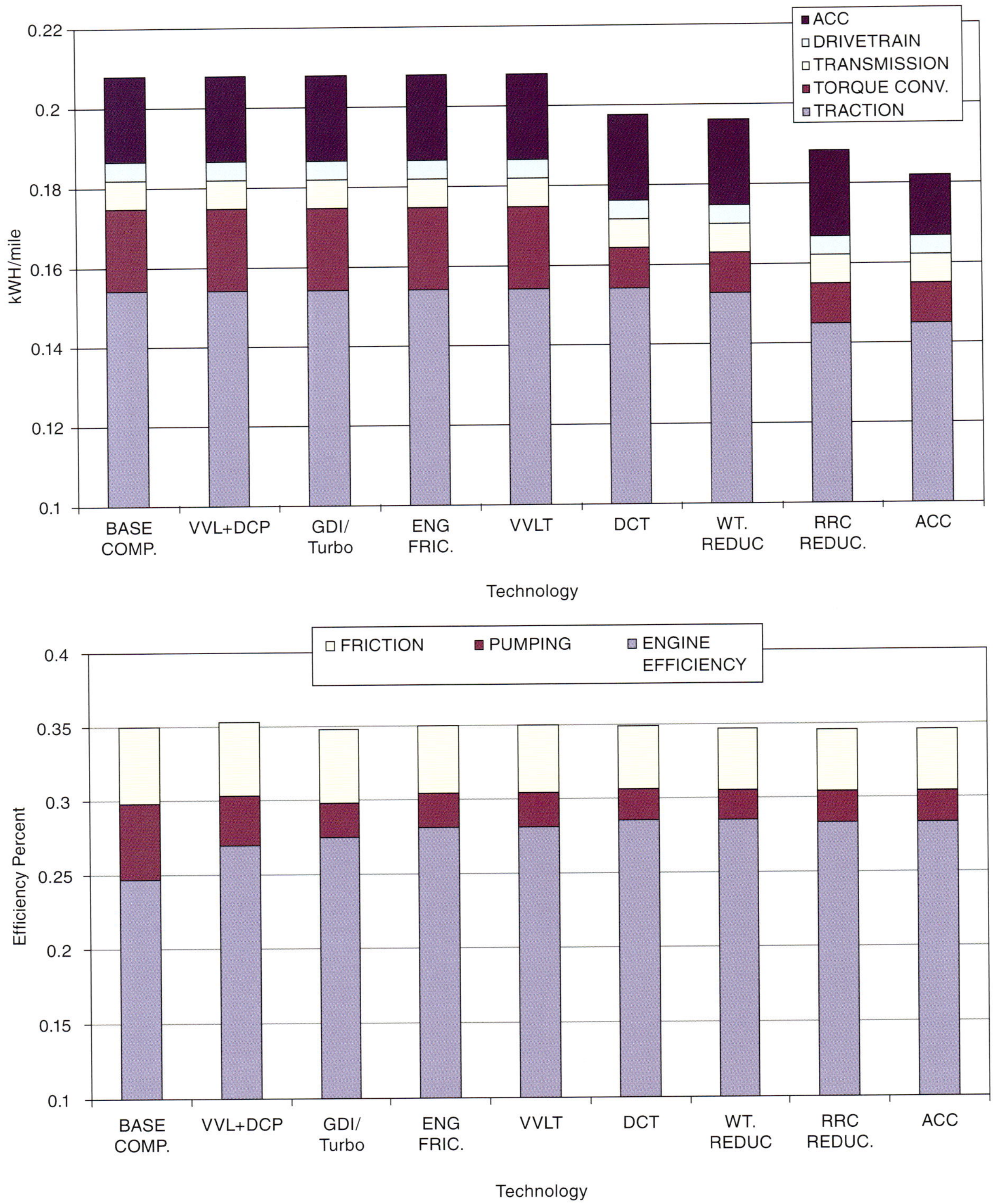

FIGURE K.4 Technology path steps and reduction in energy required to drive through the test cycle (top) and the engine efficiency (bottom), small car.

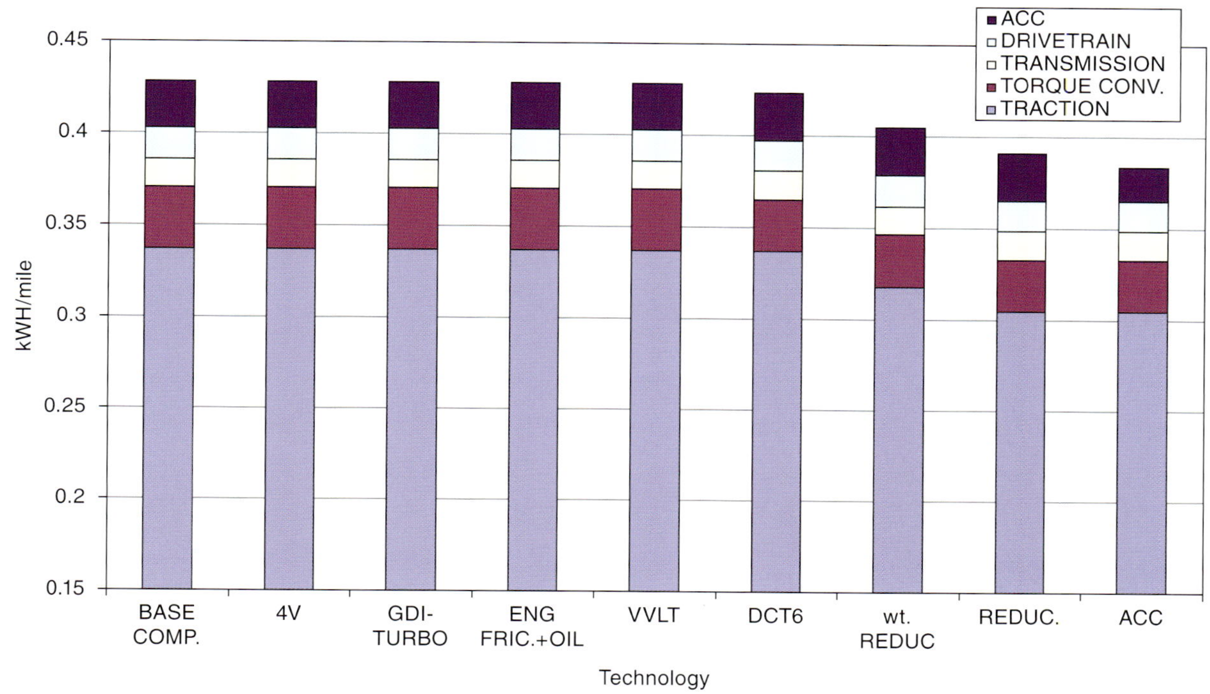

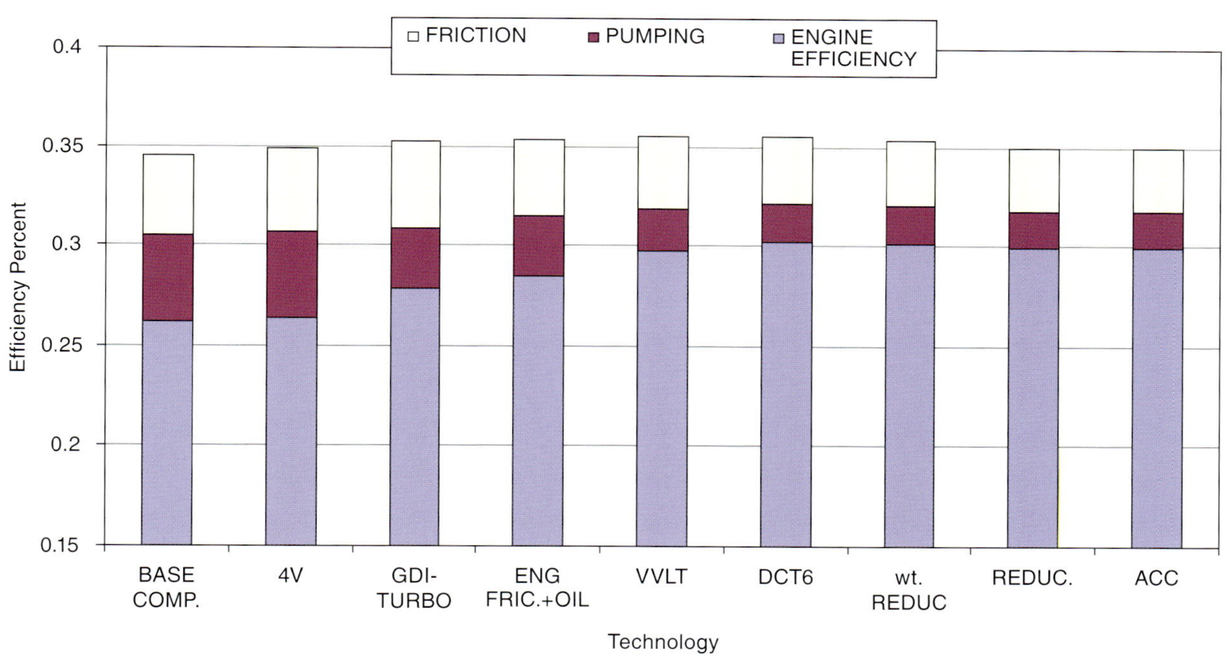

FIGURE K.5 Technology path steps and reduction in energy required to drive through the test cycle (top) and the engine efficiency (bottom), full-size truck.